QUANTUM SOCIAL SCIENCE

Written by world experts in the foundations of quantum mechanics and its applications to social science, this book shows how elementary quantum mechanical principles can be applied to decision making paradoxes in psychology, and used in modeling information in finance and economics.

The book starts with a thorough overview of some of the salient differences between classical, statistical, and quantum mechanics. It presents arguments on why quantum mechanics can be applied outside of physics and defines quantum social science. The issue of the existence of quantum probabilistic effects in psychology, economics, and finance is addressed and basic questions and answers are provided. Aimed at researchers in economics and psychology, as well as physics, basic mathematical preliminaries and elementary concepts from quantum mechanics are defined in a self-contained way.

EMMANUEL HAVEN is a Professor at the School of Management, University of Leicester, UK. He has published numerous articles in a variety of fields, such as operations research, economics, and finance.

ANDREI KHRENNIKOV is a Professor of Applied Mathematics at Linnaeus University, Sweden, and Director of the International Centre for Mathematical Modelling in Physics, Engineering, Economics and Cognitive Science.

QUANTUM SOCIAL SCIENCE

EMMANUEL HAVEN
University of Leicester

AND

ANDREI KHRENNIKOV
Linnaeus University

CAMBRIDGE
UNIVERSITY PRESS

CAMBRIDGE
UNIVERSITY PRESS

University Printing House, Cambridge CB2 8BS, United Kingdom

Published in the United States of America by Cambridge University Press, New York

Cambridge University Press is part of the University of Cambridge.

It furthers the University's mission by disseminating knowledge in the pursuit of education, learning and research at the highest international levels of excellence.

www.cambridge.org
Information on this title: www.cambridge.org/9781107012820

First published 2013

A catalogue record for this publication is available from the British Library

Library of Congress Cataloguing in Publication data
Haven, Emmanuel, 1965–
Quantum social science / Emmanuel Haven and Andrei Yu. Khrennikov.
p. cm.
Includes bibliographical references and index.
ISBN 978-1-107-01282-0 (hbk.)
1. Social sciences – Mathematical models. 2. Quantum theory.
I. Khrennikov, A. IU. (Andrei IUr'evich), 1958– II. Title.
H61.25.H38 2012
300.1'53012 – dc23 2012027126

ISBN 978-1-107-01282-0 Hardback

To our lovely Wives – Irina and Sophie

To our lovely Children – Anton, Nath, and Sam

Contents

Foreword

This new book by Emmanuel Haven and Andrei Khrennikov argues that information processing in social systems can to a degree be formalized with the mathematical apparatus of quantum mechanics. This is a novel approach. Understanding decision making is a central objective of economics and finance and the quantum like approach proposed here, is used as a tool to enrich the formalism of such decision making. Emmanuel and Andrei argue for instance that probability interference can be used to explain the violation of the law of total probability in well known paradoxes like the Ellsberg decision making paradox.

Emmanuel and Andrei's book forms one of the very first contributions in a very novel area of research. I hope this book can open the road for many new books to come. More new results are needed, especially in the area of decision making.

H. Eugene Stanley
William Fairfield Warren Distinguished Professor;
Professor of Physics; Professor of Chemistry;
Professor of Biomedical Engineering;
Professor of Physiology (School of Medicine)
Director, Center for Polymer Studies,
Department of Physics, Boston University

By chance a few days before Andrei Khrennikov and Emmanuel Haven asked me to write this Foreword to their new book *Quantum Social Science,* I was browsing the collected works of Wolfgang Pauli, *Writings on Physics and Philosophy,* eds. Charles P. Enz and Karl von Meyenn, Springer (1994). I was just coming off a busy semester, including teaching a rather advanced course on harmonic analysis and quantum physics. To those erstwhile Ph.D. students in mathematics and physics, I had found myself counseling them with utterances such as "look, all physicists need to think semi-classically or even classically," or "you have to do something, you cannot just say it is all random motion," or "Heisenberg didn't really understand mathematics, but his intuition was sufficient to guide him."

Therefore I was very pleased to see Haven and Khrennikov also going to some of Pauli's thoughts in their Preface. Pauli, one of the greatest thinkers on quantum mechanics, was often preoccupied with the interaction of experiment with observer, and in analogy with the interaction of the conscious with the unconscious. Pauli's advocacy of the coupling of objective quantum physics to the subjective, e.g. psychic, was patterned upon Bohr's fundamental notion of complementarity. Two mutually contradictory concepts, e.g. those of particle and wave, may co-exist.

Indeed, quantum mechanics has forced upon us a new reality, possessing many co-existing dualities. One has the Schrödinger picture of differential equations describing all the chemical elements upon which the universe depends, and the Heisenberg picture stressing more the probabilistic nature of scattering interactions. The two pictures were more or less reconciled by Born in 1926, with his concept of probability wave. I have reviewed the Born probability interpretation of quantum mechanics from its inception to the present in K. Gustafson, *The Born Rule*, AIP Proceedings 962 (2007) pp. 98–107. I detailed in that review how often the great pioneers of quantum theory had to resort to reasonings of classical physics. So one should not think that quantum mechanics is all " hocus-pocus." Quantum mechanics is grounded in reality.

On the other hand, it is quite important to stress that the Born interpretation places the physics into an abstract configuration space, and not in real 3d space. As a consequence, from then on one must rely on the mathematics. Quantum mechanics has generated some very powerful mathematics. Ideally, this then should be coupled with new quantum-like thinking that one will not find in classical physics. It is the authors' intention in the present book to apply these powerful new mathematical tools and the evolving new non-classical quantum intuition to social science, behavioral economics, decision theory, and financial engineering.

Both authors already have considerable experience in this endeavor. Andrei Khrennikov is the founder of the celebrated series of annual quantum physics conferences held in Växjö Sweden for the last dozen years. At those conferences Emmanuel Haven from the economics side has joined with Khrennikov in recent years to organize special sessions on the subject matter of this book. Khrennikov has previously put forth his thinking in two books, *Information Dynamics in Cognitive, Psychological and Anomalous Phenomena*, Kluwer (2004), and *Ubiquitous Quantum Structure: From Psychology to Finance,* Springer (2010). Haven brings to the present book more expertise in economics and finance.

Overall, one could describe their basic approach as that of embedding situations from the social or economic sciences into a quantum mechanical context and then using the methods of the latter to obtain new insights and results for the former.

Such approach presumes of the reader a substantial knowledge of both contexts, that of quantum mechanics, and that of the particular social field of application. That is asking a lot.

I chose to address this issue, that of more needed interdisciplinary competence in education, science, and the general public, in my recent autobiography *The Crossing of Heaven: Memoirs of a Mathematician*, Springer (2012). I have come to the conclusion that we must invoke and enforce a new term, that of Multidisciplinarity. Interdisciplinarity is a weak word. It implies that one is less than one hundred percent committed to each of the two fields. Or that one is slightly weak in one's own field and leaning on an expert from the other field, who is probably a bit weak also in his field. I have worked successfully in several fields of science and I can assure you that you should plan on becoming an expert also in "the other field," and that will take you, say, at least five years before you have a chance of becoming competitive there.

Thus a collateral message of this foreword is that of advancing the concept and indeed the cause of creating more multidisciplinarity in our future mathematicians, physicists, social scientists, and, in a more general sense, throughout the educated public. A tall order! But great opportunities will open up to those who are strong enough.

This book by Haven and Khrennikov is a move in that direction, a pioneering effort.

Karl Gustafson
Professor Of Mathematics
University of Colorado at Boulder

Preface

The current level of specialization of knowledge in a variety of fields of inquiry may make it quite challenging for a researcher to be at the same time a "developer" and a "tester" of a theory. Although a theory can exist without a necessary clear and obvious practical end goal, the ultimate test of the validity of a theory (whether it is situated in the exact or social sciences) will always be how measurement can "confirm" or dislodge a theory.

This book is largely dedicated to the *development* of a theory. We will be the very first to accept the accusation that the duo "theory-test" is widely absent in this work, and we believe it necessary to make this statement at the very beginning.

This book is about a very counter-intuitive development. We want to use a physics machinery which is meant to explain sub-atomic behavior, in a setting which is at the near opposite end of the size spectrum, i.e. the world as we know and live it through our senses. We may know about the sub-atomic world, but we do not have human experience of the sub-atomic world. Do we have credible and provable stories which can explain how the sub-atomic engages into the mechanics of the statistical macro-world? Probably not. Why do we bother then about being so exotic? The interested reader will want us to provide for a satisfactory answer to this obvious question, and we want to leave it up to him or her to decide whether we have *begun*, via the medium of this book, to convince that the level of "exoticality" (and "yes" how exotic is that word?) is sensibly less than anticipated. We can possibly give a glimmer of "hope," even at this early stage. Consider the words of one of the towering giants of physics of the twentieth century – Wolfgang Pauli. In an unpublished essay by Pauli, entitled "Modern examples of 'background physics'," which is reproduced in Meier* (pp. 179–196), we can read Pauli's words (Meier* (p. 185)): "Complementarity in physics ... has a very close analogy with the terms 'conscious' and 'unconscious' in psychology in

* Meier C. A. (2001). *Atom and Archetype: The Pauli/Jung Letters, 1932–1958.* Princeton University Press.

that any 'observation' of unconscious contents entails fundamentally indefinable repercussions of the conscious on these very contents." The words of Pauli are important. They show there is promise for a connection between "concepts" of utmost importance in two very different sciences: complementarity in quantum physics and "complementarity" between consciousness and unconsciousness in psychology.

In this book, we intend to give the reader a flavor of an intellectual development which has taken shape over several years via the usual media many academics use: conference presentations and academic articles. The theory presented here is nowhere complete but we strongly believe that it merits presentation in book form.

The models presented in this book can be called "quantum-like." They do not have a direct relation to quantum physics. We emphasize that in our approach, the quantum-like behavior of human beings is not a consequence of quantum physical processes in the brain. Our basic premise is that information processing by complex social systems can be described by the mathematical apparatus of quantum mechanics. We present quantum-like models for the financial market, behavioral economics, and decision making.

Connecting exact science with social science is not an easy endeavor. What reveals to be most difficult is to dispel an intuition that somehow there *should* exist a natural bridge between physics and the modeling of social systems. This is a very delicate issue. As we have seen above it is possible to think of "complementarity" as a concept which could bridge physics and psychology. However, in some specific areas of social systems, the "physics equivalent" of the obtained results may have very little meaning.

It is our sincere hope that with this book we can convince the brave reader that the intuition of the authors is not merely naive, but instead informative. Hence, may we suggest that "reading on" is the command of the moment? Let the neurons fire!

Acknowledgements

Luigi Accardi and A. Khrennikov and M. Ohya (2009). Quantum Markov model for data from Shafir-Tversky experiments in cognitive psychology. *Open Systems & Information Dynamics*, 16(4), 378–383. This material is reproduced with permission of World Scientific Publishing Co Pte Ltd.

Masanari Asano and M. Ohya and A. Khrennikov (2011). Quantum-like model for decision making process in two players game. *Foundations of Physics*, 41, 538–548. Foundations of Physics by Springer New York LLC. Reproduced with permission of Springer New York LLC in the format reuse in a book/textbook via Copyright Clearance Center.

Masanari Asano and M. Ohya and Y. Tanaka and I. Basieva and A. Khrennikov (2011). Quantum like model of brain's functioning: decision making from decoherence. *Journal of Theoretical Biology*, 281, 63–64. Journal of Theoretical Biology by Academic Press. Reproduced with permission of Academic Press in the format reuse in a book/textbook via Copyright Clearance Center.

Sheldon Goldstein (2010). Bohmian mechanics and quantum information. *Foundations of Physics*, 40, 335–355. © Springer Science+Business Media, LLC 2009. Quotes reproduced with kind permission from Springer Science+Business Media: Foundations of Physics.

Emmanuel Haven (2005). Analytical solutions to the backward Kolmogorov PDE via an adiabatic approximation to the Schrodinger PDE. *Journal of Mathematical Analysis and Applications*, 311, 442–443. © 2005. Reprinted with Permission from Elsevier Inc.

Emmanuel Haven (2008). Private information and the "information function": a survey of possible uses. *Theory and Decision*, 64, 200–201. © Springer 2007. Reprinted with kind permission from Springer Science+Business Media.

Emmanuel Haven (2008). The variation of financial arbitrage via the use of an information wave function. *International Journal of Theoretical Physics*, 47,

196–199. © Springer Science+Business Media, LLC 2007. Reprinted with kind permission from Springer Science+Business Media, LLC 2007.

Emmanuel Haven (2010). The Blackwell and Dubins theorem and Rényi's amount of information measure: some applications. *Acta Applicandae Mathematicae*, 109, 751–757. © Springer Science+Business Media B.V. 2008. Reprinted with kind permission from Springer Science+Business Media B.V. 2008.

Emmanuel Haven (2011). Itô's lemma with quantum calculus: some implications. *Foundations of Physics*, 41, 3, 530–536. © Springer Science+Business Media, LLC 2010. Reprinted with kind permission from Springer Science+Business Media.

Peter Holland (2000). *The Quantum Theory of Motion: An Account of the de Broglie–Bohm Causal Interpretation of Quantum Mechanics*. © Cambridge University Press 1995. Reproduced with permission.

Hiromu Ishio and E. Haven (2009). Information in asset pricing: a wave function approach. *Annalen der Physik*, 18, 1, 40–42. © 2009 Wiley-VCH Verlag GmbH & Co. KGaA, Weinheim. Reprinted with kind permission from J. Wiley.

Max Jammer (1974). *The Philosophy of Quantum Mechanics*. New York: J. Wiley. This material is reproduced with permission of John Wiley & Sons, Inc.

Andrei Khrennikov (2010). *Ubiquitous Quantum Structure: From Psychology to Finance, Prisoner's Dilemma*. © Springer-Verlag, Berlin, Heidelberg 2010. Reprinted with kind permission from Springer Science+Business Media.

Andrei Khrennikov (2011). Quantum-like model of processing of information in the brain based on classical electromagnetic field. *Biosystems*, 105, 251–259. Biosystems by Elsevier Science. Reproduced with permission of Elsevier Science.

C. A. Meier (2001). *Atom and Archetype: The Pauli/Jung Letters, 1932–1958*. © Princeton University Press 2001. Reprinted with permission.

Michael A. Morrison (1990). *Understanding Quantum Physics: A User's Manual*. Volume 1. 1st edition. © 1990. Reprinted by permission of Pearson Education Inc., Upper Saddle River, NJ.

Salih Neftci (2000). *An Introduction to the Mathematics of Financial Derivatives*, pp. 346–358. © Elsevier 2000.

Hal R. Varian (1992). *Microeconomic Analysis*. 3rd edition. © 1992, 1984, 1978 W. W. Norton and Company, Inc. Used by Permission of W. W. Norton and Company Inc.

List of symbols

Some mathematics symbols used in the book

- $\mathbb{R}$: space of real numbers
- $\mathbb{C}$: space of complex numbers
- $\rho(.,.)$: probability density function (2 dimensional)
- $\rho(t,.,.)$: time-dependent probability density function
- $L_2(\mathbb{R}^3)$: space of square integrable complex valued functions $\psi : \mathbb{R}^3 \to \mathbb{C}$
- $H = L_2(\mathbb{R}^3)$: complex Hilbert space with a scalar product
- $l.i.m$: limit in the mean square sense
- $\mathcal{P} = (\Omega, F, P)$: $\mathcal{P}$ is a probability space and points ω of Ω (which is a non-empty set) are said to be elementary events. F is a so-called σ-algebra and P is a probability measure
- $m.s.$: mean square
- $\delta(x - x_0)$: Dirac δ-"function"
- $\mathbf{P}^{b|a}$: matrix of transition probabilities
- $d_q f(x)$: q differential of a function $f(x)$
- $d_h f(x)$: h differential of a function $f(x)$

Some physics symbols used in the book

- m: mass
- a: acceleration
- f: force acting on particle
- V: real potential function
- ϕ, S: phase of a wave function
- ν: frequency
- t: time
- ∇V: gradient of the real potential function

- p: momentum
- q: position
- $\mathcal{H}(.,.)$: Hamiltonian function
- $\{f, g\}$: Poisson bracket of two functions f and g on an N particle phase space
- $\{f_1, f_2\}$: Poisson bracket for a pair of classical observables f_1, f_2
- $\phi(t, x, y, z,)$: field state at instant t of vector with coordinates x, y and z
- $E(t, x, y, z,)$: electrical field at instant t of vector with coordinates x, y and z
- $B(t, x, y, z,)$: magnetic field at instant t of vector with coordinates x, y and z
- h: Planck's constant
- $\hbar$: rationalized Planck constant
- $\Delta E_{ij} = E_i - E_j$: discrete portion of energy
- L: angular momentum of an electron
- I: intensity of the electromagnetic field
- $A = (a_{ij})$: Hermitian matrix
- $\widehat{\mathcal{H}}$: Hermitian matrix representing the energy observable (quantum Hamiltonian)
- $\hat{q}$: position operator
- $\hat{p}$: momentum operator
- σ_x: standard deviation of position
- σ_p: standard deviations of momentum
- Δ_{q_j}: Laplace operator
- $\psi(t, q)$: probability amplitude on time, t, and position, q
- Γ: phase space of hidden states
- $|\psi\rangle$: element of the Hilbert space H : a ket vector
- $\langle\phi|$: element of the dual space H^*, the space of linear continuous functionals on H: a bra vector
- $\langle\psi_1|\widehat{w}\psi_2\rangle$: Dirac braket, where ψ_1^* denotes the complex conjugate of ψ_1 and $\widehat{w}$ acts on the state function ψ_2.
- k: wave number
- $A(k)$: amplitude function of wave number k
- $\langle p\rangle$: average momentum
- Q: quantum potential
- $\mathbf{P}(.|C)$: conditional probability dependent on the context, C
- D_+: mean forward derivative
- D_-: mean backward derivative

Some economics/finance symbols used in the book

- σ: volatility
- $\alpha(\sigma)$: drift function of volatility
- $\beta(\sigma)$: diffusion function of volatility

- dX, dz, dW: Wiener process
- $\overrightarrow{q} = (q_1, q_2 \ldots q_n)$: n-dimensional price vector
- m_j: number of shares of stock j
- $T_j(t)$: market capitalization of trader j at time t
- $V(q_1, \ldots, q_n)$: interactions between traders as well as interactions from other macro-economic factors
- Π: portfolio value
- F: financial option price
- S: stock price
- $\Delta = \frac{\partial F}{\partial S}$: delta of the option
- f_u; f_d: intrinsic values of the option when the price of the asset is respectively going up and down
- $E(r)$: expected return
- $\delta\Pi$: discrete change in the value of the portfolio, Π
- μ: expected return
- dF: infinitesimal change in F (the option price)
- r_f: risk free rate of interest
- $\phi(S, t)$: part of the premium invested in the stock, S
- S_T: asset price at the expiration of the option contract
- S_0: asset price at the inception of the option contract
- $P(., .|., .)$: conditional probability distribution
- $E[S_T|I_t]$: conditional expectation of a stock price at time $T > t$, given the information you have at time t
- $E(e^{Y_t\lambda})$: moment generating function, λ is some arbitrary parameter, and Y_t follows a probability density function (pdf) with mean μt and $\sigma^2 t$
- $E^{\widetilde{P}}[., .]$: expectation with respect to a risk neutral probability measure $\widetilde{P}$
- $E^P[., .]$: expectation with respect to a probability measure P
- C_t: option call value at time t
- P_t: option put value at time t
- $\overrightarrow{\Phi} = (\Phi_1, \Phi_2, \ldots, \Phi_K)$: K-dimensional state price vector
- $\overrightarrow{D_1}, \ldots, \overrightarrow{D_K}$: security price vector at time t_1, if the market is, respectively, in state $1, \ldots, K$
- λ: Lagrangian multiplier
- $E(u(W))$: expected utility of wealth, W
- $\succ$: preference relation
- $\succeq$: weak preference relation
- β_i: CAPM - Beta of asset i

Part I

Physics concepts in social science? A discussion

1

Classical, statistical, and quantum mechanics: all in one

This chapter provides for a very short course on classical (Newtonian as well as statistical) mechanics and quantum mechanics. Readers who have not been trained in physics will be able to gain basic knowledge of the main physical theories developed during the last 400 years, with the inclusion of some of the interpretational problems of these theories.

1.1 Newtonian mechanics

We discuss one of Newton's laws, namely Newton's second law: "the product of mass and acceleration is equal to the force" or in mathematical symbols:

$$ma = f. \tag{1.1}$$

We state that m is the mass of a particle, a is its acceleration, and f is the force acting on the particle.

Newton also introduced the notion of a continuous (infinitely divisible) physical space which was used to describe the dynamics of a particle. Here we can also mention the contribution of Leibniz. However, the rigorous mathematical formalization of the real continuum was done much later, at the end of the nineteenth century. Physical space was represented by the mathematical model $\mathbb{R}^3$, the Cartesian product space $\mathbb{R} \times \mathbb{R} \times \mathbb{R}$ of three real lines. In this mathematical model, Newton's second law can be formalized in the following way. Let us introduce the following notations. Let $q = (x, y, z)$, be a three-dimensional vector, where x, y, z are the particle's coordinates:

$$v = \frac{dq}{dt} = \left(\frac{dx}{dt}, \frac{dy}{dt}, \frac{dz}{dt}\right) \tag{1.2}$$

is the vector of velocity, and, finally:

$$a = \frac{dv}{dt} = \left(\frac{d^2x}{dt^2}, \frac{d^2y}{dt^2}, \frac{d^2z}{dt^2}\right) \tag{1.3}$$

is the vector of acceleration. The dynamics of a particle, $t \to q(t)$ (its trajectory in physical space), is described by the ordinary differential equation:

$$m\frac{d^2q(t)}{dt^2} = f(t, q(t)). \tag{1.4}$$

To find the particle's dynamics, we also have to know the initial conditions:

$$q(t_0) = q_0,\ v(t_0) = \frac{dq}{dt}(t_0) = v_0, \tag{1.5}$$

where q_0 and v_0 are respectively the particle's position and velocity at the initial instant of time t_0.

An important class of forces is given by the so-called "conservative forces." We start with the one-dimensional model, a particle on the line: there is only one coordinate x. We state that a force $f(x)$ is called conservative if there exists a real valued function $V(x)$ (which is called the potential) such that:

$$f(x) = -\frac{dV}{dx}. \tag{1.6}$$

The potential $V(x)$ represents the potential energy of a particle (see for instance the next chapter for the use of such potential in a social science setting). To illustrate this notion, consider a basic example which also plays an important role in quantum mechanics. The harmonic oscillator is a well-known object in classical mechanics. The restoring force f which is proportional to the displacement of a particle from its equilibrium position is:

$$f(x) = -kx, \tag{1.7}$$

where k is a positive constant. This is a conservative force with the potential $V(x) = kx^2/2$. This is the equation of the standard parabola indicating the fact that the potential energy has its minimum at the point $x = 0$. Newton's second law for the system is:

$$m\frac{d^2x}{dt^2} = -kx. \tag{1.8}$$

Solving this ordinary differential equation, we find that the motion is described by the function:

$$x(t) = A\cos(2\pi\nu t + \phi), \tag{1.9}$$

where:

$$\nu = \frac{1}{2\pi}\sqrt{\frac{k}{m}} = \frac{1}{T}. \tag{1.10}$$

Remark that ν indicates the frequency expressed as the number of cycles per time unit. Clearly, as is intuitive, the force constant k and the mass m influence this frequency. The position $x(t)$ depends on this frequency ν but also on the amplitude A and phase ϕ. They can be found from the system of equations:

$$A\cos\phi = x_0, \quad A\sin\phi = -v_0/2\pi\nu. \tag{1.11}$$

In the case of a particle in the three-dimensional case, the force f is a vector $f = (f_x, f_y, f_z)$. It is called conservative if there exists a (real) potential $V(q)$, $q = (x, y, z)$, such that $f_x = -\frac{\partial V}{\partial x}$, $f_y = -\frac{\partial V}{\partial y}$, $f_z = -\frac{\partial V}{\partial z}$. We also recall the notion of the gradient of a function V. This is a vector composed of its partial derivatives and it is denoted as ∇V. Hence, a conservative force can be represented as the "negative gradient" of the potential:

$$f = -\nabla V. \tag{1.12}$$

Although in this book we try to minimize mathematical details as much as possible, we need to point out the theorem of the existence and uniqueness of the solution of the equation (1.4) with the initial conditions (1.5). Such a problem, i.e. an equation with initial conditions, is called the *Cauchy problem.* This is one of the basic mathematical problems of classical mechanics. The simplest version of the aforementioned theorem is that if the force is described by a smooth function f, i.e. differentiable and with continuous derivative, and the derivative is bounded, i.e. there exists a constant $c > 0$ such that, for every $q \in \mathbb{R}^3$, $|f'(q)| \leq c$, then, for any pair (q_0, v_0), a unique solution of the Cauchy problem exists (1.4), (1.5). This mathematical theorem was the main source of the *causal deterministic viewpoint* to classical mechanics: if we know the position and velocity of a particle at $t = t_0$, then we can find them at any instant of time $t > t_0 : q = q(t), v = v(t) = \frac{dq(t)}{dt}$.

Consider the following quote by Laplace [1]:

> We ought to regard the present state of the universe as the effect of its antecedent state and as the cause of the state that is to follow. An intelligence knowing all the forces acting in nature at a given instant, as well as the momentary positions of all things in the universe, would be able to comprehend in one single formula the motions of the largest bodies

as well as the lightest atoms in the world; provided that its intellect were sufficiently powerful to subject all data to analysis; to it nothing would be uncertain, the future as well as the past would be present to its eyes.

Later interpretations of quantum mechanics also leave the theoretical possibility of such a super intellect contested.

This is a good example of how pure mathematics generates fundamental philosophic principles. As it often happens in science, it is not easy to change philosophic principles which have been established on the basis of some special mathematical results and models. During Laplace's lifetime, the theory of differential equations had not yet been well developed. Nowadays, it is well known that the Cauchy problem (1.4), (1.5) may have a non-unique solution even for continuous forces. If f is smooth, then the solution is unique only locally, i.e. for a small neighborhood of the point (t_0, x_0). However, globally it can be non-unique. Hence, modern mathematics does not imply determinism even in classical mechanics (see [2] for usage of this argument in classical non-deterministic biological dynamics). We also remark that if the dynamics of a particle is even deterministic, but unstable, then a small disturbance of initial conditions, can change crucially the trajectory of such a particle. In such a case, although the principle of determinism is formally valid, it has no usage in real practice, since it is impossible to determine initial conditions with infinite precision. This argument against the uncontrollable usage of the principle of determinism in classical mechanics was presented by Blohinzev [3] in his comparison of classical and quantum mechanics. In conclusion, we can see from the above that Laplace's causal determinism is indeed a mere prejudice.

Besides Laplace's prejudice, we can also mention the Kantian prejudice which says that physical space has to be identified with its Euclidean model [4]. This prejudice was based on two-thousand years of Euclidean geometry. The first blow to the Kantian views of physical space was given by Lobachevsky. However, the genius of Einstein was needed to establish modern views of the geometry of physical space.

The above discussion raises a reasonable recommendation: the reader may want to veer close to mathematics and instead steer away from general physical, metaphysical, and philosophic principles.

1.2 References

[1] See http://plato.stanford.edu/entries/determinism-causal and the reference contained therein, cited as: Laplace, P. S. (1820). *Essai philosophique sur les probabilités* forming the introduction to his *Théorie analytique des probabilités*, Paris: V. Courcier repr. Truscott, F. W. and Emory, F. L. (1951). *A Philosophical Essay on Probabilities.* Dover, New York.

[2] Conte, E., Federici, A., Khrennikov, A., and Zbilut, J. P. (2004). Is determinism the basic rule in dynamics of biological matter? *Proceedings of Quantum Theory: Reconsideration of Foundations. Series in Mathematical Modelling in Physics, Engineering and Cognitive Science*, 10, 639–678, Växjö University Press, Sweden.
[3] Blochinzev, D. I. (1976). *Elementary Quantum Mechanics.* Nauka, Moscow (in Russian).
[4] Kant, I. (2008). *The Critique of Pure Reason.* Penguin Classics, London.

1.3 The Hamiltonian formalism

To proceed from classical to quantum mechanics, one typically uses the Hamiltonian formalism for the description of the motion of classical particles. As usual, let us introduce the momentum $p = mv$ of a particle and consider *phase space* with coordinates (q, p), where q is position. Points of the phase space are interpreted as states of classical particles. We state again that, by Newton's second law, to determine the trajectory of a particle it is necessary to know both initial position q_0 and the velocity v_0. In particular, knowledge of only position is not sufficient. Therefore, it is natural to define the *particle's state* as the pair (q, v). By scaling the velocity by the particle's mass, we introduce its momentum, p, and equivalently we represent the particle's state as a pair (q, p).

We remark that the momentum's definition can be expressed in the form of an ordinary differential equation:

$$\frac{dq}{dt} = \frac{p}{m}. \tag{1.13}$$

Hence, Newton's second law, (1.4), can be written as:

$$\frac{dp}{dt} = -\frac{dV}{dq}. \tag{1.14}$$

Let us introduce the following function on the phase space:

$$\mathcal{H}(q, p) = \frac{p^2}{2m} + V(q). \tag{1.15}$$

$\mathcal{H}(.,.)$ is called the *Hamiltonian function*. This is the total energy of a particle which moves under the action of the force induced by the potential V and the kinetic energy $\frac{p^2}{2m}$. The system of equations (1.13), (1.14) can be written as:

$$\frac{dq}{dt} = \frac{\partial \mathcal{H}}{\partial p}, \; \frac{dp}{dt} = -\frac{\partial \mathcal{H}}{\partial q}. \tag{1.16}$$

This is the system of Hamiltonian equations. It is easy to prove that the energy is preserved in the process of motion:

$$\mathcal{H}(q(t), p(t)) = \mathcal{H}(q(t_0), p(t_0)). \tag{1.17}$$

To prove this important fact (the law of energy conservation), it is sufficient to use the basic rule for the differentiation of a composition of functions and then to apply this rule to the system of Hamiltonian equations.

By using the Hamiltonian formalism, we can formulate a feature of classical mechanics, which can be called *locality.* Let us consider a system consisting of N particles with the three-dimensional coordinates $q_j = (x_j, y_j, z_j)$, $j = 1, \ldots, N$ and corresponding momenta p_j. The Hamiltonian function of a system of N particles with masses m_j moving in the potential $V(q_1, \ldots, q_N)$ has the form:

$$\mathcal{H}(q, p) = \sum_{j=1}^{N} \frac{p_j^2}{2m_j} + V(q), \tag{1.18}$$

where $q = (q_1, \ldots, q_N)$, $p = (p_1, \ldots, p_N)$. The above Hamiltonian gives the total energy of this system composed of N particles. The system of Hamiltonian equations describing the dynamics of this composite system can be written as:

$$\frac{dq_j}{dt} = \frac{\partial \mathcal{H}(q, p)}{\partial p_j}, \quad \frac{dp_j}{dt} = -\frac{\partial \mathcal{H}(q, p)}{\partial q_j}, \quad j = 1, \ldots, N. \tag{1.19}$$

Within the potential V, the interaction between different particles is described by terms containing coordinates of a few particles. We can consider the interaction between particles by writing for instance terms of the form $q_i \ldots q_N$ (various products of different coordinates). But let us consider now a potential which does *not* contain interaction terms, $V(q) = V_1(q_1) + \cdots + V_N(q_N)$. The corresponding system of Hamiltonian equations is:

$$\frac{dq_j}{dt} = \frac{p_j}{m_j}, \quad \frac{dp_j}{dt} = -\frac{\partial V_j}{\partial q_j}, \quad j = 1, \ldots, N. \tag{1.20}$$

This is a system of N-independent equations.

Hence, an important principle emerges from our discussion so far: *Hamiltonian mechanics is local, i.e. in the absence of interaction between particles, such particles move independently of each other.*

We remark that non-local motion, as is the case with for instance Bohmian mechanics (see Chapter 6), has the following (paradoxical from the viewpoint of our classical intuition) feature. In the absence of interaction, even for $V \equiv 0$, the dynamics of different particles are *dependent* on each other. Changing the state of one particle (q_j, p_j) induces changing the states (q_i, p_i), $i \neq j$, of other particles. In the classical world, we have never seen such a behavior of physical systems.

Let us introduce a mathematical tool which has a key role in the Hamiltonian formalism. The *Poisson bracket* of two functions on the N-particle phase space,

$f(q, p), g(q, p)$, is defined as:

$$\{f, g\} = \sum_{j=1}^{N} \Big(\frac{\partial f(q, p)}{\partial q_j} \frac{\partial g(q, p)}{\partial p_j} - \frac{\partial f(q, p)}{\partial p_j} \frac{\partial g(q, p)}{\partial q_j} \Big). \tag{1.21}$$

As an example, consider functions $f(q, p) = q_j, g(q, p) = p_j$. Then:

$$\{q_j, p_j\} = 1, \ \{q_j, p_k\} = 0, \ j \neq k. \tag{1.22}$$

$$\{q_j, q_k\} = 0, \ \{p_j, p_k\} = 0. \tag{1.23}$$

By using the Poisson bracket, we rewrite the system of Hamiltonian equations as:

$$\frac{dq_j}{dt} = \{q_j, \mathcal{H}\}, \ \frac{dp_j}{dt} = \{p_j, \mathcal{H}\}. \tag{1.24}$$

This form of the Hamiltonian dynamics will be used to proceed from classical Hamiltonian mechanics to quantum mechanics.

1.4 Statistical mechanics and the Liouville equation

In studying the dynamics of an ensemble of a huge number, say N particles, the presence of the system of Hamiltonian equations plays merely a methodological role. From as early as the nineteenth century until the 1960s, it was simply impossible to solve this system for large N and non-trivial potentials. Nowadays in principle one can solve it numerically and obtain millions of trajectories in the phase space. However, it is not clear how one can use or visualize the results of such computations. Already in the nineteenth century it was proposed that instead of studying the trajectories of individual particles, it would be better to consider the probability to find a particle in some domain, say W, of the phase space. Such an approach meant in effect a move away from the deterministic description of mechanics to a statistical description. Hence, the name *statistical mechanics* was coined to denote this particular area of study.

Let us consider the phase space of the system of N particles, $\mathbb{R}^{2N}$, with points (q, p), where $q = (q_1, \ldots, q_N), p = (p_1, \ldots, p_N)$. What is the probability density function which indicates the probability to find the first particle at point q_1 with momentum p_1, the second particle at point q_2 with momentum $p_2, \ldots,$ the Nth particle at q_N with momentum p_N? Since momenta are mass scalings of velocities, the question can be reformulated as: "What is the probability density function of the first particle at point q_1 with velocity v_1, the second particle at point q_2 with velocity $v_2, \ldots,$ the Nth particle at q_N with velocity v_N?" We state that mathematically a probability density is a function $\rho(q, p)$ which is non-negative and normalized by 1:

$$\int_{\mathbb{R}^{2N}} \rho(q, p) dq dp = 1. \tag{1.25}$$

The probability to find a particles at a point (q, p) from the domain W of the phase space is calculated with the aid of the probability density function:

$$P((q, p) \in W) = \int_W \rho(q, p) dq dp. \tag{1.26}$$

A fundamental problem now consists to describe the dynamics of the (time-dependent) probability density, $t \to \rho(t, q, p)$. Fortunately, the *Liouville equation* gives us the answer:

$$\frac{\partial \rho(t,q,p)}{\partial t} = \{\mathcal{H}(q, p), \rho(t, q, p)\}, \tag{1.27}$$

$$\rho(t_0, q, p) = \rho_0(q, p). \tag{1.28}$$

Here, we remark:

$$\{\mathcal{H}, \rho\} = \sum_{j=1}^{N} \Big(\frac{\partial \mathcal{H}}{\partial q_j} \frac{\partial \rho}{\partial p_j} - \frac{\partial \mathcal{H}}{\partial p_j} \frac{\partial \rho}{\partial q_j} \Big). \tag{1.29}$$

Hence, if the initial probability density function is known, then (by solving the Liouville equation) we can find the density at any instant of time. We remark that the usage of the probabilistic description of an ensemble of N particles does not contradict the existence of the deterministic Hamiltonian dynamics.

We do not provide the formal derivation of the Liouville equation. Instead of this, we present physical arguments which lie behind this equation. Consider the probability density function on a trajectory $(q(t), p(t))$ in the phase space $\rho(t, q(t), p(t))$. We calculate its total derivative with respect to time:

$$\frac{d\rho}{dt} = \frac{\partial \rho}{\partial t} + \sum_{j=1}^{N} \left(\frac{\partial \rho}{\partial q_j} \frac{dq_j}{dt} + \frac{\partial \rho}{\partial p_j} \frac{dp_j}{dt} \right). \tag{1.30}$$

We now use the system of Hamiltonian equations (1.19) (this describes the trajectory in the phase space) to express the time derivatives of coordinates and momenta and obtain:

$$\frac{d\rho}{dt} = \frac{\partial \rho}{\partial t} + \sum_{j=1}^{N} \Big(\frac{\partial \rho}{\partial q_j} \frac{\partial \mathcal{H}(q, p)}{\partial p_j} - \frac{\partial \rho}{\partial p_j} \frac{\partial \mathcal{H}(q, p)}{\partial q_j} \Big).$$

Then the Liouville equation is equivalent to the following statement: the distribution function is constant along any trajectory in phase space, i.e. $\frac{d\rho}{dt} = 0$.

1.5 The classical kingdom . . .

At the end of the nineteenth century, it was widely believed that *all* physical processes can (at least in principle) be described with the aid of classical statistical mechanics.[1] The two fundamental principles of classical mechanics, namely *determinism* and *locality*, had never been doubted. No experimental data were available to disprove whether these principles actually existed.

The elaboration of a very consistent picture of natural phenomena was emerging. Here are some examples. A previous state determines coming states. A change of a state could happen only through the action of some force. For a composite system consisting of a few particles and where interaction is absent, a change of the state of any particle could not change the states of other particles. Such *philosophical* views of natural phenomena are known under the term of *local realism.* Here "realism" denotes the realism of states of physical systems, i.e. their positions and momenta. Their dynamics are described by transitions from one state to another.

Let us speculate on a physical world in which local realism is violated. Start with a world in which the principle of locality is violated, but the principle of determinism holds true ("non-local realism"). Consider two cars moving, e.g. in Moscow and New York. The motion of each car is deterministic. Each car has a well-defined position in space and a velocity. However, if we were to consider there exists *action at a distance,* then if, say, the car in Moscow stops near a traffic light, the car in New York may feel the effect of this stop instantaneously! Note that instantaneously means at least faster than the speed of transmission (at the speed of light) of a signal between Moscow and New York.[2] If indeed such instantaneous action were possible, then a state formed by the Moscow car stopping could affect the state of the New York car.

Consider now a local world in which the principle of determinism is violated, i.e. the previous state of a system does not determine uniquely successive states. A car moving in Moscow can appear with some probability in any place of the world. In all cases, the initial position and velocity of this car are the same. Hence, the initial state does not determine the car's dynamics.

We remark that physical determinism influenced some social doctrines. For example, the philosophical basis of Marxism–Leninism is known under the name of "historical materialism." This says that one social system will inevitably change

[1] When Max Planck finished highschool, he applied to university to study physics. A professor interviewing applicants explained to young Max that it was totally meaningless to start an education in physics. The professor was sure nothing more could be studied in physics. All was now known! Young people would only destroy their lives by starting careers in physics. Max was very talented in music and the professor recommended him to start a career in music. See [1].

[2] The optic fiber signal speed can be close to the speed of light. Therefore, in the case of financial transactions for instance, transmission speeds can attain close to speed of light transmission.

another, capitalism → socialism → communism. Even Sigmund Freud's psychological determinism was created under the influence of classical mechanics. In [2], we can read about Freud's psychoanalysis: "What is attractive about the theory, even to the layman, is that it seems to offer us long sought-after and much-needed causal explanations for conditions which have been a source of a great deal of human misery. The thesis that neuroses are caused by unconscious conflicts buried deep in the unconscious mind in the form of repressed libidinal energy would appear to offer us, at last, an insight in the causal mechanism underlying these abnormal psychological conditions as they are expressed in human behavior, and further show us how they are related to the psychology of the 'normal' person" (see also [3]). Furthermore, [3] mentions that "In psychology, those, like Freud, who believe in psychic determination in psychiatry, assume that all mental events have causes. Freud believed that the existence of unconscious forces proved psychic determinism to be a fact of mental life... he regarded psychoanalysis as a science based on causal-deterministic assumptions." See [4] [5] for an attempt to combine Freudian psychological determinism and free will through the Bohmian quantum model. See also Chapter 6 for the Bohmian mechanics model.

1.6 References

[1] See http://backreaction.blogspot.com/2011/12/advent-calender-9-prof-jollys-advice.html, and the reference contained therein: Planck, M. (1933). *Wege zur physikalischen Erkenntnis* (S. Hirzel), p. 128.

[2] Internet *Encyclopedia of Philosophy*. University of Tennessee. See http://www.iep.utm.edu/freud/ (under b. The coherence of the theory).

[3] *The Encyclopedia69 Dictionary*. Please see at: http://www.encyclopedia69.com/eng/d/determinism/determinism.htm.

[4] Khrennikov, A. (2002). *Classical and Quantum Mental Models and Freud's Theory of Unconscious Mind. Series in Mathematical Modelling in Physics, Engineering and Cognitive Sciences*. Växjö University Press, Sweden.

[5] Khrennikov A. (2004). *Information Dynamics in Cognitive, Psychological, Social, and Anomalous Phenomena. Series: Fundamental Theories of Physics*. Kluwer, Dordrecht.

1.7 Classical fields

The completion of the theory of classical mechanics of particles was performed via the introduction of the notion of the classical mechanics of fields, i.e. *classical field theory* – the local deterministic theory of the electromagnetic field. The field's state (at the instant of time t) is given by the vector:

$$\phi(t, x, y, z)) = (E(t, x, y, z), B(t, x, y, z)), \tag{1.31}$$

where:

$$E(t,x,y,z) = (E_1(t,x,y,z), E_2(t,x,y,z), E_3(t,x,y,z)) \tag{1.32}$$

and:

$$B(t,x,y,z) = (B_1(t,x,y,z), B_2(t,x,y,z), B_3(t,x,y,z)) \tag{1.33}$$

are electric and magnetic fields, respectively. The dynamics of the electromagnetic field also can be described by the Cauchy problem (i.e., a dynamical equation combined with initial conditions):

$$\frac{\partial \phi(t,x,y,z)}{\partial t} = L(\phi(t,x,y,z)),\ \phi(t_0,x,y,z) = \phi_0(t,x,y,z), \tag{1.34}$$

where L is a differential operator of the first order with partial derivatives with respect to coordinates. At the moment, its form is not important. This operator was found by Maxwell. We remark that *the system of Maxwell equation (1.34) can be written as a Hamiltonian system with respect to field components* [1]. The electric component plays a role of position, $q(t,x,y,z) \equiv E(t,x,y,z)$, and the magnetic component plays a role of momentum, $p(t,x,y,z) \equiv B(t,x,y,z)$. What is the phase space of this field system? The energy of the electromagnetic field (at the fixed instant of time) is given by the integral:

$$\mathcal{E}(E,B) = \int_{\mathbb{R}^3} (E^2(x,y,z) + B^2(x,y,z))dxdydz. \tag{1.35}$$

Since this integral is finite, the field's position $q(x,y,z) = E(x,y,z)$ and the field's momentum $p(x,y,z) = B(x,y,z)$ have to be square integrable (integrals of squared functions are less than infinity). By using mathematical analogy, we can see E^2 relates to a real potential – B^2 relates to kinetic energy – in the non-field setting. Denote the space of square integrable functions by L_2. Thus, the field's phase space is the Cartesian product of two L_2-spaces: $L_2 \times L_2$.

The electric and magnetic components can be combined in a single complex valued field $\phi = E + iB$. This is well known (in classical signal theory) as the *Riemann–Silberstein* representation of the classical electromagnetic field. This representation induces the complex structure on the phase space, $\phi = q + ip$, where all functions depend on spatial coordinates.[3] This is equivalent to the consideration of the space H of square integrable complex valued functions:

$$H = \left\{\phi : \int |\phi(x,y,z)|^2 dxdydz < \infty\right\}. \tag{1.36}$$

[3] This complex representation of the classical electromagnetic field was in usage before the creation of quantum mechanics. Nowadays, little attention is paid to this historical fact. The fathers of quantum mechanics, including Schrödinger, were well aware of the Riemann–Silberstein representation. Originally, the complex wave function was invented merely as an analogue of the complex (classical) electromagnetic field.

1.8 Reference

[1] Khrennikov, A. (2008). Quantum-like model for classical random electromagnetic field. *Journal of Modern Optics*, 55, 2257–2267.

1.9 The Born–Sommerfeld quantization

Contrary to what could be expected, the first steps towards edifying quantum mechanics were not revolutionary at all! We note that those first steps were not at all accompanied by a fundamental change in the philosophical foundations of science. The story started with a graph representing experimental data, and it contained a spike, which from the viewpoint of classical mechanics was quite difficult to explain. The experimental graph turned out to be about *black body radiation* and it was Max Planck who found that the spike could be explained in a classical statistical mechanics framework, but it required one novel assumption: *radiation is emitted not continuously, but by discrete portions.* Purely formally the energy space was decomposed into cells. The size of a cell depends on the frequency ν of the oscillations of the electromagnetic field. Max Planck postulated that dependence on the frequency is linear and the coefficient does not depend on the frequency:

$$\Delta E_\nu = h\nu. \tag{1.37}$$

This coefficient of proportionality was later called *Planck's constant.* Since the frequency has dimension l/time, the Planck constant is expressed in units of "energy × time." This is the dimension of the classical physical variable called *action.* This constant was measured with very good precision:

$$h \approx 6.6260693(11) \times 10^{-34} \text{ J} \times \text{sec}. \tag{1.38}$$

The decomposition of the energy space into small cells and the summation over cells is similar to the standard procedure of forming Riemann sums.[4] To calculate the Riemann integral, the size of cells has to go to zero. However, Planck did not make this last step and wrote the answer using cells of a finite size (proportional to h). We note that in quantum folklore, one can find a story that Max Planck obtained the correct answer, because he simply did not know that in order to calculate Riemann's integral one has to consider that the limit of the cell's size goes to zero.

We remark that the tool of discretization of the energy space was not as novel as it is typically presented in textbooks on quantum mechanics. It was actually rather standard in classical statistical mechanics! In particular, Boltzmann used

[4] These sums are used in the construction of the Riemann integral.

discretization $\epsilon, 2\epsilon, \ldots, n\epsilon, \ldots$, where ϵ was a "minimal quant" of energy. This is maybe the reason why the work of Planck was very welcome in the classical statistical community: nobody considered the introduction of a parameter of energy discretization as an attack against classical statistical mechanics!

The discretization parameter h ceased to be merely a parameter only after Einstein's work [1] (1905). In his work, Einstein claimed that $\Delta E_\nu = h\nu$ is not just a minimal portion of energy which can be transmitted for the frequency ν, but that even in the absence of interaction of the electromagnetic field and matter, the field is "quantized," i.e. it is split into a collection of quanta of lights. Later these quanta were called *photons*. Thus, in opposition to classical field theory, the electromagnetic field has to be decomposed into an ensemble of photons, i.e. it has corpuscular features.[5]

The next step towards quantum theory was performed on the basis of *Bohr's quantization condition.* We want to explicitly state that in classical Hamiltonian mechanics, the energy is preserved on each trajectory (see (1.17)). Suppose now that there exist constraints (of an unknown nature) which forbid some motions and some trajectories, and that the system can move only via a discrete set of trajectories. Denote those possible trajectories (consistent with the constraints) by:

$$\gamma_1, \ldots, \gamma_n, \ldots \tag{1.39}$$

Since the energy is constant on each of them, we obtain a discrete set of possible energies:

$$E_1 = E(\gamma_1), E_2 = E(\gamma_2), \ldots, E_n = E(\gamma_n), \ldots \tag{1.40}$$

This idea was explored by Niels Bohr in his model of the atom. It was known from experiments that atoms can emit and absorb energy only by quantized portions. Bohr proposed a model describing this feature of atoms. In this model, the electron is a classical-like particle which moves around the nucleus. However, in such a motion a purely classical charged particle would continuously emit radiation and lose energy. Finally, it would fall onto the nucleus. This was one of the main unsolved problems of classical electrodynamics. Bohr postulated that an electron can move only on a discrete set of orbits (1.39) and hence its energy can take only a discrete series of values (1.40). Since an electron can only jump from one orbit to another, the atom can emit (absorb) only discrete portions of energy $\Delta E_{ij} = E_i - E_j$. To match experimental data, Bohr postulated that the frequency

[5] There is a piece of irony in this story. Although Albert Einstein introduced quanta of light and hence in this way he made the first step towards modern quantum theory, later (in the 1920s) he gave up and until the end of his life he worked to create a classical field theory which would describe quantum phenomena.

ν of emitted (absorbed) electromagnetic radiation is determined by Planck's law:

$$\Delta E_{ij} = E_i - E_j = h\nu. \tag{1.41}$$

Since these frequencies were known from experiment, he could find energy spacing in the atom. Bohr was also able to "derive" energy spacing even theoretically and he obtained a key result which indicates that the angular momentum L of an electron is to be an integer multiple of $\hbar$

$$L = n\frac{h}{2\pi} = n\hbar, \tag{1.42}$$

where $n = 1, 2, 3, \ldots$ is the quantum number, and $\hbar = h/2\pi$.

Bohr's quantization condition (1.42) determining the electron's orbits in the atom was generalized to the famous Bohr–Sommerfeld quantization rule (which also had been postulated):

$$\int_{\gamma_n} p dq = hn, \tag{1.43}$$

where γ_n is the permitted orbit corresponding to a natural number n. In the original Bohr model, only circular orbits were permitted; in the Bohr–Sommerfeld model orbits can be elliptic.

Thus, the first step towards quantum theory was the recognition that some physical quantities, first of all energy, which were considered as continuous in classical mechanics, are *fundamentally discrete*. Discreteness by itself is less fascinating. However, the concept is fascinating, and even mystical, when it is combined with the wave features of the systems under consideration (see below). Photons are mystical not because they have corpuscular features,[6] but because these features are combined with wave behavior. As was shown in the famous experiment on the interference of quantum light (see Chapter 5, Section 5.3), photons did not lose their wave features. They interfere as usual waves. "Quantumness" was exhibited by the detection procedure. As a consequence of the discreteness of energy in experiments on quantum interference, one registers not intensities of signals and interference of these intensities, but rather clicks of detectors which are "eating" discrete portions of energy. The probability density of the number of clicks presents the interference picture similar to the ordinary wave interference.

[6] Already Newton invented corpuscles of light.

1.10 Reference

[1] Einstein, A. (1905). Über einen die Erzeugung und Verwandlung des Lichtes betreffenden heuristischen Gesichtspunkt. *Annalen der Physik*, 4, 17, 132–148.

1.11 Theory of quantum waves

The next step of great importance was performed by Louis de Broglie who *assigned the wave length to any quantum particle*.[7] Thus, not only a photon has a wave length (and hence a frequency),[8] but even an electron or a neutron, i.e. a massive particle. Later, it was shown experimentally that de Broglie's conjecture was correct and even massive micro-systems can demonstrate non-trivial interference. As in the case of photons, this is interference of the number of discrete counts of detectors, i.e. interference encoded in the probability density.

The natural question arises, e.g. [1], how big (in size and mass) can systems be to exhibit wave properties, e.g. interference? The theoretical formulation of quantum mechanics cannot provide the definite answer to this question. Bohr emphasized that the laws of quantum mechanics are valid only at scales related to the fundamental quant of action – the Planck constant. He had in high esteem the role of the *correspondence principle*, which established the coupling between classical and quantum mechanics. This principle is discussed in detail in Chapter 5, Section 5.15. This principle (which, in fact, was and still remains very vaguely formulated) says that if $h \to 0$, i.e. the Planck constant is negligibly small compared with the action scale (time$\times$energy) of systems under consideration, then quantum features (such as interference) disappear and classical mechanics becomes applicable. However, Bohr's correspondence principle has never been really justified. Although an advanced mathematical formalism related to this principle was developed, e.g. [2], its physical meaning is still the subject of many discussions. Therefore, it should not come as a surprise to witness the fact that high-level researchers in quantum mechanics have claimed that wave features can be exhibited by macroscopic massive bodies. For instance, Nobel prize winner A. Leggett expects the creation of Schrödinger's cat-like states – i.e. macroscopic superpositions. This came after a discussion held following a talk from one of the authors of this book.[9] A. Zeilinger performed famous experiments on the interference of

[7] de Broglie interpreted "quantum waves" as physical waves. Later he elaborated the double solution theory by which quantum particles appeared as singularities in quantum waves. The former were driven by the latter. De Broglie's double solution theory was the first step towards modern pilot wave theory and Bohmian mechanics. This is a theory which combines corpuscular and wave features of quantum systems in a realistic manner. See also Chapter 6 where we discuss the basics of Bohmian mechanics.

[8] This was not surprising.

[9] This refers to the talk given by A. Khrennikov on May 3, 2004, at the Beckman Institute of the University of Illinois, Urbana Champain.

macro-molecules [3]. Recently, De Martini created macroscopic photonic clouds in states of superposition and, moreover, he created entangled macroscopic clouds of photons [4]. However, all these studies are far from fully clarifying! Indeed, they suffer from interpretational problems. We need to indicate that superposition is also a fundamental feature of *classical* macroscopic waves. Thus, superposition by itself is not a unique feature of quantum mechanics. We note that the superposition attains a quantum character when it is combined with the discreteness of measurement results. Interference of continuous quantities, such as intensities of classical radio signals, cannot be considered as a quantum feature. However, the interference of discrete clicks of detectors is really a quantum phenomenon. To check the property of "quantum superpositions" is a difficult problem. For example, in experiments of De Martini [4] discreteness is created by hand, through threshold type measurements of the intensity of the electromagnetic field: if $I > \epsilon$ the result is $+1$, if $I < \epsilon$ it is -1. However, as was shown by Adenier and one of the authors of this book [5], such "threshold generated" combination of discreteness and interference can be obtained even for classical electromagnetic signals.

Finally, quantum theory obtained a solid mathematical base via Schrödinger's dynamical equation (see also Section 1.20 and Chapter 5). In a similar way as de Broglie, Schrödinger considered waves which were associated with micro-systems as real physical waves. He derived an evolutionary equation for the wave associated with a massive particle, e.g. a non-relativistic electron, moving under the action of an arbitrary potential $V(q)$.

1.12 References

[1] Khrennikov, A. (2009). *Contextual Approach to Quantum Formalism*. Springer Verlag, Berlin, Heidelberg.
[2] Khrennikov, A. (1999). *Superanalysis*. Kluwer, Dordrecht.
[3] Zeilinger, A. (2010). *Dance of the Photons: From Einstein to Quantum Teleportation*. Macmillan; Farrar, Straus & Giroux, New York.
[4] De Martini, F., Sciarrino, F., and Vitelli, Ch. (2008). Schrödinger cat: entanglement test in a micro-macroscopic system. *Physical Review Letters*, 100, 253601.
[5] Adenier, G. and Khrennikov, A. (2007). Is the fair sampling assumption supported by EPR experiments? *Journal of Physics B: Atomic, Molecular and Optical Physics*, 40, 131–141.

1.13 Heisenberg's symbolic calculus

Somewhat before Schrödinger's work, Heisenberg started a new trend in physical science which can be called *symbolism*. He made a mathematical observation which was trivial by its very mathematical nature, but it turned out to play a fundamental role in the further development of quantum theory.

Consider the notion of the *Hermitian matrix* $A = (a_{ij})$: its elements satisfy the condition $a_{ij} = \bar{a}_{ij}$, where, for a complex number $a = a_1 + ia_2$, $\bar{a} = a_1 - ia_2$ denotes its conjugate. Heisenberg identified discrete values of physical observables with eigenvalues of Hermitian matrices. In Heisenberg's approach, all physical observables have to be represented by Hermitian matrices.

As is well known from linear algebra,[10] any Hermitian matrix of finite size, $n \times n$, can be diagonalized in the basis consisting of eigenvectors corresponding to its eigenvalues (eigenvalues are real numbers and eigenvectors are orthogonal). However, matrices in quantum theory are of infinite size. We shall explain later in the book why one cannot proceed with matrices of finite size. In such a case, some Hermitian matrices cannot be diagonalized. Besides eigenvalues, their spectra can contain a non-discrete part. It can even happen that there are no eigenvalues at all, and then such spectra are called continuous. This mathematical formalism matches the physical situation: some physical observables, such as the particle's position and momentum, are still continuous (as it is in classical mechanics).

1.13.1 Canonical commutation relations

Heisenberg performed a formal translation of classical Hamiltonian mechanics (in which observables were given by functions on the phase space) into a new type of mechanics (quantum mechanics), in which observables are represented by Hermitian matrices. He correctly noted the crucial role the Poisson brackets could play in classical formalism. They are defined for any pair of classical observables, f_1, f_2. We remark that Poisson brackets are antisymmetric $\{f_1, f_2\} = -\{f_2, f_1\}$. A natural operation on the set of Hermitian matrices corresponding to Poisson brackets is the commutator of two matrices:

$$[A_1, A_2] = A_1 A_2 - A_2 A_1, \tag{1.44}$$

defined with the aid of standard matrix multiplication. Note that the commutator is anti-symmetric as well. The transformation of the usual multiplication of functions into matrix multiplication was the great contribution of Heisenberg towards the creation of quantum formalism. Starting with the classical position and momentum observables q_j, p_k (coordinates on the phase space) and the equalities for their Poisson brackets, see (1.22), (1.23), he postulated that corresponding quantum observables denoted by $\hat{q}_j, \hat{p}_k$ (hats are used to distinguish classical and quantum

[10] Please see Chapter 4 where various linear algebra concepts are dealt with in more detail.

observables, functions and matrices) have to satisfy the following commutation relations:

$$[\hat{q}_j, \hat{p}_j] = ihI, \ [\hat{q}_j, \hat{p}_k] = 0, \ j \neq k; \tag{1.45}$$

$$[\hat{q}_j, \hat{q}_k] = 0, \ [\hat{p}_j, \hat{p}_k] = 0, \tag{1.46}$$

where I is the unit matrix. These commutation relations were called the *canonical commutation relations.* The appearance of i in the non-trivial commutator is not surprising. One cannot simply borrow the relation (1.22) and put the unit matrix I, instead of the constant function $f \equiv 1$, on the phase space. Take two Hermitian matrices A_1 and A_2 and form their commutator $B = [A_1, A_2]$. The latter is a skew-Hermitian, i.e. its elements satisfy the condition $b_{ij} = -\bar{b}_{ji}$. However, the unit matrix is simply Hermitian; by multiplying it by i, we obtain a skew-Hermitian matrix iI. The Planck constant h in (1.45) plays a role of scaling factor (the scale of energy for micro-systems under consideration). One can say that Heisenberg introduced a *non-commutative phase space.*

1.13.2 Schrödinger's representation

Later Schrödinger found a concrete representation for the quantum observables of position and momentum, and nowadays authors of textbooks typically simply start with this (Schrödinger) representation. However, this may give the impression that Schrödinger's concrete choice of the operators of position and momentum $\hat{q}_j, \hat{p}_j$ played a primary role in the derivation of the canonical commutation relations. This was not the case. As was pointed out, Heisenberg really started with classical Poisson brackets on the phase space variables. As was mentioned in Section 1.11, Schrödinger considered the wave function of a quantum system as a real physical wave. Therefore, we work in the space $L_2(\mathbb{R}^3)$ of square integrable complex valued functions $\psi : \mathbb{R}^3 \to \mathbb{C}$, i.e.:

$$\int_{\mathbb{R}^3} |\psi(q)|^2 dq < \infty. \tag{1.47}$$

For a shorter notation, one can set $H = L_2(\mathbb{R}^3)$. This is a complex Hilbert space with the scalar product:

$$\langle \psi_1, \psi_2 \rangle (\equiv \langle \psi_1 | \psi_2 \rangle) = \int_{\mathbb{R}^3} \overline{\psi_1(q)} \psi_2(q) dq. \tag{1.48}$$

We note that the Hilbert space is defined in Chapter 4, Section 4.3. Contrary to Heisenberg, Schrödinger worked with operators (and not matrices), i.e. he considered a more general framework. For simplicity, we consider here the one-dimensional case, i.e. of a particle moving on the real line. Here $H = L_2(\mathbb{R})$. He introduced the operators of position and momentum:

$$\hat{q}\psi(q) = q\psi(q), \; \hat{p}\psi(q) = -i\hbar\frac{\partial\psi(q)}{\partial q}. \tag{1.49}$$

The result of the action of the position operator on the square integrable function $\psi(q)$ (vector in H) leads to another function, $q \to q\psi(q)$. In the same way, the momentum operator $\hat{p}$ transforms $\psi(q)$ into its derivative, up to the constant factor $-i\hbar$. Formally, we can write:

$$\hat{q} = q, \; \hat{p} = -i\hbar\frac{\partial}{\partial q}. \tag{1.50}$$

We remark that H was considered by Schrödinger as the space of classical fields. See also Section 1.7 and especially (1.36). Thus, originally the appearance of the complex Hilbert space to represent operators of position and momentum was just the operator reformulation of the theory of classical fields and signals. Of course, this is correct only with respect to views relative to Schrödinger, but not at all relative to Heisenberg or Bohr.

1.13.3 Heisenberg's dynamics

Finally, Heisenberg put operators satisfying the canonical commutation relations in the system of the Hamiltonian equation (1.24), instead of the classical phase space variables. In this way, he derived the basic equations of quantum dynamics:

$$\frac{d\hat{q}_j}{dt} = \frac{i}{h}[\hat{q}_j, \widehat{\mathcal{H}}], \; \frac{d\hat{p}_j}{dt} = \frac{i}{h}[\hat{p}_j, \widehat{\mathcal{H}}], \tag{1.51}$$

where $\widehat{\mathcal{H}}$ is a Hermitian matrix representing the energy observable. It is called the *quantum Hamiltonian.* Thus, classical and quantum dynamics have the same form, cf. (1.24) and (1.51). However, the variables have different physical meanings and mathematical representations. In (1.24), the "position" and "momentum" are real valued functions; in particular they commute. In (1.51), the "position" and "momentum" are Hermitian operators (which can be represented by infinite Hermitian matrices); they satisfy the canonical commutation relations. In classical mechanics, the position and momentum are interpreted as objective properties of systems. In quantum mechanics, these are observables which cannot be considered as objective properties of quantum systems.

1.13.4 Quantization procedure

How can one find such a Hamiltonian? Consider the classical Hamilton function of the form (see also Section 1.3):

$$\mathcal{H}(q, p) = \frac{p^2}{2m} + V(q), \tag{1.52}$$

where V is a polynomial function, e.g. $V(q) = kq^2$ and k is a real constant. Then one can construct the corresponding Hermitian matrix, i.e the quantum observable, by formally putting matrices $\hat{q}$ and $\hat{p}$ in the classical Hamiltonian function, instead of classical variables. For the aforementioned potential, we obtain:

$$\widehat{\mathcal{H}} = \frac{1}{2m}\hat{p}^2 + k\hat{q}^2. \tag{1.53}$$

If the potential is not a polynomial, then the mathematics are more complicated. The operator theory and Schrödinger's representation of the canonical commutation relations have to be used, see Section 1.13.1. The main problem arises for classical observables, functions on the phase space, which contain products of position and momentum variables, e.g. $f(q, p) = qp$. In principle, we can form a family of matrix expressions corresponding to this function, $\hat{f} = \alpha\hat{q}\hat{p} + \beta\hat{p}\hat{q}$, where α, β are real numbers, $\alpha + \beta = 1$. However, we obtain the Hermitian matrix only for $\alpha = \beta = 1/2$, $\hat{f} = (\hat{q}\hat{p} + \hat{p}\hat{q})/2$. This rule is known as the *Weyl quantization*.[11]

If one uses the Schrödinger's representation of the canonical commutation relations, i.e. (1.50), the quantum Hamiltonian, (1.52), can be written as:

$$\widehat{\mathcal{H}} = -\frac{\hbar^2}{2m}\frac{\partial^2}{\partial q^2} + V(q),$$

where $V(q)$ is the operator of multiplication by the function $V(q)$ in the space of square integrable functions, namely $\psi(q) \to V(q)\psi(q)$. Thus, in Schrödinger's representation all Hamiltonians are simply partial differential operators.

In general, quantization has the meaning of a transition from functions on the phase space, $f(q, p)$, to Hermitian matrices or more generally operators, by using operators of position and momentum, instead of corresponding classical variables. It is important to stress that there is no "reasonable explanation" for such a formal procedure which is required when transiting from classical physical quantities to quantum physical quantities. However, it works well.

We remark that one of the problems in the application of quantum formalism to social sciences (such as economics, finance, psychology, and cognitive science) is that, roughly speaking, we do not have classical (such as Newtonian or Hamiltonian)

[11] Let $f(q, p) = q^2 p$. Find the corresponding matrix representation.

mechanics. In other words, we do not have classical quantities which we can automatically quantize. Therefore, the majority of quantum-like quantities used in the aforementioned domains of science are phenomenological. They are invented by analogy with quantum theory. Hence, and we need to emphasize this, we do not start with a classical model and then quantize it, but we directly mimic the quantum approach. This forms one of the important problems of the quantum-like approach. One possible solution of this problem consists in using phenomenological Hamiltonians.

1.14 Heisenbergian symbolism in physics: a version of symbolism in art

Heisenberg's approach to micro-phenomena was really symbolic (operational). Heisenberg was not able to present physical reasons (in the classical meaning) for the introduction of matrices, instead of functions on the phase space. His calculus was useful to encode observed energy levels in the spectra of Hermitian matrices, but nothing more.[12] Nevertheless, this symbolic approach has been very fruitful and it played a key role. The history of the creation of quantum mechanics is an interesting subject for some social scientists. The main positive impact of Heisenberg's symbolic approach was the novelty in the description of physical phenomena. In fact, this was not a detailed and realistic description as in classical physics, but instead a fuzzy (operational) description of results of measurements. We should also mention Bohr's contribution who emphasized the role of the so-called *experimental context*. For him, it was meaningless to speak about an object outside of the concrete measurement context. The main negative impact was the aggressive anti-classical attitude. From the very beginning, Heisenberg claimed that his symbolic (operational) description of experimental data for micro-systems could not be derived on the basis of a finer classical-like model of micro-reality. Moreover, he and Bohr strongly advertised the viewpoint Mach held that it is meaningless even to try to create such models, since such an activity belongs to the domain of metaphysics and not real physics.[13] One could say that Heisenberg and Bohr were not correct, since a number of "prequantum" (classical-like) models reproducing results of quantum experiments have been created, for example Bohmian mechanics (see Chapter 6) and stochastic electrodynamics. However, it is clear that Heisenberg and Bohr would not agree with such a statement, as for them "prequantum models" are metaphysical.

The behavior and writings of Heisenberg and especially Bohr remind us very much of the manifestos of symbolism and futurism. Please see [1] [2].

[12] We note that to find theoretically these levels, one has to use Schrödinger's approach.

[13] This is a good place to recall that Mach intensively attacked Boltzmann by claiming that, since molecules are not observable (as was the case at that time), they are metaphysical creatures and hence they have to be excluded from real physical theory. Mach's attacks against Boltzmann's realism may have played a role in Boltzmann's tragic death.

We also mention works of Plotnitsky [3] [4] who analyzed similarities between "romanticism" in literature and "romanticism" in Heisenberg's and Bohr's writing see also Khrennikov [5] [6]. Plotnitsky also gave talks on Beckett and quantum mechanics and on Blake and quantum epistemology. For example, he recently presented his views at an interdisciplinary workshop (philosophy, finance, psychology, quantum physics) at the University of Leicester, UK, November 2011.

1.15 References

[1] Balakian, A. (1967). *The Symbolist Movement: A Critical Appraisal.* Random House, New York.
[2] Martin, M. W. (1978). *Futurist Art and Theory.* Hacker Art Books, New York.
[3] Plotnitsky, A. (1995). All shapes of light: the quantum mechanical Shelley. In *Shelley: Poet and Legislator of the World.* Eds. Curran, S. and Bennett, B., Johns Hopkins University Press, Baltimore, USA.
[4] Plotnitsky, A. (2004). In principle observable: Werner Heisenberg's discovery of quantum mechanics and romantic imagination. *Parallax*, 10, N3, 20–35.
[5] Khrennikov, A. Yu. (2002). *Växjö Interpretation of Quantum Mechanics.* Please see: http://arxiv.org/abs/quant-ph/0202107 and also http://arxiv.org/abs/quant-ph/0401072, Cornell University.
[6] Khrennikov, A. Yu. (2002). On foundations of quantum theory. In *Quantum Theory: Reconsideration of Foundations.* Ed. Khrennikov, A. *Series in Mathematical Modelling*, 2, 163–196, Växjö University Press, Växjö, Sweden.

1.16 Completeness of quantum mechanics and a possibility to apply quantum mechanics outside of physics

Heisenberg was the first who claimed that a finer description of micro-phenomena than that given by the matrix representation of physical observables is in principle impossible. His claim was supported by Bohr who later elaborated the principle of *completeness of quantum mechanics*. The main Heisenberg argument was based on the uncertainty relation:

$$\Delta x \Delta p \geq h, \tag{1.54}$$

which is related to the first equality in the canonical commutation relations (1.45). Heisenberg was sure that, because of the uncertainty relation, the precise position and momentum cannot be jointly assigned to a quantum system, e.g. to an electron. Therefore, it is impossible to create a finer description of quantum phenomena, e.g. based on the classical phase space description.

Later this interpretation of Heisenberg's uncertainty relation was criticized by Margenau [1] and Ballentine [2]. To understand their point, we remark that Heisenberg established the expression (1.54) as the minimum amount of unavoidable momentum disturbance caused by any position measurement. However, he did not give a precise definition for the uncertainties Δx and Δp. It was Kennard in 1927 who first proved the modern inequality:

$$\sigma_x \sigma_p \geq \frac{\hbar}{2}, \tag{1.55}$$

where $\hbar = h/2\pi$, and σ_x, σ_p are the standard deviations of position and momentum, respectively. The crucial point is that the standard deviations are statistical quantities. They can be defined only through measurements of either position or momentum for huge ensembles of particles. Therefore the Heisenberg uncertainty relation (1.55) has nothing to do with the joint determination of position and momentum for a single particle. This is a statistical (dispersion) relation applicable to an ensemble of systems and describing the interrelation between standard deviations of positions and momenta of systems in such an ensemble. However, at the beginning of the twentieth century, the Margenau–Ballentine argument was not yet known. Unfortunately, this argument against the "naive interpretation" of Heisenberg's uncertainty relation is until today practically unknown, and, if it is known, it is simply discarded in the modern quantum community. By the "naive interpretation" we understand this to be a rather common viewpoint to Heisenberg's uncertainty relation: it is impossible to assign to a quantum system both position and momentum, since, by (1.54), increasing the precision of determining the position should automatically decrease the precision of determining the momentum and vice versa. A closely related interpretation is that it is impossible to perform a joint measurement of position and momentum with arbitrary high precision. Neither of these interpretations could be justified by (1.54), because in reality one does not operate with (1.54) (whose meaning is rather unclear), but with (1.55). The latter has nothing to do with position and momentum of an individual particle. Therefore, in spite of (1.54), we have Bohmian mechanics (see Chapter 6) in which position and momentum of a quantum particle are well defined.

Niels Bohr inspired by Heisenberg's discovery concluded that quantum mechanics is complete. Even the grandchildren of our grandchildren will not be able to find a finer description of micro-phenomena... At the time of Bohr's writing, the "finer description" had the meaning "space time description." Later the orthodox "Copenhagenists"[14] generalized this principle of impossibility to any kind of

[14] The word "Copenhagenists" comes from the Bohr–Heisenberg Copenhagen interpretation of quantum mechanics. By this interpretation, the wave function of a quantum system provides the most complete description of its state, i.e. no finer description could be created.

additional variables which might improve the quantum description. Nowadays, such variables are known under the name *hidden variables*. When this term was coined, the idea was that position and momentum were known. What was hidden was only the *pair* of those variables, i.e. a point of the phase space. The completeness of quantum mechanics implies that determinism cannot be recovered through a finer description. Thus, instead of the symbolic dynamics of Hermitian matrices, a more detailed dynamics, e.g. in space time, cannot be constructed.

We state again that one has to distinguish determinism as a general philosophic principle from its applications to real phenomena. As was remarked, even in classical mechanics determinism can be violated by dynamics with some (continuous) forces $f(q)$. Moreover, the initial conditions (position and velocity) cannot be determined with infinite precision. If the dynamics are unstable, a small perturbation can change the trajectory crucially. In this case, the principle of determinism has merely a theoretical value.[15] For an ensemble consisting of a huge number of particles, in practice we can operate only with probabilities and the dynamics will be given by the Liouville equation. The fact of the existence of the underlying Hamiltonian mechanics has merely a metaphysical value: positions and momenta can be really assigned to individual particles and these quantities can be imagined as evolving in space independently of our measurements. By the Bohr–Heisenberg approach, it is sufficient to operate only with probabilities for results of measurements.

A reader may be curious and ask the following question: "*Why do we need all these philosophic considerations on the completeness of quantum mechanics?*" The main problem is that we want to use quantum mathematics without having to share the views of Bohr and Heisenberg, i.e. the so-called Copenhagen interpretation of quantum mechanics. Our position is that we consider Heisenberg's discovery as merely a discovery of a *new mathematical formalism* describing results of measurements for systems characterized by a high sensitivity to external influences. The reader can hopefully understand our dilemma: we want to use the total power of the quantum operational (symbolic) approach and at the same time we do not want to give up *realism*. The latter is very important for us. The very subject of this book is tied to this position we adopt. We plan to apply quantum mathematics to social and cognitive phenomena. We cannot forget (even if we wished) that these phenomena are based on *classical* physical processors of information, the brains (individual and collective).

[15] We remark that classical mechanical determinism is rigidly coupled to the mathematical model of space-time, namely to the real continuum. The states of classical systems are given by pairs of real numbers (or by pairs of real vectors). The infinite divisibility of this space plays an important role, cf. with p-adic spaces which have been recently used in theoretical physics [3].

We are well aware of attempts to create models of a so-called *quantum brain* (see also Chapter 14). In such models, quantum processes at the micro-level induce our mental activity, see e.g. Penrose and Hameroff [4] [5]. However, we do not want to couple our modeling to this very special model of the brain's activity. In particular, we do not want to give up the *neuronal paradigm* of neurophysiology and cognitive science. Neurons are classical processors of electrical signals. Therefore, we try to convince the reader that the arguments of Bohr and Heisenberg about the completeness of quantum mechanics were not fully justified. The structure of quantum mathematics does not exclude a possibility of a peaceful coexistence, of e.g. classical signal processing of information in the brain (performed by neurons) and the quantum probabilistic description of the results of (self-)measurements performed by the brain! Please see again Chapter 14 for more details. A similar viewpoint was presented by de Barros and Suppes [6]: classical oscillatory processes in the brain can in principle induce information processing which has to be described by the mathematical formalism of quantum mechanics. We can also mention Nelson's stochastic quantum mechanics [7] in which quantum probabilistic behavior can be reduced to classical randomness. See also Chapter 13 for more details on this approach.

When considering a quantum-like model of the functioning of the brain proposed in [8], the brain processes concrete mental images by using a classical representation of information. However, abstract mental images, such as *concepts,* are processed on the basis of the quantum-like representation of information. A physical mechanism of creation of the quantum-like representation of classical signals is presented. Quantum-like images are encoded by *covariance matrices* of classical signals. In the quantum terminology, these are *density matrices.* Thus concepts are represented by density matrices (which are in fact classical covariance matrices). The same model can be applied to "collective brains" and thus social systems.

Thus, we do not merely reject the Bohr–Heisenberg thesis about completeness of quantum mechanics. We explore the incompleteness of the quantum (and more general quantum-like) description. Our main claim is that such a description is useful precisely in situations in which the complete information is not available or even if it is available it is impossible to process it in the complete setting.[16] A cognitive system must make a cut-off in the information space. But such a cut-off has to be done in a consistent way. This can be done by using quantum information theory. In our approach, the latter is simply a version of classical information theory adjusted for the processing of incomplete information.

[16] For example, a cognitive (social) system does not have time or computational resources to analyse the complete data about a problem.

For applications to social systems, it is important to know that such systems can process information in a quantum-like way just as it occurs in macroscopic classical systems. Thus, instead of all the quantum mysteries, we have "only" special formal schemes of processing of incomplete information. Our main thesis is that social systems developed the ability to use such quantum-like scheme of information processing and decision making.

1.17 References

[1] Margenau, H. (1950). *The Nature of Physical Reality*. McGraw-Hill, New York.
[2] Ballentine, L. E. (1989). *Quantum Mechanics*. Prentice Hall, Englewood Cliffs, NJ.
[3] Khrennikov, A. Yu. (1997). *Non-Archimedean analysis: Quantum Paradoxes, Dynamical Systems and Biological Models*. Kluwer, Dordrecht.
[4] Penrose, R. (1994). *Shadows of the Mind*. Oxford University Press.
[5] Hameroff, S. (1994). Quantum coherence in microtubules: a neural basis for emergent consciousness? *Journal of Consciousness Studies*, 1, 91–118.
[6] de Barros, J. A. and Suppes, P. (2009). Quantum mechanics, interference, and the brain. *Journal of Mathematical Psychology*, 53, 306–313.
[7] Nelson, E. (1985). *Quantum Fluctuations*. Princeton University Press, NJ.
[8] Khrennikov, A. Yu. (2010). On the physical basis of the theory of "mental waves." *Neuroquantology*, 8, 4 (Suppl. 1), S71–S80.

1.18 Brownian motion

The creation of a mathematical model of so-called "Brownian motion" was a crucial step towards the probabilistic description of physical reality. Maybe we can claim it was not less important than the creation of the quantum formalism?[17] We would like to compare the probabilistic dynamics of the Brownian motion (also known as the Smoluchowski equation) with the probabilistic dynamics of quantum particles (given by the Schrödinger equation). The consideration of the probabilistic dynamics given by the Smoluchowski equation and not by the Liouville equation has an important philosophic impact. By creating a mathematical model of the Brownian motion, physicists understood that the Hamiltonian phase space model cannot be directly applied to this (classical) phenomenon. The basic problem is that in general the velocity of a Brownian particle is not well defined; its trajectory, although continuous, is not differentiable. Therefore, instead of the probability density $\rho(q, p)$ defined on the phase space, one has to use the probability density of spatial coordinates $\rho(q)$.

The q-space is often called *configuration space.*

[17] In this chapter we are interested in physics, but it has to be remarked that the first mathematical construction of the corresponding stochastic process was done in finance; in the doctoral thesis of Louis Bachelier [1] in 1900, see also Section 2.1.

In phase space mechanics, it is always possible to obtain the configuration space density from the phase space:

$$\rho(t, q) = \int \rho(t, q, p)dp. \tag{1.56}$$

The Liouville dynamics on the phase space induce the dynamics of the probability density on the configuration space.

Suppose now that $q(t)$ is continuous, but not differentiable. Then the velocity (and, hence, the momentum) is not well defined. Nevertheless, the density on the configuration space is used to find the probability that the particle's coordinate q belongs to a domain V of the configuration space:

$$P(q(t) \in V) = \int_V \rho(t, q)dq. \tag{1.57}$$

The problem of the derivation of the corresponding dynamical equation for $\rho(t, q)$ now arises.

The simplest probabilistic dynamics of this type correspond to the Brownian motion. In a few words, we can say that a particle moves in a linear fashion colliding with a molecule which changes the direction of motion, and again the particle resumes a same linear based movement. Note that at the points of collision the velocity is not well defined. Einstein and Smoluchowski found the form of this probability density to be of the Gaussian form:

$$\rho(t, q) = \frac{1}{\sqrt{2\pi t}} e^{-q^2/2t}, \tag{1.58}$$

and then by calculating its derivatives they found that this function satisfies the partial differential equation:

$$\frac{\partial \rho(t, q)}{\partial t} = \frac{1}{2} \frac{\partial^2 \rho(t, q)}{\partial q^2}. \tag{1.59}$$

This partial differential equation formally (i.e. from a mathematics point of view) coincides with the Fokker–Planck partial differential equation. See Chapter 2. However, physically they are different: the Smoluchowski's equation describes the probability density of position and the Fokker–Planck's equation is the equation for the probability density of velocity. Equation (1.59) describes the dynamics of the probability density on the configuration space. If one knows the initial probability density $\rho(t_0, q) = \rho_0(q)$, it is possible to find the density at any instant of time $t > t_0$.

However, the knowledge of the initial position $q(t_0) = q_0$ of a particle does provide a possibility to determine uniquely its state $q(t), t > t_0$. The dynamics depend on a chance parameter, say ω, $q \equiv q(t, \omega)$. Formally, if ω was known in advance, it would be possible to recover determinism along the trajectory $t \to q(t, \omega)$. However, ω is never known. For example, for a Brownian particle, ω includes the knowledge of all possible collisions which will occur for this particle! Hence, the *principle of determinism has no practical value for classical stochastic processes.*

One natural way to recover determinism is to use a coarser description of reality, namely to define states of particles not as points of the configuration space, but as probability densities on this space. The dynamics (1.58) emerging from such an approach are deterministic. We shall speculate that something similar was done in quantum mechanics. However, this "recovery of determinism" for the Brownian motion has an unexpected consequence and that is *"non-locality" of this model!* This is a rather unexpected statement: non-locality of classical dynamics! To illustrate this viewpoint, we proceed to the multi-dimensional case.

In the multi-dimensional case, the Cauchy problem[18] for Smoluchowski's equation has the form:

$$\frac{\partial \rho(t, q_1, \ldots, q_N)}{\partial t} = \frac{1}{2} \sum_{j=1}^{N} \Delta_{q_j} \rho(t, q_1, \ldots, q_N), \tag{1.60}$$

$$\rho(t_0, q_1, \ldots, q_N) = \rho_0(q_1, \ldots, q_N), \tag{1.61}$$

where Laplace operators Δ_{q_j} are defined as:

$$\Delta_{q_j} \rho = \frac{\partial^2 \rho}{\partial x_j^2} + \frac{\partial^2 \rho}{\partial y_j^2} + \frac{\partial^2 \rho}{\partial z_j^2},$$

and $q_j = (x_j, y_j, z_j)$, $j = 1, \ldots, N$. If the initial density is factorizable:

$$\rho_0(q_1, \ldots, q_N) = \rho_{01}(q_1) \ldots \rho_{0N}(q_N), \tag{1.62}$$

i.e. at the initial instant of time $t = t_0$ particles are independent, then the solution of equation (1.60) can be found in the form:

$$\rho(t, q_1, \ldots, q_N) = \rho_1(t, q_1) \ldots \rho_N(t, q_N), \tag{1.63}$$

with:

$$\rho_j(t_0, q_j) = \rho_{0j}(q_j), \; j = 1, \ldots, N. \tag{1.64}$$

[18] We state again that this is a dynamical equation with initial conditions.

Thus, the Cauchy problem (1.60), (1.61) can be separated into a system of independent problems:

$$\frac{\partial \rho_j(t, q_j)}{\partial t} = \frac{1}{2} \Delta_{q_j} \rho_j(t, q_j), \tag{1.65}$$

$$\rho_j(t_0, q_j) = \rho_0(q_j). \tag{1.66}$$

Such a dynamics can be called local: the dynamics of the state (in the sense of the probability density) of the jth particle has no influence on the dynamics of the ith particle, $i \neq j$. We state again that equation (1.60) describes the motion of non-interacting Brownian particles.

However, if the initial density is not factorizable, i.e. at $t = t_0$ particles are not independent, they will be correlated (e.g. as the result of the common preparation procedure), and then we cannot use the ansatz (1.63) and split the Cauchy problem for Smoluchowski's equation, see (1.60), (1.61).

Thus, in the very classical situation, the Brownian motion, the determinism (realism) may be incompatible with locality. An attempt to recover the former through the usage of a coarser mathematical description, via the interpretation of a probability density as a state of a system, may lead to the impossibility to treat independently the dynamics of N non-interacting Brownian particles.

We remark that little attention is paid to this trivial mathematical fact: initial correlations can have an impact on further dynamics. We have presented here this "determinism–locality" analysis of the Brownian motion to keep closer to typical discussions of quantum foundations. We shall see that the quantum situation will remind us very much of our presentation of the "determinism–locality" problem in the classical theory of random processes, see [2] (in Russian) for more detail.

1.19 References

[1] Bachelier, L. (1900). Théorie de la Spéculation. *Annales scientifiques de l'Ecole Normale Supérieure*, 21–86.

[2] Khrennikov, A. (2008). *Introduction to Quantum Information Theory*. Nauka, Fizmatlit, Moscow (in Russian).

1.20 The Schrödinger equation

It is clear that the dynamics of a quantum particle cannot be simpler than the dynamics of a Brownian particle: a quantum particle collides not only with other particles, but also with fields. Moreover, it is evident that the picture of the vacuum as the totally empty space does not match physical reality: the random background is present everywhere.[19] Therefore, one could not expect that it would be possible to construct the phase space dynamics of the Hamiltonian type for quantum particles. The dynamics of $\rho(t, q)$ were needed. In principle, we might refer to Heisenberg's uncertainty relation with the original Heisenberg interpretation: position and momentum cannot be jointly determined. However, we prefer to use purely classical arguments.

Roughly speaking, in quantum mechanics the following ansatz was used:

$$\rho(t, q) = \psi^2(t, q). \tag{1.67}$$

So, it was proposed to search for a "density amplitude" $\psi(t, q)$ (wave function) and not directly for a density. At the beginning (in the pioneering studies of Schrödinger), this amplitude was real. Later it was proposed to extend this ansatz and consider complex amplitudes, i.e. to look for a density which is represented as:

$$\rho(t, q) = |\psi(t, q)|^2, \tag{1.68}$$

where $\psi(t, q)$ takes values in the field of complex numbers $\mathbb{C}$. The correct equation for this probability amplitude is the famous Schrödinger equation. We write it directly for an N-particle system:

$$i\hbar \frac{\partial \psi(t, q_1, \dots, q_N)}{\partial t} = -\sum_{j=1}^{N} \frac{\hbar^2}{2m_j} \Delta_{q_j} \psi(t, q_1, \dots, q_N)$$
$$+ V(q_1, \dots, q_N)\psi(t, q_1, \dots, q_N), \tag{1.69}$$

$$\psi(t_0, q_1, \dots, q_N) = \psi_0(q_1, \dots, q_N), \tag{1.70}$$

where m_j is the mass of the jth particle and V is a potential. In the modern presentation of quantum mechanics, this dynamical equation is simply postulated. There are no reasonably justified derivations of Schrödinger's equation in conventional quantum theory. The original Schrödinger derivation was heuristic and it had little to do with modern quantum theory. When quantum theory was being developed, Schrödinger considered the ψ-function (at least for a single particle) as a real physical field.[20]

[19] Although Copenhagenists (see Section 1.16) had never accepted Planck's viewpoint of the spontaneous emission of light by atoms, in modern quantum theory, especially quantum field theory, the vacuum is also considered as a highly complex structure which can produce experimentally observable effects, e.g. the Casimir effect. Although formally classical field and quantum viewpoints of the vacuum are totally different, the resulting picture of the vacuum as the actively acting environment is practically the same.

[20] Feynman claimed that he derived Schrödinger's equation by using the formalism of the path integral, e.g. [1]. However, since he treated the path integral formally, merely as a symbol, his "derivation" cannot be considered as rigorous. In a series of papers of one of the authors [2]–[4], it was shown that by representing the ψ-function in the form $\psi(t, q) = Q(t, q) + iP(t, q)$, where Q and P are the real and imaginary parts of ψ, Schrödinger's equation can be written as a system of Hamiltonian equations with respect to the field variables Q and P. Thus, the Schrödinger equation is a special case of the Hamiltonian dynamics for the field

If one assumes that quantum mechanics is complete, i.e. the state of a quantum system is given by the wave function ψ (and any finer description is forbidden), then the theory is deterministic: the knowledge of the initial condition, (1.70), provides the knowledge of the wave function at any instant of time. Under natural conditions, the Cauchy problem (1.69), (1.70) has the unique solution. Of course, the price to be paid for obtaining such determinism is to allow for the probabilistic nature of a quantum state. In the same way as for the Brownian motion, we are also equally aversive by the fundamental non-locality of quantum dynamics.

Let us now present the quantum analogue of the consideration which was presented for the Smoluchowski equation for a system of classical particles.

Suppose that there are no mutual interactions between different particles, i.e.:

$$V(q_1, \ldots, q_N) = V_1(q_1) + \cdots + V_N(q_N). \tag{1.71}$$

If the initial wave function is factorizable:

$$\psi_0(q_1, \ldots, q_N) = \psi_{01}(q_1) \ldots \psi_{0N}(q_N), \tag{1.72}$$

then the solution of equation (1.69) can be found in the form:

$$\psi(t, q_1, \ldots, q_N) = \psi_1(t, q_1) \ldots \psi_N(t, q_N), \tag{1.73}$$

with:

$$\psi_j(t_0, q_j) = \psi_{0j}(q_j),\ j = 1, \ldots, N. \tag{1.74}$$

Thus the Cauchy problem (1.69), (1.70) can be separated into a system of independent problems:

$$i\hbar \frac{\partial \psi_j(t, q_j)}{\partial t} = -\frac{\hbar^2}{2m_j} \Delta_{q_j} \psi_j(t, q_j) + V_j(q_j) \psi_j(t, q_j), \tag{1.75}$$

$$\psi_j(t_0, q_j) = \psi_0(q_j). \tag{1.76}$$

Such dynamics are local: the dynamics of the state (in the sense of the probability amplitude) of the jth particle has no influence on the dynamics of the state of the ith particle, $i \neq j$.

We remark that the factorization of the wave function implies factorization of the probability density, and hence independence of different particles. However,

phase space (we remark that such phase space is infinitely dimensional). We state again that the system of Maxwell equations for the classical electromagnetic field can also be represented as the Hamiltonian system with respect to the field variables. Therefore, Schrödinger's equation can be considered as the generalization of the Maxwell equation for the case of massive particles. However, such an interpretation is incompatible with the conventional interpretation of the wave function.

the analogy between classical Brownian motion and the dynamics of a system of quantum particles is not complete. The factorizability of the density, i.e. the probabilistic independence of particles, does not imply factorization of the probability amplitude. For example, take:

$$\psi_0(q_1, q_2) = e^{iqp}\psi_{01}(q_1)\psi_{02}(q_2). \tag{1.77}$$

Then:

$$\rho_0(q_1, q_2) = |\psi_0(q_1, q_2)|^2 = |\psi_{01}(q_1)|^2|\psi_{02}(q_2)|^2 = \rho_{01}(q_1)\rho_{02}(q_1). \tag{1.78}$$

For such an initial ψ-function, we are not able to split the Cauchy problem (1.69), (1.70) into the system of independent problems (1.75), (1.76). Thus, classical probabilistic independence of initial conditions does not provide independence of dynamics. This is the essence of quantum non-locality. Within the conventional (Copenhagen) approach, this observation is difficult to comprehend. Pilot wave theory was one of the first theories which took this observation seriously (Bohmian mechanics is an improvement of the de Broglie's double solution model). See also Chapter 6 for more details on Bohmian mechanics. This theory was one of the first attempts to go beyond quantum mechanics and create a finer (in fact, space time) description of the micro-phenomena.

The so-called *prequantum classical statistical field theory* (PCSFT) is a promising new approach which may go beyond quantum physics [3]. In this model, quantum non-locality has been reduced to classical correlations in the background field.

Our analysis of quantum non-locality differs from conventional considerations. In the conventional presentation, quantum non-locality is a consequence of quantum entanglement, i.e. the existence of wave functions of the form:

$$\psi(q_1, q_2) = \psi_1(q_1)\psi_2(q_2) + \psi_2(q_1)\psi_1(q_2), \tag{1.79}$$

where $\psi_1 \neq \psi_2$. From a (mathematical) point of view, it is impossible to factorize the wave function of the composite system $S = (S_1, S_2)$, where S_1 and S_2 are two systems. Factorization can be written as:

$$\psi(q_1, q_2) = \psi_1(q_1)\psi_2(q_2). \tag{1.80}$$

Typically, one may not always pay attention to the following trivial mathematical fact. If the corresponding probability density is not factorizable:

$$\rho(q_1, q_2) \neq \rho_1(q_1)\rho_2(q_2), \tag{1.81}$$

i.e.:

$$|\psi(q_1, q_2)|^2 \neq |\psi_1(q_1)|^2|\psi_2(q_2)|^2. \tag{1.82}$$

The impossibility to split the Cauchy problem for the Schrödinger equation has the purely classical probabilistic explanation: initial correlations will affect the future dynamics. Thus, the correlation of *phases* (and *not the coordinates*) is the essence of quantum entanglement.

Thus, the quantum multi-particle correlations have a lot in common with multi-particle correlations of classical particles (see [5] for a deep analogy between quantum entanglement and correlation of classical Brownian motions). However, quantum correlations cannot be totally reduced to the classical. Besides the correlations of particle coordinates, the correlations of phases, encoded in the wave function (complex probability amplitude), play an important role.

1.21 References

[1] Feynman, R. and Hibbs, A. (1965). *Quantum Mechanics and Path Integrals*. McGraw-Hill, New York.

[2] Khrennikov, A. (2006). Prequantum classical statistical field theory: complex representation, Hamilton–Schrödinger equation, and interpretation of stationary states. *Foundations of Physics Letters*, 19, 299–319.

[3] Khrennikov, A. (2005). A pre-quantum classical statistical model with infinite-dimensional phase space. *Journal of Physics A: Mathematical and General*, 38, 9051–9073.

[4] Khrennikov, A. (2009). Entanglement's dynamics from classical stochastic process. *Europhysics Letters*, 88, 40005.1–6.

[5] Allahverdyan, A., Khrennikov, A., and Nieuwenhuizen, Th. M. (2005). Brownian entanglement. *Physical Review A*, 71, 032102-1–032102-14.

1.22 Do not be afraid of no-go theorems!

The so-called "no-go theorems" (see von Neumann [1] [2], Bell [3], and Kochen-Specker [4]) are often considered as a barrier on the way towards applications of quantum formalism, outside physics. Since the mathematical statements made about the completeness of quantum mechanics are taken as mathematical proofs, there is an impossibility to combine quantum formalism and realism. The majority of scientists are sure that quantum phenomena cannot be emergent from some classical phenomena. As was already emphasized, such a scientific ideology is not acceptable by the authors of this book. Therefore, we have to present arguments against the common interpretation of no-go theorems. A detailed analysis of no-go ideology was performed in [1]. No-go theorems are statements about the impossibility to construct "prequantum models" reproducing the probabilistic predictions of quantum mechanics and operating with classical probabilities. The main point is that physics cannot tell us anything about the features of such models and their coupling with the quantum model. How can one hope to prove that something does not exist without any knowledge (or at least not much) about this "something"? Therefore, each author of a no-go theorem, in fact, formulates a list of features of a "prequantum model" and its correspondence with the quantum model. Then the author proves that his list of assumptions is in contradiction with quantum formalism. Hence, instead of the standard conclusion that the classical probabilistic description is incompatible with the quantum description, we can say that the list of assumptions underlying a no-go theorem is not properly formulated. Therefore, the contradiction with quantum formalism is not surprising. We remark that the situation is made more complicated by the vague formulations of no-go theorems. The authors of these theorems, e.g. von Neumann and Bohr, use some "hidden assumptions" which play an important role in "derivations" (see [1] for an analysis of hidden assumptions in Bell's theorem). We cannot say it better than Bell [3] himself: "long may Louis de Broglie continue to inspire those who suspect that what is proved, by impossibility proofs, is lack of imagination."[21]

Hence, we recommend to researchers applying quantum formalism outside quantum physics: do not be afraid of no-go theorems! However, one has to use the lessons of the no-go activity. It seems that no-go statements work against classical mechanics of particles, but not against classical wave models. We claim that classical waves and oscillatory processes can produce quantum (or more general quantum-like) representations of information, see [4], and Chapter 14.

1.23 References

[1] von Neumann, J. (1932). *Mathematische Grundlagen der Quantenmechanik*. Springer Verlag, Berlin.

[2] von Neumann, J. (1955). *Mathematical Foundations of Quantum Mechanics*. Princeton University Press, New Jersey.

[3] Bell, J. (1987). *Speakable and Unspeakable in Quantum Mechanics*. Cambridge University Press.

[4] Kochen, S. and Specker, E. (1967). The problem of hidden variables in quantum mechanical systems. *Journal of Mathematics and Mechanics*, 17, 59–87.

[21] We recall that de Broglie also advertised the viewpoint that one has not yet found a proper prequantum model.

2

Econophysics: statistical physics and social science

2.1 Science and social science: econophysics?

Economics and finance have for long used concepts drawn from the exact sciences. Famous economists such as Vilfredo Pareto, Nobel Prize winners Paul Samuelson and Maurice Allais, and many others, all used concepts from the exact sciences in order to aid them in modeling economic phenomena.

As we mentioned already in a footnote in Section 1.18 of Chapter 1, Louis Bachelier [1] submitted in 1900 a doctoral thesis in mathematics on the use of a Brownian motion (of the arithmetic type) as a general descriptor of an asset price process. Paul Samuelson [2] proposed, 60 years later, to use the *geometric* Brownian motion so as to allow for the fact that asset prices cannot be negative. This Brownian motion becomes then a major ingredient in the famous Black–Scholes option pricing theory [3].

The famous American mathematician, John von Neumann[1] (with economist O. Morgenstern), wrote the basics of game theory. Their book [4], entitled *Theory of Games and Economic Behavior*, which first appeared in 1947, is the "classic" of any game theorist's library.

Important contributions to theoretical economics were also proposed by Gerard Debreu [5], a Nobel Prize winner in economics (and a mathematician by training), on the development of the so-called core of the economy.

Leonard Savage [6], a famous statistician, developed the Savage expected utility[2] model (1954), where subjective probabilities are used in the calculation of expected utility. Robert Aumann [7], also a mathematician by training and a recent economics Nobel Prize laureate, developed the so-called Anscombe–Aumann approach (1963), where expected utility is measured with a mixture of subjective and objective probabilities.

[1] See for instance reference to von Neumann in Section 1.22 of Chapter 1 in the context of no-go theorems.
[2] We refer the interested reader to Chapter 7, where we expand more on the meaning of utility functions.

The Savage expected utility model mentioned above was found to have inconsistencies in its axiomatic content. As an example, the Ellsberg paradox [8], which we also discuss in much more detail in Chapter 8, indicates that the famous sure-thing principle, a key assumption of the Savage model, can be refuted on experimental grounds. The theoretical economics community has attempted to answer the shortcomings laid bare by the Ellsberg paradox.

In light of this very brief overview, how can we understand the birth of the so-called "econophysics" movement? What does "econophysics" stand for?

From a symptomatic perspective, the econophysics community is mainly composed of physicists (but we can also count some social scientists in that community) who are interested in explicitly using physics techniques and concepts in the social sciences. Econophysics does not only target economics or finance. One can find applications of physics concepts in sociology and even religious studies. We note that the econophysics movement is *not* characterized by social scientists attempting to apply social science concepts in physics.

It is very much debatable how the econophysics movement came into being. The book by Nikolae Georgescu-Roegen [9], a Harvard economist, could possibly be seen as a starting point for the movement. Bertrand Roehner [10] in a survey paper on the econophysics movement, indicates that the term "econophysics" was coined for the very first time in a talk delivered by Eugene Stanley at a statistical physics meeting held in August 1995 in Xiamen (China). In effect, Eugene Stanley may well be called the founding father of the movement. As editor-in-chief of the journal *Physica A*, numerous articles on the topic of econophysics have been published in that journal. Rosario Mantegna and Stanley [11] published a successful handbook on some of the basics of econophysics as early as 1999. Thomas Lux and Marchesi [12], succeeded in publishing an economics article in the famous journal *Nature* in 1999. In 2006, economist Xavier Gabaix joint with Eugene Stanley (and others) [13] published a highly cited econophysics article in the top-notch economics journal, *Quarterly Journal of Economics*.

Can we now come to a clearer definition of what econophysics is? Not really! Most of the literature in econophysics will be characterized by the use of mainly *statistical physics* techniques to social science areas of application. A frequently used concept in econophysics is the so-called power law. A lot of important econophysics work has indeed attempted to find the existence of such power laws in economic data. Please consult for instance Mantegna and Stanley [11]. We will show some other examples of simple econophysics applications later in this chapter.

Maybe we can try to define the econophysics movement by what it is not. Econophysics does not seem to be concerned with applying quantum physical concepts to social science problems. It is in that sense that econophysics is distinct from what we would dare to call "quantum social science." For a much more extensive discussion, please see Chapter 3, Section 3.1.

As the reader may have noticed, this book is thus not about econophysics. Instead this book is very much pre-occupied with attempting to convince the reader that simple quantum mechanical concepts can be employed to good use in a variety of social science settings. In Chapter 3, we will be focussing on explaining, in substantial detail, what so-called "quantum social science" is all about. Part III of the book will examine quantum probabilistic effects in psychology, while Part IV will consider quantum probabilistic effects in economics, finance, and brain sciences. The classical stochastic model of a financial process presented in this chapter (see Sections 2.3 and 2.7) plays an important role in the transition to quantum models, which we propose in Part IV of this book (notably Chapters 11, 12, and 13).

From the outset, we need to mention that one of the main problems which emerge when considering applications of the quantum formalism to the social sciences (particularly in finance and economics) is that we cannot perform the so-called "quantization procedure," i.e. the transition from classical to quantum-like quantities, by simply using operators instead of classical financial variables. Surprisingly, the problem is not in the formalism of the transition (i.e. quantization) by itself, but in the absence of a financial analogue of classical statistical mechanics, cf. Chapter 1. To proceed to quantum-like financial modeling we need to operate with classical quantities such as financial energy, momentum, and so on. As was pointed out already, current financial mathematics does not provide us with such quantities. Nevertheless, some analogs and motivations can be extracted from classical finance. One of the crucial points when considering the coupling of classical and quantum finance is the usage of the basic classical financial model which refers to the fundamental equation of classical statistical mechanics, i.e. the Fokker–Planck equation.[3] Therefore, we present below in brief detail applications of this equation in classical finance. From these considerations, we also get the motivation to introduce financial analogs of potential, energy, and momentum – see Section 2.5.

As we already indicated above, the theory of classical stochastic processes is actively used in Sections 2.3 and 2.7. We have tried to present the fundamentals of classical stochastic finance in a simple and brief way. Nevertheless, readers who have never used stochastic calculus may encounter resistance to proceed easily through these sections. In principle, these sections can be omitted (at least during a first reading). We remark that a brief introduction to the theory of classical stochastic processes is presented in Chapter 4, Section 4.18.

In the next section, we consider a simple economics application of the forward Kolmogorov partial differential equation, also known as the "Fokker–Planck"

[3] Mathematically, the Fokker–Planck equation coincides with the Smoluchowski equation, see Chapter 1, Section 1.18 (on the probabilistic description of the trajectories of Brownian motion).

partial differential equation. In the next section (Section 2.5), we briefly discuss potential and kinetic energy equivalents in economics. In Section 2.7, we show an example of a very famous backward Kolmogorov partial differential equation: the Black–Scholes partial differential equation. The Black–Scholes model aids in pricing financial derivative products. In the penultimate Section 2.9, we give a flavor of the wonderfully "weird" world of finance, where fake probabilities are high up the agenda to solving asset pricing problems. We finish the chapter with a very short section, which has the sole aim to "whet the appetite" on how quantum mechanics can be connected to basic option pricing. No quantum mechanics knowledge is required at this stage.

2.2 References

[1] Bachelier, L. (1900). Théorie de la spéculation. *Annales scientifiques de l'Ecole Normale Supérieure*, 21–86.
[2] Samuelson, P. (1965). Proof that properly anticipated prices fluctuate randomly. *Industrial Management Review*, 4, 41–50.
[3] Black, F. and Scholes, M. (1973). The pricing of options and corporate liabilities. *Journal of Political Economy*, 81, 637–659.
[4] von Neumann, J. and Morgenstern, O. (1947). *Theory of Games and Economic Behavior*. Princeton University Press, New Jersey.
[5] Debreu, G. (1959). *Theory of Value: An Axiomatic Analysis of Economic Equilibrium*. Yale University Press, New Haven, CT.
[6] Savage, L. J. (1954). *The Foundations of Statistics*. J. Wiley, New York.
[7] Anscombe, F. and Aumann, R. (1963). A definition of subjective probability. *Annals of Mathematical Statistics*, 34, 199–205.
[8] Ellsberg, D. (1961). Risk, ambiguity and Savage axioms. *Quarterly Journal of Economics*, 75, 643–669.
[9] Georgescu-Roegen, N. (1999). *The Entropy Law and the Economic Process*. Harvard University Press, Cambridge, MA.
[10] Roehner, B. (2010). Fifteen years of econophysics: worries, hopes and prospects. Please see at: http://arxiv.org/abs/1004.3229, Cornell University.
[11] Mantegna, R. and Stanley, H. E. (1999). *An Introduction to Econophysics: Correlations and Complexity in Finance.* Cambridge University Press.
[12] Lux, Th. and Marchesi, M. (1999). Scaling and criticality in a stochastic multi-agent model of a financial market. *Nature*, 397, 498–500.
[13] Gabaix, X., Gopikrishnan, P., Plerou, V., and Stanley, H. E. (2006). The trades of large institutional investors in illiquid markets explain many patterns in the extreme behavior of returns and trading volume. *Quarterly Journal of Economics*, 121, 2, 461–504.

2.3 The physics-based "Fokker–Planck" PDE in economics

We have already discussed the Fokker–Planck partial differential equation (PDE) as describing the time evolution of a probability density function of velocity. We have already mentioned this equation in Section 1.18 (please see under equation (1.59)).

Many textbooks in physics will deal with the background and construction of the Fokker–Planck PDE. One area where such an equation can be used to beneficial effect is in the estimation of stochastic volatility. The concept of volatility has wide bearing in financial asset pricing. Following Hull [1] (p. 206), the volatility of the price of an asset is a measure of the uncertainty about the returns of a stock. It is universally defined as the standard deviation of the return of an asset (a stock for instance). If stock prices are assumed to be lognormally distributed,[4] then we can say that the percentage changes in the stock price are normally distributed. The finance community often uses the concept of "volatility" and this is the square root of the variance of the return of an asset. Recall that Brownian motion has already been discussed in Section 1.18 of Chapter 1. We can imagine that the volatility of the price of an asset moves over time following a Brownian motion. Hence, suppose we have the stochastic differential equation for volatility, σ, to be:

$$d\sigma = \alpha(\sigma)\,dt + \beta(\sigma)\,dz, \tag{2.1}$$

where $d\sigma$ denotes the infinitesimal change in the value of volatility, σ; $\alpha(\sigma)$ is some drift function of volatility; dt denotes the infinitesimal change in time; $\beta(\sigma)$ is some diffusion function of volatility; dz is a Wiener process. Imagine, following Wilmott [2] (p. 319) that from empirical data you can estimate: $d\sigma = \alpha(\sigma)\,dt + \upsilon\sigma^{\lambda}dz$; with $\upsilon = 0.88$ and $\lambda = 1.05$. One can estimate $\alpha(\sigma)$ using the Fokker–Planck PDE approach, by using $\frac{\partial f}{\partial t} - \frac{1}{2}\frac{\partial^2 f}{\partial \sigma^2}\beta^2 + \frac{\partial f}{\partial \sigma}(\alpha) = 0$. As per Wilmott [2] (p. 320), impose time independency and assume that we know the steady-state distribution of $f : f_\infty$. One can write $\frac{1}{2}\frac{\partial^2 f_\infty}{\partial \sigma^2}\beta^2 = \frac{\partial f_\infty}{\partial \sigma}(\alpha)$ and integrating once (towards σ), Wilmott [2] (p. 320) obtains $\alpha f_\infty = \frac{1}{2}\frac{\partial f_\infty}{\partial \sigma}\beta^2$ from which $\alpha = \frac{1}{2 f_\infty}\frac{\partial f_\infty}{\partial \sigma}\beta^2$ and where the constant is assumed zero. Hence, if one knows f_∞, one can find the drift, $\alpha(\sigma)$. We note that Wilmott [2] (p. 320), indicates that f_∞ could be a lognormal probability density function (pdf) on volatility.

2.4 References

[1] Hull, J. (2006). *Options, Futures and Other Derivatives*. 6th edition. Prentice Hall.

[2] Wilmott, P. (1998). *Derivatives: The Theory and Practice of Financial Engineering*. J. Wiley, New York.

2.5 Potential and kinetic energy in social science

We may wonder how potential and kinetic energy connect to economics? Khrennikov [1] [2] and Choustova [3] [4] [5] [6] have shown that many concepts in basic physics can have a social science equivalent (including potential energy).

[4] Please note that other distributions can certainly be assumed too! The 2008 financial crisis showed us that so-called fat-tailed distributions would be much more appropriate.

The model presented in this section can be considered as a "classical mechanical model" for finance. In principle, it can be used as the basis for what we could call "financial quantization," i.e. a transition from the functional to the operator representation of financial variables. The main problem is that at this level of the "classical mechanical model" of the financial market we do not know which "precise" financial Hamiltonian function describes the financial analogue of energy. In effect, we thus "play" around with Hamiltonian functions borrowed from (classical) mechanics. Quantization will lead to the use of quantum financial Hamiltonians.

Let us follow closely, in the development below, Khrennikov [7] (pp. 155–157).

Consider the most simple concept: the phase space. Assume there exists a configuration space $Q = \mathbb{R}^n$ of price vectors $\overrightarrow{q} = (q_1, q_2, \ldots, q_n)$, where q_j is the price of the share of the jth corporation. The dynamics of prices can then be described by a trajectory $\overrightarrow{q}(t) = (q_1(t), q_2(t), \ldots, q_n(t))$ in Q. One can define $\delta q_j(t) = q_j(t + \Delta t) - q_j(t)$ which symbolizes a discrete price change. A continuous price change $v_j(t) = \dot{q_j}(t) = \lim_{\Delta t \to 0} \frac{q_j(t+\Delta t)-q_j(t)}{\Delta t}$ can then be defined. Note that one can consider time scales such as $\Delta t, 2\Delta t, \ldots$

There exists a phase space $Q \times V$, where $V \equiv \mathbb{R}^n$ and $\overrightarrow{v} = (v_1, v_2, \ldots, v_n) \in V$. A state (q, v) is called a classical state.

The analogue of physical mass can also be introduced, i.e. the number of shares of stock j: m_j. The market capitalization of trader j (firm j) is then $T_j(t) = m_j q_j(t)$.

Total energy is the sum of kinetic and potential energy. Kinetic energy is defined in physics as $\frac{1}{2}\sum_{j=1}^{n} m_j v_j^2$ and the interpretation with the former economics-based analogies is now immediate. Potential energy is denoted as usual $V(q_1, \ldots, q_n)$. It describes interactions between traders as well as interactions from other factors, such as macro-economic factors. What is its functional form? An example of a simple financial potential would be $(q_i - q_j)^2$. Classical price dynamics are then defined by a price momentum $p_j = m_j v_j$. We then have $m_j \left(\lim_{\Delta t \to 0} \frac{v_j(t+\Delta t)-v_j(t)}{\Delta t}\right) = \frac{-\partial V}{\partial q_j}$ and here we find the well-known Newtonian equation.

The development above can also be found in other work. Any of the references mentioned above will give a similar treatment (i.e. Khrennikov, [1] [2] and Choustova [3] [4] [5] [6]).

2.6 References

[1] Khrennikov, A. (2004). *Information Dynamics in Cognitive, Psychological and Anomalous Phenomena*. Series *Fundamental Theories of Physics*, vol. 138. Kluwer, Dordrecht.

[2] Khrennikov, A. Yu. (1999). Classical and quantum mechanics on information spaces with applications to cognitive, psychological, social and anomalous phenomena. *Foundations of Physics*, 29, 1065–1098.

[3] Choustova, O. (2004). Bohmian mechanics for financial processes. *Journal of Modern Optics*, 51, 6/7, 1111.
[4] Choustova, O. (2006). Quantum Bohmian model for financial market. *Physica A*, 374, 304–314.
[5] Choustova, O. (2007). Quantum modeling of nonlinear dynamics of prices of shares: Bohmian approach. *Theoretical and Mathematical Physics*, 152, 7, 1213–1222.
[6] Choustova, O. (2007). Quantum Bohmian model for financial market. Department of Mathematics and System Engineering, International Center for Mathematical Modeling, Växjö University, Sweden.
[7] Khrennikov, A. (2010). *Ubiquitous Quantum Structure: From Psychology to Finance*. Springer Verlag, Berlin.

2.7 The backward Kolmogorov PDE in finance

2.7.1 What is an option?

One of the most famous statistical physics equations in finance is the so-called Black–Scholes PDE. This PDE, when solved, yields the price of a so-called "option" contract. There do exist two basic option contracts: the call option and the put option.

In Bailey [1] (p. 439), a call option is defined as:

Definition 1 *A security that gives its owner the right, but not the obligation, to purchase a specified asset for a specified price, known as the exercise price or the strike price.*

In Bailey (p. 439), the put option is defined as:

Definition 2 *A security that gives its owner the right, but not the obligation, to sell a specified asset for a specified price, known as the exercise price or the strike price.*

When you go "long" on an option, you pay a premium (this is the option price). The writer of the option has the so-called "short" position.

2.7.2 The Black–Scholes portfolio and preferences for risk

The typical Black–Scholes PDE derivation is as follows. The interested reader may want to consult Black and Scholes [2]. Assume a collection (also called a portfolio) of two assets: a stock and an option. The portfolio value can be called, Π. It can be written as follows:

$$\Pi = -F + \frac{\partial F}{\partial S} S, \tag{2.2}$$

where F is the financial option price and S is the stock price. We remark that the financial option price is the price of a call or a put option (please see Definitions 1 and 2 above). This option price is typically a function of two variables: time and the value of the underlying asset. Option pricing can be applied to a myriad of underlying assets, such as bonds, stocks, and other options. In the derivation we provide here, we will assume that the stock (or also "share") is the underlying asset. The practical significance of the option price is that it is the price of the option contract. This price is also known under the name of "premium." As an example, if we were to think of a put contract as an insurance contract against a certain contingency, then the put price, F, would be nothing else than the price of the insurance contract.

We note the quite abstract form of this postulated portfolio: the minus sign which precedes the financial option F indicates we must sell an option. The $+\frac{\partial F}{\partial S}$ indicates the quantity we need to buy of the underlying asset, S. Often the question is raised how the inventors of the option pricing formula got the "intuition" to propose such an abstract portfolio formula. We may want to take a step back now. Consider the so-called binomial option pricing model. The binomial option pricing model can be shown to converge to the Black–Scholes option pricing model. The key paper is by Cox *et al.* [3]. In the binomial option pricing model, the portfolio used is quite similar to the portfolio presented in the above equation (2.2). As any text book on option pricing will show, the finding of a quantity Δ (which is similar – albeit not the same! – to the partial derivative of F towards S) makes that the portfolio in the binomial option pricing model has the same value, no matter whether the price of the underlying asset S goes up or down. Following Hull [4] (p. 243), the portfolios in the binomial option pricing model can be written as $S.u.\Delta - f_u = S.d.\Delta - f_d$, where $S.u$ is the price of the underlying asset when it goes "up," while $S.d$ is the price of the underlying asset when it goes down; f_u and f_d are the so-called intrinsic values of the option when the price of the asset is going up or down, respectively. Following the above two definitions (Definitions 1 and 2), the intrinsic value of an option, when it is a (long) call (Definition 1), would be $\max\{S.u - X, 0\}$ when considering $S.u$, where X is the strike price. Using Definition 2, we would have as intrinsic value for a (long) put, $\max\{X - S.d, 0\}$ when considering $S.d$. From the imposed equality, $S.u.\Delta - f_u = S.d.\Delta - f_d$, one can find $\Delta = \frac{f_u - f_d}{S.u - S.d}$. This Δ is essential as it allows the use of the so-called "risk free rate of interest" as a return parameter on the portfolio. Such a rate of return is in financial economics a highly desirable variable, since it is independent of preferences for risk. But the novice reader may now wonder about those two new terms: "risk free rate of interest" and "preferences for risk." The risk free rate of interest is an interest rate which is relevant for assets which do not bear any risk. As an example, bonds issued by a government would in theory need to yield a risk free rate of interest if those bonds are guaranteed to be repaid no matter the financial circumstance. The concept of "preference for risk" is quite easy to understand. As individual investors, we all have different levels of tolerance towards risk. Some investors are much more sensitive towards the perceived risk position of a certain asset, while others do perceive the level of risk as quite acceptable.

Although it is very intuitive to understand that we have levels of tolerance for risk *vis-à-vis* a certain financial asset, it is quite a much more daunting task to formulate, in a rigorous way, such levels of tolerance for risk. Economists, have for long used the tool of utility functions to "reflect" our attitude towards risk (more on this function in Chapter 7).

As an example, we could for instance think of projecting various scenarios of asset returns in the future. Typically, we assume the existence of states of nature (a state of nature such as "an expanding economy" or also a state of nature such as "a collapsing economy"). If we were to attach probabilities to the likelihood of such states occurring and if we had some idea what return each asset would carry in each of the different states of nature, then we could calculate the mean and the variance of the returns of each asset across the different states of nature. Assets could then be classified according to their risk/return profile. However, this is not fault free. The probabilities of states and estimated levels of returns are mostly subjective assessments. Furthermore, one can easily be confronted with the situation where asset A has a higher expected return than asset B, but asset A's volatility is also higher compared to asset B's volatility. As Danthine and Donaldson [5] (p. 23) remark: "what decrease in expected return is a [decision maker] willing to accept for a 1% decrease in standard deviation of returns?" Answering this question requires the notion of risk tolerance.

Clearly, if we can dispense with the use of any considerations towards the formulation of attitudes towards risk, then we have in some sense a much more tractable theory. The very objective of the portfolio as formulated in equation (2.2) above (and we gave some background information as to how we can see a similar portfolio in the binomial option pricing formula) is thus to form a portfolio which allows for the application of the use of the risk free rate of return for a judicious choice of the Δ. Indeed, if equation (2.2) were to be rewritten as $-F + \Delta' S$, where we used Δ' to distinguish it clearly from the Δ in the binomial option pricing model, then if we replace $\Delta' = \frac{\partial F}{\partial S}$ in the derivation of the Black–Scholes model (please see below – especially equation (2.8)), then we can obtain a risk free portfolio. The attentive reader will of course have spotted that the relation between Δ' and Δ is mathematically not that remote. For more on the background of choice under (un)certainty and the various models of expected utility, please see Chapter 7, Sections 7.1 and 7.3.

2.7.3 A Taylor expansion in a stochastic environment?

The discrete change in the value of the above Black–Scholes portfolio (equation (2.2)), which we denote as $\delta\Pi$, is:

$$\delta\Pi = -\delta F + \frac{\partial F}{\partial S}\delta S. \tag{2.3}$$

It is an assumption that the stock price process follows a geometric Brownian motion (see Samuelson [6]):

$$dS = \mu S dt + \sigma S dz, \tag{2.4}$$

where μ is the expected return; σ is volatility of the stock price S; dz is a Wiener process; t is time. Remark that the expected return, μ, includes preferences for risk. It is precisely one of the great achievements of the Black–Scholes model that, by virtue of its portfolio construction, this expected return is fully absent from the final partial differential equation (i.e. the Black–Scholes partial differential equation), which is (2.12). Assume one wants to know how to write the infinitesimal change in F (the option price (put or call)), denoted as dF, when the derivative F is a function of the stock price and time.

We cover in some detail in Chapter 4, Section 4.18.4 the so-called Itô Lemma [7], which we will use now. Employing the Itô Lemma in the context presented here involves the use of equations (2.5) and (2.6) below. As per Hull [4] (p. 291), the Itô Lemma indicates that:

$$dF = \left(\frac{\partial F}{\partial S}\mu S + \frac{\partial F}{\partial t} + \frac{1}{2}\frac{\partial^2 F}{\partial S^2}\sigma^2 S^2\right)dt + \left(\frac{\partial F}{\partial S}\right)\sigma S dz, \tag{2.5}$$

and we can put this into a discrete format as follows:

$$\delta F = \left(\frac{\partial F}{\partial S}\mu S + \frac{\partial F}{\partial t} + \frac{1}{2}\frac{\partial^2 F}{\partial S^2}\sigma^2 S^2\right)\delta t + \left(\frac{\partial F}{\partial S}\right)\sigma S \delta z. \tag{2.6}$$

INTERMEZZO. The reader will have noticed our "casual" switching from continuous to discrete changes. Such a change is not to be taken lightly. In a stochastic sense, does it make sense to write that $E(X'(t)) = E\left[l.i.m_{\varepsilon\to 0}\frac{x(t+\varepsilon)-x(t)}{\varepsilon}\right] = \lim_{\varepsilon\to 0} E\left[\frac{x(t+\varepsilon)-x(t)}{\varepsilon}\right]$, where $l.i.m$ is the limit in the mean square sense (and this is a very different notion from the ordinary limit definition)? We will assume standard Hilbert spaces $L^1 \equiv L^1(\Omega, \mathcal{F}, \mathbf{P})$ and $L^2 \equiv L^2(\Omega, \mathcal{F}, \mathbf{P})$, which are defined as $L^1 = \{Y : E[|Y|]\} < \infty\}$ and $L^2\left\{Y : E\left[Y^2\right]\right\} < \infty\}$. Note that $L^2 \subset L^1$. Here is an interesting proof. The proof was provided by Alexander [8].

Proof. We can write: $\left|E(X'(t)) - E(\frac{x(t+\varepsilon)-x(t)}{\varepsilon})\right| = \left|E(X'(t)) - (\frac{x(t+\varepsilon)-x(t)}{\varepsilon}))\right|$. Square both sides and use Jensen's inequality then we obtain:

1. $0 \leq \left|E(X'(t)) - E(\frac{x(t+\varepsilon)-x(t)}{\varepsilon})\right|^2 = \left|E(X'(t)) - (\frac{x(t+\varepsilon)-x(t)}{\varepsilon})\right|^2$
2. $\left|E(X'(t)) - (\frac{x(t+\varepsilon)-x(t)}{\varepsilon})\right|^2 \leq E\left\{(X'(t) - (\frac{x(t+\varepsilon)-x(t)}{\varepsilon}))^2\right\}$ (which is Jensen's inequality)

Now we know that $\lim_{\varepsilon\to 0} E\left[\left(\frac{x(t+\varepsilon)-x(t)}{\varepsilon} - X'(t)\right)^2\right] = 0$ since this is the definition of $l.i.m$. Therefore, we can write (using 1 and 2 and the definition of $l.i.m$): $0 \leq \lim_{\varepsilon\to 0}\left|E(X'(t)) - E(\frac{x(t+\varepsilon)-x(t)}{\varepsilon})\right|^2 = \left|E(X'(t)) - (\frac{x(t+\varepsilon)-x(t)}{\varepsilon})\right|^2 \leq \lim_{\varepsilon\to 0} E\left[(X'(t) - (\frac{x(t+\varepsilon)-x(t)}{\varepsilon}))^2\right] = 0$ and therefore $\lim_{\varepsilon\to 0}\left|E(X'(t)) - E(\frac{x(t+\varepsilon)-x(t)}{\varepsilon})\right|^2 = 0$ which of course implies that $\lim_{\varepsilon\to 0}\left|E(X'(t)) - E(\frac{x(t+\varepsilon)-x(t)}{\varepsilon})\right| = 0$ and which therefore proves that $E(X'(t)) = \lim_{\varepsilon\to 0} E\left[\frac{x(t+\varepsilon)-x(t)}{\varepsilon}\right]$. ∎

After this brief intermezzo, let us put the geometric Brownian motion in discrete format:

$$\delta S = \mu S \delta t + \sigma S \delta z. \tag{2.7}$$

As per Hull [4] (p. 291), if we substitute the δF equation (see above) and the δS equation (see above) into $\delta\Pi = -\delta F + \frac{\partial F}{\partial S}\delta S$, then we get:

$$\begin{aligned}\delta\Pi = &-\left(\left(\frac{\partial F}{\partial S}\mu S + \frac{\partial F}{\partial t} + \frac{1}{2}\frac{\partial^2 F}{\partial S^2}\sigma^2 S^2\right)\delta t + \left(\frac{\partial F}{\partial S}\right)\sigma S \delta z\right) \\ &+ \frac{\partial F}{\partial S}\left(\mu S \delta t + \sigma S \delta z\right).\end{aligned} \tag{2.8}$$

You will observe that the discrete change in the value of the portfolio will be written as:

$$\delta\Pi = \left(-\frac{\partial F}{\partial t} - \frac{1}{2}\frac{\partial^2 F}{\partial S^2}\sigma^2 S^2\right)\delta t. \tag{2.9}$$

A crucial argument (see also Hull [4] (p. 292)) now consists in claiming that since we do not have the Wiener process we can say that the return of the portfolio $(\frac{\delta\Pi}{\Pi})$ per unit of time $\frac{1}{\delta t}$ is equal (under no arbitrage) to the risk free rate of interest:

$$\frac{\delta\Pi}{\Pi}\frac{1}{\delta t} = r_f, \tag{2.10}$$

where r_f is the risk free rate. Now remark that this risklessness of the portfolio is only valid for a brief instant of time.

Finally, all that now needs to be done is to substitute $\delta\Pi$ and Π into the above equations, to get:

$$\left(-\frac{\partial F}{\partial t} - \frac{1}{2}\frac{\partial^2 F}{\partial S^2}\sigma^2 S^2\right)\delta t = r_f\left(-F + \frac{\partial F}{\partial S}S\right)\delta t. \tag{2.11}$$

Cancel both δt on the left-hand side and the right-hand side of the equation and you obtain the Black–Scholes equation (i.e. the partial differential equation):

$$\frac{\partial F}{\partial t} + r_f S \frac{\partial F}{\partial S} + \frac{1}{2}\sigma^2 S^2 \frac{\partial^2 F}{\partial S^2} = r_f F. \tag{2.12}$$

The Black–Scholes PDE is a famous example of a so-called backward Kolmogorov PDE. There is a lot of literature on this. The backward Kolmogorov PDE can be written as (see Wilmott [9] (p. 144)) $\frac{\partial P}{\partial t} + \frac{1}{2}B(y,t)\frac{\partial^2 P}{\partial y^2} + A(y,t)\frac{\partial P}{\partial y} = 0$, where P is a pdf; $B(y,t)$ is some diffusion function; $A(y,t)$ is some drift function; y is some position variable; t is time.

It is easy to show that the Black–Scholes PDE is a backward Kolmogorov PDE. Following Wilmott [9] (pp. 147–148), using the format above and substituting for $B(y,t) = B(S,t) = \sigma^2 S^2$ and $A(S,t) = \mu S$: $\frac{\partial P}{\partial t} + \frac{1}{2}\sigma^2 S^2 \frac{\partial^2 P}{\partial y^2} + \mu S \frac{\partial P}{\partial S} = 0$. Imposing $P(S,t) = (\exp(r_f(T-t))F(S,t)$ and taking the necessary derivatives, one gets:

- $\frac{\partial P}{\partial t} = (\exp(r_f(T-t))F(S,t)'_t = -r_f e^{r_f(T-t)}F(S,t) + \frac{\partial F(S,t)}{\partial t}e^{r_f(T-t)}$,
- $\frac{\partial P}{\partial S} = e^{r_f(T-t)}\frac{\partial F}{\partial S}$,
- $\frac{\partial^2 P}{\partial S^2} = e^{r(T-t)}\frac{\partial^2 F}{\partial S^2}$.

Substituting all this into $\frac{\partial P}{\partial t} + \frac{1}{2}\sigma^2 S^2 \frac{\partial^2 P}{\partial S^2} + \mu S \frac{\partial P}{\partial S} = 0$ and dividing by $e^{r_f(T-t)}$, one obtains: $-r_f F + \frac{\partial F}{\partial t} + \frac{1}{2}\sigma^2 S^2 \frac{\partial^2 F}{\partial S^2} + \mu S \frac{\partial F}{\partial S} = 0$.

Bouchaud [10], Bouchaud and Potters [11], Bouchaud and Sornette [12], Schweizer [13], and Schäl [14] provide for an alternative and very elegant way of deriving the option price. We expand on this approach in Chapter 13, Section 13.20.

2.8 References

[1] Bailey, R. (2005). *The Economics of Financial Markets*. Cambridge University Press.
[2] Black, F. and Scholes, M. (1973). The pricing of options and corporate liabilities. *Journal of Political Economy*, 81, 637–659.
[3] Cox, J., Ross, S., and Rubinstein, M. (1979). Option pricing: a simplified approach. *Journal of Financial Economics*, 7, 229–263.
[4] Hull, J. (2006). *Options, Futures and Other Derivatives*. 6th edition. Prentice Hall.
[5] Danthine, J. P. and Donaldson, J. B. (2005). *Intermediate Financial Theory*. Academic Press Advanced Finance Series.
[6] Samuelson, P. A. (1965b). Rational theory of warrant pricing. *Industrial Management Review*, 6, 13–39.
[7] Øksendal, B. (1992). *Stochastic Differential Equations*. Springer Verlag, Berlin.
[8] Alexander, D. (2003). Private communication.
[9] Wilmott, P. (1998). *Derivatives: The Theory and Practice of Financial Engineering*. J. Wiley.

[10] Bouchaud, J. P. (2002). An introduction to statistical finance. *Physica A*, 313, 238–251.
[11] Bouchaud, J. P. and Potters, M. (2000). *Theory of Financial Risks*. Cambridge University Press.
[12] Bouchaud, J. P. and Sornette, D. (1994). The Black–Scholes option pricing problem in mathematical finance: generalization and extensions for a large class of stochastic processes. *Journal de Physique I*, 4, 863–881.
[13] Schweizer, M. (1994b). Approximating random variables by stochastic integrals. *Annals of Probability*, 22, 1536–1575.
[14] Schäl, M. (1994). On quadratic cost criteria for option hedging. *Mathematics of Operations Research*, 19, 121–131.

2.9 What a weird world! Martingales and fake probabilities

How can one solve this backward Kolmogorov PDE? One approach is to use martingales to solve such PDE. Martingales are a well-known mathematical instrument in finance. A heuristic (and in this case incomplete view) of the martingale is very simple: the conditional expectation of say a stock price at time $T > t$, given the information you have at time $t : I_t$ is $E[S_T|I_t] = S_t$, for all $t < T$. For an excellent introductory book on this topic, see Neftci [1]. More advanced material can be found in the excellent text by Øksendal [2].

One possible approach to solving the backward Kolmogorov PDE (Black–Scholes PDE) is provided for in Neftci [1]. We now closely follow Neftci [1] and we summarize[5] below the development[6] contained in Neftci [1] (pp. 346–358). Consider a stock price $S_t = S_0 \exp(Y_t)$ and let Y_t be following a normal pdf with mean μt and variance $\sigma^2 t$. The moment generating function, defined as $E(e^{Y_t \lambda})$, where λ is some arbitrary parameter can be found as being (assume that Y_t follows a pdf with mean μt and $\sigma^2 t$):

$$E(e^{Y_t \lambda}) = \exp\left(\lambda \mu t + \frac{1}{2}\sigma^2 t \lambda^2\right). \tag{2.13}$$

The first moment can then be found as $\frac{dE(e^{Y_t \lambda})}{d\lambda} = t\mu$ when this derivative is evaluated at $\lambda = 0$. We can write the change in Y_t as ΔY_t and this then follows $N(\mu(t-s), \sigma^2(t-s))$. The moment generating function is now:

$$E(e^{\Delta Y_t \lambda}) = \exp\left(\lambda \mu (t-s) + \frac{1}{2}\sigma^2 (t-s) \lambda^2\right). \tag{2.14}$$

Therefore, $E(e^{\Delta Y_t}) = \exp(\mu(t-s) + \frac{1}{2}\sigma^2(t-s))$, where we set $\lambda = 1$. One can write, $E\left[\frac{S_t}{S_u} | S_u, u < t\right] = E[\exp(\Delta Y_t)|S_u]$. Please note that the notation says we

[5] The summary is based on work from Neftci, S., *An Introduction to the Mathematics of Financial Derivatives*, pp. 346–358, Elsevier (2000).
[6] Up to Section 2.10.

obtain the expectation of the division of S_t over S_u for fixed u such that $u < t$. We can write, $E[S_t|S_u, u < t] = S_u \exp(\mu(t-u) + \frac{1}{2}\sigma^2(t-u))$, and where S_u is treated as non-random at time u (and hence can be taken out of the expectation operator).

We would like to introduce a new probability measure $\widetilde{P}$, which will help us to transfer S_t, which is not a martingale, into an S_t, which will be a martingale, and that up to a deterministic process. This approach describes a mean shift induced with the presence of the risk free (and constant) rate of interest.

Hence, let us now write the following, $E^{\widetilde{P}}[(\exp(-r_f t)S_t|S_u, u < t] = \exp(-r_f u)S_u$, which in words says that the expectation of the discounted value of a risky asset can be discounted with the risk free rate (r_f), as long as we use the probability measure $\widetilde{P}$. This is very useful from a finance perspective! The risk free rate of interest is not linked to preferences for risk. Hence, from a natural science perspective, the risk free rate of interest can almost be considered as a (time-dependent...) constant. We have already alluded to the usefulness of the risk free rate of interest (and its importance) in Section 2.7.2 above.

The key issue in the above development is to find an expression for $\widetilde{P}$. Associate that new probability to a normal density with mean ρt, and the same variance as before $\sigma^2 t$. But what is ρ? Sure we can write, in analogy with $E^P[S_t|S_u, u < t]$, that $E^{\widetilde{P}}[S_t|S_u, u < t] = S_u \exp(\rho(t-u) + \frac{1}{2}\sigma^2(t-u))$. But that does not bring us much further.

We can augment the above expression, $E^{\widetilde{P}}[S_t|S_u, u < t]$, a little more by considering a discount factor and we could write $E^{\widetilde{P}}\left[e^{-r_f(t-u)}S_t|S_u, u < t\right] = S_u e^{-r_f(t-u)} \exp(\rho(t-u) + \frac{1}{2}\sigma^2(t-u))$. To say something sensible about the at-present undefined $\widetilde{P}$, one could think of imposing that the expectation $E^{\widetilde{P}}\left[e^{-r_f(t-u)}S_t|S_u, u < t\right]$ obeys the martingale. This would then mean that $E^{\widetilde{P}}\left[e^{-r_f(t-u)}S_t|S_u, u < t\right] = S_u$. This is reasonable since this is equivalent to saying that $E^{\widetilde{P}}\left[e^{-r_f t}S_t|S_u, u < t\right] = e^{-r_f u}S_u$. Therefore, impose on $S_u e^{-r_f(t-u)} \exp(\rho(t-u) + \frac{1}{2}\sigma^2(t-u))$ so that it is equal to S_u.

Therefore, one needs that $e^{-r_f(t-u)} \exp(\rho(t-u) + \frac{1}{2}\sigma^2(t-u)) = e^0$. One can easily see this now by just writing $e^{((\frac{1}{2}\sigma^2 - r_f)(t-u)) + \rho(t-u)} = e^0$ and this is only the case when $\rho = r_f - \frac{1}{2}\sigma^2$. Thus, the probability $\widetilde{P}$ associated with the normal density $N(\rho t, \sigma^2 t)$ can now be made more precise (by using the martingale condition) $N(\rho t, \sigma^2 t) = N\left(\left(r_f - \frac{1}{2}\sigma^2\right)t, \sigma^2 t\right)$.

What one can see clearly is that the transition from P to $\widetilde{P}$ has shifted the mean. This leads to a very nice analogy. If the payoffs in a lottery dice game are to be changed, how could one proceed? The mean of the value of that lottery can be altered by changing the payoffs you get from each dice outcome. However, the mean value of the lottery can also be altered by changing the probabilities associated to each dice outcome! This is indeed a seriously weird solution: if the dice is a fair dice, then we know what the objective probabilities are of each dice's outcome: 1/6. Thus, in analogy with the dice example, you can think of the change from P to $\widetilde{P}$ as a change of probability, from the real probability P to some "fake" probability $\widetilde{P}$!

If one writes that the call value at time $t = 0$:

$$C_0 = E^{\widetilde{P}}\left[e^{-r_f T} \max\left[S_T - X, 0\right]\right], \tag{2.15}$$

where X is the strike price (i.e. the price at which we can exercise the option) and S_T is the terminal stock price, then the main query consists in asking how to work out the expectation $E^{\widetilde{P}}$?

We have now an answer. By imposing the martingale, one could find that the fake probability $\widetilde{P}$ can be associated with $N(\rho t, \sigma^2 t)$, with $\rho = r_f - \frac{1}{2}\sigma^2$. Let us use the integral to calculate the expectation. We remind ourselves that the probabilities we want to use in this expected value calculation must be drawn from $N(\rho t, \sigma^2 t)$. We write:

$$C_0 = \int_{-\infty}^{\infty} e^{-r_f T} \max[S_T - X, 0] \frac{1}{\sqrt{2\pi\sigma^2 T}} e^{-\frac{1}{2\sigma^2 T}(Y_T - (r_f - \frac{1}{2}\sigma^2)T)^2} dY_T. \tag{2.16}$$

We are interested in the "ideas" behind this integral. We have the following important parts:

- $\frac{1}{\sqrt{2\pi\sigma^2 T}} e^{-\frac{1}{2\sigma^2 T}(Y_T - (r_f - \frac{1}{2}\sigma^2)T)^2}$: this is the functional form of the Normal density associated with $N(\rho t, \sigma^2 t)$, with $\rho = r_f - \frac{1}{2}\sigma^2$.
- $(Y_T - (r_f - \frac{1}{2}\sigma^2)T)^2$: notice the presence of the mean here, $(r - \frac{1}{2}\sigma^2)$.
- dY_T: notice that we clearly integrate towards Y_T here.

We can get rid of the max-operator by imposing $S_T = S_0 e^{Y_T} \geq X$ and this can be rewritten as:

$$Y_T \geq \ln\left(\frac{X}{S_0}\right) : \int_{\ln\left(\frac{X}{S_0}\right)}^{\infty} e^{-r_f T}\left(S_0 e^{Y_T} - X\right) \frac{1}{\sqrt{2\pi\sigma^2 T}} e^{-\frac{1}{2\sigma^2 T}(Y_T - (r_f - \frac{1}{2}\sigma^2)T)^2} dY_T. \tag{2.17}$$

This integral can then be split into two parts:

- $S_0 \int_{\ln\left(\frac{X}{S_0}\right)}^{\infty} e^{-r_f T} e^{Y_T} \frac{1}{\sqrt{2\pi\sigma^2 T}} e^{-\frac{1}{2\sigma^2 T}(Y_T - (r_f - \frac{1}{2}\sigma^2)T)^2} dY_T = \textit{Part A}$
- $-Xe^{-r_f T} \int_{\ln\left(\frac{X}{S_0}\right)}^{\infty} \frac{1}{\sqrt{2\pi\sigma^2 T}} e^{-\frac{1}{2\sigma^2 T}(Y_T - (r_f - \frac{1}{2}\sigma^2)T)^2} dY_T = \textit{Part B}.$

A handy way of solving Part B is to set $Z = \frac{Y_T - (r_f - \frac{1}{2}\sigma^2)T}{\sigma\sqrt{T}}$, which allows one to write $-Xe^{-r_f T} \int_{\frac{\ln\left(\frac{X}{S_0}\right) - \left(r_f - \frac{1}{2}\sigma^2\right)T}{\sigma\sqrt{T}}}^{\infty} \frac{1}{\sqrt{2\pi}} e^{-\frac{1}{2}(Z)^2} dZ$. Remark here, integration now happens towards Z. In the lower bound, one reads $\ln\left(\frac{X}{S_0}\right)$. One can rewrite this as $\ln\left(\frac{X}{S_0}\right) = \ln\left(\frac{S_0}{X}\right)^{-1} = -\ln\left(\frac{S_0}{X}\right)$. The lower bound of the integral can be rewritten as $\frac{\ln\left(\frac{X}{S_0}\right)-(r_f-\frac{1}{2}\sigma^2)T}{\sigma\sqrt{T}} = \frac{-\ln\left(\frac{S_0}{X}\right)-(r_f-\frac{1}{2}\sigma^2)T}{\sigma\sqrt{T}} = -\frac{\ln\left(\frac{S_0}{X}\right)+(r_f-\frac{1}{2}\sigma^2)T}{\sigma\sqrt{T}} = -d_2$. Because of symmetry of the normal density, $-Xe^{-r_fT}\int_{-\infty}^{d_2}\frac{1}{\sqrt{2\pi}}e^{-\frac{1}{2}(Z)^2}dZ = -Xe^{-r_fT}N(d_2)$. This latter part is "one-half" of the Black–Scholes solution.

Part A can be solved likewise, with a little more trickery, to obtain $S_0\int_{-\infty}^{d_2}\frac{1}{\sqrt{2\pi}}e^{-\frac{1}{2}\left(Z+\sigma\sqrt{T}\right)^2}dZ$; $Z = \frac{Y_T-(r_f-\frac{1}{2}\sigma^2)T}{\sigma\sqrt{T}}$. Introduce $H = Z+\sigma\sqrt{T}$: $S_0\int_{-\infty}^{d_2+\sigma\sqrt{T}}\frac{1}{\sqrt{2\pi}}e^{-\frac{1}{2}(H)^2}dH = SN(d_1)$, where $d_1 = d_2+\sigma\sqrt{T}$. The full Black–Scholes solution for a European call is now available, taken from parts A and B.

The development above which thus deals with a solution technique to solving the Black–Scholes PDE can also be found in other sources. Books on the topic of partial differential equations will often carry solution techniques for certain types of PDEs. As we already remarked in Section 2.1 above, the Black–Scholes PDE is a backward Kolmogorov PDE.

2.10 References

[1] Neftci, S. (2000). *An Introduction to the Mathematics of Financial Derivatives.* Academic Press.

[2] Øksendal, B. (1992). *Stochastic Differential Equations.* Springer Verlag, Berlin.

2.11 Whetting the quantum appetite

Li and Zhang [1] show in an interesting paper how option pricing can also be modeled via the use of the most fundamental quantum mechanics partial differential equation: the Schrödinger PDE. Without going into any depth (please see Chapter 13 for a more detailed treatment of this approach), the authors show [2] (p. 2) that the necessary transformations to achieve such a pricing approach are that the call function (which is a function of the stock price and time), $Call(S,t) = \sqrt{\sigma(S)}g(x,t)$, where $g(x,t)$ is the quantum mechanical wave function (which we will discuss in much more detail in the next chapter), and $\sigma(S)$ is some volatility function of the asset price S. Furthermore, the authors also show in the same paper (p. 2) that $x(S) = \delta\int\frac{1}{\sigma(S)}dS$, with $\delta = +/-1$. The authors then show that one can use the Schrödinger PDE to solve for $g(x,t)$. Interesting potential functions are considered in a fully financial environment. The concept of the potential function has already been discussed in Sections 1.1, 1.3, 1.13.4, 1.20 in Chapter 1 and 2.5 in this chapter.

Haven [3] shows that when using the so-called WKB approximation method, the Schrödinger PDE can be used to obtain the option price. We will discuss the WKB and the Li and Zhang approaches in more depth in Chapter 13.

In summary, we have covered some simple examples which show how elementary physics equations such as the Fokker–Planck PDE and the backward Kolmogorov PDE can be successfully used in economics. This chapter has introduced the reader to the concept of mean changes via the use of a fake probability. This approach is beneficial, from a finance perspective, as the risk free rate of interest can then be employed. Such interest rates can be seen as the near equivalent of a natural science constant.

2.12 References

[1] Li, Y. and Zhang, J. E. (2004). Option pricing with Weyl–Titschmarsh theory. *Quantitative Finance*, 4, 457–464.

[2] Zhang, J. E. and Li, Y. (2010). New analytical option pricing models with Weyl–Titschmarsh theory. *Quantitative Finance*, DOI: 10. 1080/14697688.2010.503659.

[3] Haven, E. (2005). Analytical solutions to the backward Kolmogorov PDE via an adiabatic approximation to the Schrödinger PDE. *Journal of Mathematical Analysis and Applications*, 311, 439–444.

3

Quantum social science: a non-mathematical motivation

3.1 What is quantum social science?

After a short discursion into the field of "econophysics" we now arrive at a much more exotic stop. However, readers who have perused Chapter 1, will possibly be less astonished to find that instead of making reference to social science within a statistical mechanics environment, we now make reference to social science as it could be embedded in a quantum mechanical context. Or is "quantum social science" denoting something altogether different? The term is certainly fraught with a huge potential for misinterpretation. A common, but maybe not so erroneous, interpretation of the term often made by newcomers to this "field" can be expressed via the following question: "is quantum social science about reformulating social science with quantum mechanical tools?" Such a question brings us closer to the goal of attempting to explain quantum social science, but right here and now, i.e. at this very point in this book, we have the opportunity to make a serious attempt to be really precise on what is meant with this new term.

Therefore, let us from the outset indicate what it is *not*. Quantum social science is *not* about reformulating social science on a quantum size scale. The macroscopic world does not operate at the Planck scale. On prima facie, this obvious statement does not need any further explanation. However, if the human experience is centered around consciousness and the functioning of the brain, then we are much less sure to claim there is a total absence of quantum processes. Roger Penrose has proposed the idea that the brain can operate according to quantum mechanics. The key references to note are by Penrose [1] and Penrose [2]. Roger Penrose and Stuart Hameroff [3] edited in May of 2011 a special issue of the *Journal of Cosmology* on consciousness and the universe. The special issue contains numerous articles on the topic of the relationship between quantum physics and consciousness. See also Hameroff and Penrose [4] and Hagan *et al.* [5]. We are unsure whether to subscribe to those points of view in our book. We will have more to say about this topic in Chapter 14.

Besides the brain and consciousness, anything else which is part of the human experience certainly does not operate on a scale comparable to the Planck scale.

Clearly, we certainly use the word "quantum" in quantum social science. Here is our attempt to be precise in what we mean. We even encapsulate it in a definition.

Definition 3 *Quantum social science has as goal to investigate problems within the wide remit of the social sciences – be it economics, finance, psychology, sociology, or other domains of inquiry – with the help of formal models and concepts used in quantum physics.*

Formal models from quantum physics are many and plentiful. Chapter 1 gave us already a rich flavor. Chapter 6 will give an example of a *formal model*, Bohmian mechanics, which can be used to some extent within a social science setting. An example of the use of a *formal concept* from quantum mechanics in a social science setting can be the idea of employing probability interference in psychology. This is a concept which we will cover in Chapters 5 and 8.

Two immediate questions emerge: (i) How can we argue for the use of concepts and models from quantum physics in such a macroscopic setting? (ii) What problems have been solved? We hope Chapters 1 and 2 have already given some idea of the usefulness to consider why concepts from physics (for instance statistical physics) have found inroads in social science. It is impossible to make an assessment of the achievements of "econophysics" contributions to social science, but there is little doubt that we can make the argument that physics and to a wider extent, engineering concepts, have penetrated many finance models. We think specifically of derivative pricing models. The proprietary finance models developed by research departments of international investment houses rely very heavily on concepts developed in the physical sciences. A simple example is the intensive use of Monte-Carlo simulation in finance. This simulation technique, to the best of our knowledge, was popularized in finance via the great efforts of an outstanding finance academic (with a physics training), Phelim Boyle [6]. See also Ma [7] for more on financial asset pricing.

The attentive reader has spotted that we still have not answered the first and second immediate questions we mentioned above: Why quantum physical concepts in social science? The econophysics movement which is roughly about 12 years old may have seen the emergence of a parallel movement, i.e. the one we attempted to describe in the above definition. It is difficult to pinpoint what the precise research papers were which jump-started the quantum social science movement. However, some trust can be put in the assertion that the *Foundations of Physics* paper by Andrei Khrennikov, which appeared in 1999 [8], may have had a very important impact on the new movement. We believe that the central objective of that paper and subsequent work (for instance the 2004 Kluwer book by Andrei Khrennikov [9]) was essentially asking the question: How can we model *information* to a reasonable standard in a social science setting? We do not want the reader to jump to fortuitous conclusions! Quantum social science is not *only* about modeling information in social science. We believe it is an important component but by no means the only component. Surely, there do exist important models in the social sciences, most notably in economics, where information can form an active part of for instance a pricing model. However, there is surely room for more novelty in this area. Let us consider the last sentence of the abstract of Sheldon Goldstein's paper which appeared in *Foundations of Physics* [10] (p. 335): "I would like to explore the hypothesis that the idea that information plays a special role in physics naturally emerges in a Bohmian universe."[1] Goldstein answers the question regarding the role of the central object (although by no means the unique characteristic) of quantum mechanics: the wave function. This is indeed an important question, also for our purposes, since there is a close connection, we believe, between the wave function and information. Before delving a little deeper between the connection of the wave function and information, we feel it to be quite appropriate to mention that Anton Zeilinger elaborated a purely informational interpretation of quantum mechanics. See Bouwmeester and Zeilinger [11] (Chapter 1). His theory is neither about particles nor is it about waves. It is about information as a fundamental, irreducible objective entity. Although Zeilinger may have never spoken about such a possibility, the reader may conclude that in Zeilinger's interpretation quantum theory can be applied everywhere where information plays a fundamental and active role. We also would like to mention the contributions of Christopher Fuchs [12] [13] who recently presented a novel viewpoint to quantum probabilities (and hence entropy and information). He interpreted quantum probability as subjective probability: it represents the experimenters' subjective expectations of the results of experiments. The Fuchsian interpretation of the quantum wave function is still questioned by a majority of physicists. On the other hand, in social science this interpretation can be fruitfully used.

We can possibly have a sense of discomfort which maybe generated by the thought that "the wave function does not have to be quantum mechanical to establish a link with information." This is indeed correct. However, to our knowledge, there is a specific model in quantum mechanics (which we first discuss in Chapter 6 and then in Chapters 11 and following) which provides for a usable model (in a social science setting) of linking information with the quantum mechanical wave

[1] Bohmian mechanics is covered in Chapter 6.

function. Goldstein proposes that the quantum mechanical wave function is a puzzling object. He proposes several questions that can be asked as to its existence (see [10] (p. 336)):

- "Does it merely represent information or does it describe an observer-independent reality?"
- "If it is objective, does it represent a concrete material sort of reality, or does it somehow have an entirely different and perhaps novel nature?"

To come back to the point we made about the non-unique characteristic the wave function carries in quantum physics, we hint back to Goldstein's paper [10] where he queries that the wave function is (p. 336) (i) everything, (ii) something, or (iii) nothing. For our purposes, we need to exclude (iii). But as Goldstein also indicates, the second option is the one which leaves room for the use of so-called "hidden" variables. The motivation for hidden variables can be traced back – even before quantum mechanics existed. Jammer [14] (p. 259) argues that Hertz had a theory of "verborgene Koordinaten," which actually informed so-called hidden motions and masses. Jammer [14] presents two definitions of hidden variables. The first definition we cover much later in the book, in Chapter 13. Let us consider his second definition (p. 262). We leave out the more technical parts (i.e. parts (3) and (4) of that definition, which in essence refer to the existence of probability densities): "(1) each individual quantum system described by the ... state function ψ is characterized by additional hidden states labeled by a parameter λ; the totality of all hidden states is the phase space Γ of hidden states; ψ and λ determine the result of measuring any observable on the system. (2) Each state function ψ is associated with a probability measure $\rho_\psi(\Lambda)$ on Γ such that if Λ is a measurable subset of Γ then $\rho_\psi(\Lambda)$ is the probability that the state, defined by ψ and λ, lies in Λ." What is of interest here (having thus only cited aspects (1) and (2) of this definition) is that λ are called hidden variables. As Jammer [14] also remarks (p. 262), if λ's value is known for an individual system in state ψ, then, with certainty, one can predict the result of any measurement on it. A very telling (and early) interpretation of hidden variables was in fact given by no less than Born. Jammer [14] cites him as follows (pp. 263–264): "The classical theory introduces the microscopic coordinates which determine the individual processes, only to eliminate them because of ignorance by averaging over their values; whereas the new theory gets the same results without introducing them at all. Of course it not forbidden to believe in the existence of these coordinates; but they will only be of physical significance when methods have been devised for their experimental observation" (see Born [15]). Of course, as Jammer also reports, hidden variables became much more of a "no-no" once the work of von Neumann seemed to establish the impossibility of hidden variables. The main argument[2] (see Jammer [14] (p. 270)) for the refutation of hidden variables is the argument that if hidden variables were acceptable, then no dispersive ensembles could be homogeneous. But von Neumann's proof shows that every ensemble is dispersive and no ensemble can be homogeneous. However, homogeneous ensembles do exist. This then refutes the existence of hidden variables. To put it in words which can be understandable for the non-physicist, Jammer writes [14] (pp. 270–271): "If experiments force us to base the formalism of quantum mechanics on postulates as formulated above,[3] any conception of hidden variables, designed to embed, even only conceptually, the theory into a deterministic scheme is inconsistent with these postulates." History also shows us that it was David Bohm who actually called into being again the possible existence of hidden variables. It is very interesting to note that Bohm indeed did write in his 1951 book [16] that hidden variables were an impossibility. However, the two key papers by Bohm, supporting hidden variables appeared very quickly after the book [17] [18]. We will have to say much more about the Bohmian approach in Chapter 6 of this book. Later work by Jauch and Piron will however show again that quantum mechanics would not be able to be interpreted with hidden variables. See Jammer [14] (pp. 300–302) and Jauch and Piron [19], but also Kochen and Specker [20]. We note that Bohm and Bub [21] refuted the Jauch and Piron proof.

This very short description on the evolution of hidden variables could absolutely not be complete without mention of the most important contribution of Bell [22] [23]. Bell's theorem states that [14] (p. 306): "a local hidden variable theory can not reproduce all statistical predictions of quantum mechanics."

We need to emphasize that we presented here in detail the conventional viewpoint on hidden variables. Roughly speaking, those variables cannot be introduced in a consistent way (at least locally). However, as it was stressed in Section 1.22, the aforementioned argument against hidden variables cannot be considered as the final word! In spite of all the no-go theorems, some form of a local hidden variable theory might be invented in the future. We refer the reader to the citation of Bell on the views of de Broglie in Section 1.22.

Goldstein also queries [10] (p. 337) what information the wave function represents. He mentions (p. 337) that if the wave function is about information, then "it is presumably concerned directly either with mental events or, more likely, with the behavior of macroscopic variables." But Goldstein also raises the query that (p. 337) "the very form of the Hamiltonian and wave function strongly points to a microscopic level of description." We have already hinted to the notion of a Hamiltonian in Chapter 1. The Hamiltonian in a quantum mechanical setting is

[2] A sketch of von Neumann's proof is provided for on pp. 267–270 in Jammer [14].
[3] Those are postulates P.I–P.IV in Jammer [14] (pp. 267–268).

operator derived and in that sense is very different from the classical mechanics-based Hamiltonian. There is also evidence from important work performed by Belal Baaquie [24] that the Hamiltonians which are usable in for instance an interest rate modeling environment exhibit traits which, from a physics point of view (quantum mechanical or classical mechanical), are very difficult to interpret.

Let us come back to the wave function. We mentioned in Section 1.11 that de Broglie interpreted quantum waves as physical waves. Any undergraduate student in a physical sciences curriculum will know that there is a very famous relationship, proposed by no less than Louis de Broglie many many years ago, between a wave property (i.e. the wave number[4]) and the momentum of a particle. This is important to keep in mind, especially in light of our earlier statement that wave functions are not a unique characteristic of quantum mechanics. As we will see more in detail in Chapter 6, the wave function can be embedded in a physics model such that it very clearly affects the behavior of a particle (see [10] (p. 348, Section 6)). We believe it is in that sense that the wave function takes on the role of information. The simple algebraic detail of this statement will be shown in Chapter 6, and the transposition of that important idea within a social science setting was first proposed by Andrei Khrennikov (see [8] and [9]).

Hence, so far we seem to come to the hesitating conclusion that quantum social science seems to have something to do with (i) wave functions, (ii) information connected to such wave functions, and (iii) a very peculiar model which seems to connect particles, wave functions, and information. We used the word "hesitating" as indeed quantum social science as a new field does not revolve around solely those vague (we hope it all gets clearer in Chapter 6) references.

Therefore, let us propose three other facets which may possibly also be included in this new field, quantum social science. The list is clearly non-exhaustive. We will first discuss the functional approach by Volovich. A second facet will briefly treat the issue of consciousness and quantum mechanics. The third facet will consider a "bridge" model between classical and quantum mechanics.

Volovich, in a *Foundations of Physics* paper [25], argues that a trajectory in the Newtonian approach does not have physical real meaning since it embeds the use of real numbers, while our observations require the use of rational numbers. He proposes in that paper a so-called functional approach, where the Liouville equation, which we have already discussed in Section 1.4, is now the starting point. The functional approach is applied to both classical and quantum mechanics. Interestingly, Volovich shows that with the functional approach the densities of the distribution functions (for a Gaussian wave packet (see also Chapter 5 for the idea of such packet)) are the same as the densities used in classical mechanics,

[4] See Chapter 5, section 5.9, where we discuss the wave number.

as long as a condition is met that equates the Planck constant with the product of two distribution parameters (this is equation (17) in that paper). We remark in Chapter 13, where we will discuss the so-called universal Brownian motion, that the treatment of the stochastic equivalent of so-called Hamilton–Jacobi equations will involve the setting of equalities between the Planck constant and other parameters.

Another possible facet of quantum social science deals with the treatment of consciousness and quantum physics. We already alluded to the work of Roger Penrose. But we also believe that this is the correct place in the book to discuss in some detail the approach Carl Jung, founder of analytical psychology, followed by attempting to connect psychology with exact science, in particular quantum physics. Wolfraim [26] describes how Pauli, one of the most eminent quantum physicists of the twentieth century, was in contact with Jung for many years to discuss how quantum mechanics could possibly be connected to psychology. This led to the publication of a book, jointly authored by Jung, Pauli, and Hull [27] in 1955 entitled *The Interpretation of Nature and the Psyche.* Wolfraim [26] writes that "Because mathematics reflects the order of the Unus Mundus, it solves the profound mystery of how it is that mathematics, which is a phenomenon of the mind, should prove so remarkably effective in representing phenomena occurring in the physical world." The meaning of "Unus Mundus" seems to be what Wolfraim [26] describes as the "deeper levels of the collective unconscious." There may be here some connection with one of the proposals Edward Nelson [28] once made about the nature of the so-called "formalism" approach to mathematics. Says Nelson [28] (p. 7) "what is real in mathematics is the notation, not an imagined denotation." This would then, as we argued in Haven [29], rule out the existence of so-called "models" in this approach to mathematics. We based our argument on the proposal Verhulst [30] (p. 33) made that "A model . . . is a metaphor with something added, it has not only qualitative but also quantitative aspects which adds to the precision of description." If there are metaphors which, we could propose, are equivalent to the denotations; if indeed mathematics is a representation of the mind which at the same time is also superbly effective in describing natural phenomena, then it must be that there should be very little "human construct" in mathematics. In other words, one could for instance subscribe to the formalism approach Nelson proposes.

An important point is the possible fact that we may not be able to, as Wolfraim says [26], "directly perceive an objective world but only experience a mental representation of it constructed from the brain's interpretation of incoming sense-data." Of course, if there exists an accurate representation of the external world, it must be, as per Wolfraim [26], that the images "must be governed by the same environmental constraints and functional properties that we normally attribute to physical objects, such as spatial depth, motion and rotation as well as environmental forces such as gravity as well." This statement is closely linked to what Wolfraim [26] calls the "Environmental Invariants Hypothesis." This hypothesis comes from the work of Shepard [31] and it is worth noting what this hypothesis stands for. As Wolfraim [26] says: "The Environmental Invariants Hypothesis claims that governing principles inherent in the laws of physics, such as spatiality, momentum, gravity, friction and centripetal force etc., found in the physical environment, are also actively present in the mental imagery comprising our perception of an experiential world. That is, our mental representation of the physical world is regulated by the very same governing principles that apply to the physical environment." Shepard [31] says the following (p. 1318): "Analogously in psychology, a law that is invariant across perceptual dimensions, modalities, individuals, and species may be attainable only by formulating that law with respect to the appropriate abstract psychological space."

Baaquie and Martin [32] describe the psyche of an individual as (p. 11) "being represented by a state vector, denoted as $|P\rangle$ whose elements are all possible states of an individual's psyche." Ideas can be expressed also as state vectors and if they are independent from each other, Baaquie and Martin [32] then indicate that we can write (in general) (see p. 12): $\langle \text{Idea } 1 | \text{Idea } 2 \rangle = 0$. The fact that the product of those ideas is zero embeds the notion of orthogonality. As is also remarked in their paper, ideas can be seen as superposition of (non-)independent basic ideas. As an example, watching television may be done with a level of attention which is affected by the mixing up of other ideas, i.e. one may be thinking of the kids playing in the garden while watching the television. Thus, as in Baaquie and Martin [32] (see again p. 12, but also p. 13), the individual's psyche is smeared out amongst many different states (some dependent, some independent) and they write (equations (5) and (6), p. 13): $|a\rangle = c_1|\text{Idea } 1\rangle + c_2|\text{Idea } 2\rangle + c_3|\text{Idea } 3\rangle + \ldots c_N|\text{Idea } N\rangle$, where c_i are complex numbers and it is the square modulus $|c_i|^2$ which can be seen as the weight or probability of each of the ideas in the superposed thought $|a\rangle$. When the state is normalizable we have that $\sum_{i=1}^{N} |c_i|^2 = 1$ and the state space which consists of normalizable states is called a Hilbert space. See also Chapter 5 for more definitions on this topic.

Baaquie and Martin [32] (p. 16), remark that " when the mind is observed either externally or by itself, the density matrix undergoes decoherence . . . (and one can write): $\rho = \sum_{i=1}^{N} |c_i|^2 \, |\text{Idea } i\rangle\langle \text{Idea } i|$." Free will is an important concept in all of human decision making. The way that Baaquie and Martin [32] (p. 16) define free will is interesting: it "operates in the freedom of choosing to actualize a particular state."

The stance Baaquie and Martin [32] (see p. 19) take towards the classification of the human psyche is an interesting one. The paper discusses in detail how a Fermion and Boson field can be used to express the state of the human psyche.

Baaquie and Martin [32] also provide for a beautiful idea: the definition of self-consciousness is seen as the rate of change of consciousness (see p. 27). They introduce a time variable which has the peculiarity that it can be seen as the psychological time of an individual. Baaquie has used this interpretation of time in some of his other work. Equation (17) (p. 27) defines the time-dependent change of the creation operator as $\frac{\partial a'(x,t)}{\partial t_s} = \frac{\partial t}{\partial t_s}\frac{\partial a'}{\partial t}$, and it is remarked that the operator $\frac{\partial a'}{\partial t}$ and the operator a' do not commute. As Baaquie and Martin [32] (p. 27) remark, it is not allowed that one is at the same time in the state of consciousness AND in the state of self-consciousness. The absence of precise measurement on position and momentum we will also argue for in later chapters, notably Chapter 13.

Baaquie and Martin [32] also propose the idea (p. 36) that one can probe an electron at different scales and this scale-dependent probing thus gives different results in terms of energy measurements. They propose that this scale dependency can also be applied to the field of the human psyche.

Let us consider (very briefly) the third facet to quantum social science: a "bridge" model between classical and quantum mechanics. This model proposed by Aerts and Durt appeared first in *Foundations of Physics* [33]. The authors propose a sphere model where three major situations are distinguished: (i) a situation where the fluctuations on the experimental apparatus is maximal; (ii) a situation where the fluctuations on the experimental apparatus is zero; and (iii) an intermediate situation. The fluctuations have nothing to do with measurement error due to the apparatus. Different types of probabilities (Kolmogorovian or not Kolmogorovian) are found in the three models.

Another bridge model, which we will discuss in much more detail in Chapter 13, is the model which considers the stochastic equivalent of the Hamilton–Jacobi equations.

Are we any further in our elucidation of what quantum social science now is? The aware reader can sense that this is a very multifaceted movement, but it resolutely uses a quantum mechanical tool kit. But why? We hope to convince the reader that the use of this quantum mechanical toolkit has given interesting insights into a variety of areas of the social sciences. We thus turn in the next sub-section to highlight some of the past and current achievements.

3.2 Key findings from quantum social science

The title of this section is somewhat ambitious. We will attempt to provide a brief overview of some of the key findings. Clearly, we cannot provide much detail on any of those key findings as much more detail will follow in later chapters. It is our intention to give an appetite-inducing flavor of some of the achievements to date in the field.

Quantum social science has been working on multiple problems, which we can divide into the following groups:

- financial asset pricing,
- decision making,
- quantum game theory,
- new social science concepts.

A reference, which treats all of the above groups in detail, is by Khrennikov [34].

In the area of financial asset pricing, research in quantum social science may have started as early as 1998, with a paper by Segal and Segal [35] on a reformulation of the Black–Scholes option pricing equation within a quantum mechanical context. Related work on the topic of such specific embedding was also performed by Accardi and Boukas [36]. But there was also much earlier work by Hudson and Parthasarathy [37] who brought out a paper on the formulation of a quantum version of the Itô formula. Belal Baaquie's work on using a non-stochastic approach towards financial derivative pricing can be found in his 2005 book [38]. But Baaquie went beyond a mere reformulation of derivative pricing. He extended his analysis, intensively using path integration,[5] to other areas of finance, notably to interest rate modeling [24]. He also provided for a world primer on using the Wilson expansion in finance. Baaquie also is currently developing very interesting Hamiltonians, which for the most part are quite devoid of any physics-based interpretation. This also shows quite clearly that a simple "import" of physics principles into social science may well be much less evident than previously anticipated. Brandenburger, a celebrated economist, together with Yanofsky [40] produced a paper where a classification scheme is proposed, which aids in better understanding the hidden variable theory interpretation of quantum physics. Bohmian mechanics and its applications to economics and finance, as we have already mentioned at the beginning of this chapter (and in other chapters), was first introduced by Khrennikov [8] [9]. See also Choustova [41] and Haven [42]. In Haven [43], the Wentzel–Kramers–Brillouin approximation in option pricing theory is considered.

In the area of decision making and psychology, we can mention the early work of Belal Baaquie [44] on how quantum field theory can be used to model consciousness.[6] See also Deutsch [45]. We also mention the work of Khrennikov [46] [47] on the pilot wave (Bohmian-like) model of dynamical processing of conscious and unconscious information by the brain. Here the classical configuration

[5] Path integration was developed by Feynman, and following Vardarajan [39] it can be seen as an equivalent model to the von Neumann approach when wanting to construct probability models to explain complementarity.

[6] We also mentioned some of the related work in the first section of this chapter. See [32].

space is given, not by the real continuum, but by fields of p-adic numbers (discrete hierarchic trees). Brandenburger and La Mura [48] investigate how decision makers when given access to quantum signals can improve their decision making. For instance, memory limitations can be overcome via the setting up of so-called quantum signals.

Busemeyer *et al.* [49] and Busemeyer and Wang [50] have shown that the Ellsberg paradox (a decision making paradox well known in psychology and economics) can be explained with the help of probability interference. See also Khrennikov and Haven [51]. The interference concept will be explained in Chapter 5 and it will be shown how to apply it to decision making in Chapter 8 (see for this Asano *et al.* [52]; Asano *et al.* [53]; Accardi *et al.* [54]).

Danilov and Lambert-Mogiliansky [55] develop non-expected utility with the help of quantum mechanical concepts. Finally, La Mura [56] shows how quantum mechanical concepts can solve (amongst others) the Rabin paradox. The use of experiments to uncover quantum mechanical behavior at the level of decision making are discussed in Conte *et al.* [57] and in Khrennikov [59][58]. See also Bordley and Kadane [60].

In regards to quantum game theory, influential work is by Eisert *et al.* [61], Brandenburger [62], and La Mura [63]. Quantum games do allow for another consideration of equilibriae when qubits are considered. Broekaert *et al.* [64] generalize the famous liar paradox with the aid of quantum mechanics.

Finally, new social science concepts have been contributed mainly in the area of devising a relevant social science-based uncertainty principle. Martin Shubik [65], as early as 1999, did mention that "modern finance . . . has not yet provided us with either the appropriate concepts or measures for the bounds on the minimal overall uncertainty that have to be present in an economy." Baaquie [38] discusses a pragmatic approach to a social-science-based-version of the Heisenberg uncertainty principle. See also Bordley [66].

3.3 References

[1] Penrose, R. (1989). *The Emperor's New Mind: Concerning Computers, Minds and the Laws of Physics.* Oxford University Press.

[2] Penrose, R. (1994). *Shadows of the Mind: A Search for the Missing Science of Consciousness.* Oxford University Press.

[3] Penrose, R. and Hameroff, S. (2011). Editors of the Special Issue: Consciousness and the universe. *Journal of Cosmology*, 14.

[4] Hameroff, S. and Penrose, R. (1996). Orchestrated reduction of quantum coherence in brain microtubules: a model for consciousness? In *Toward a Science of Consciousness: The First Tucson Discussions and Debates.* Eds. Hameroff, S., Kaszniak, A., and Scott, A., MIT Press, pp. 507–540.

[5] Hagan, S., Hameroff, S., and Tuszynski, J. (2002). Quantum computation in brain microtubules? Decoherence and biological feasibility. *Physical Review E*, 65, 061901.

[6] Boyle, P. (1977). Options: a Monte Carlo approach. *Journal of Financial Economics*, 4, 4, 323–338.

[7] Ma, C. (2010). *Advanced Asset Pricing Theory.* Imperial College Press.

[8] Khrennikov, A. Yu. (1999). Classical and quantum mechanics on information spaces with applications to cognitive, psychological, social and anomalous phenomena. *Foundations of Physics*, 29, 1065–1098.

[9] Khrennikov, A. Yu. (2004). Information dynamics in cognitive, psychological and anomalous phenomena. *Series in the Fundamental Theories of Physics*. Kluwer, Dordrecht.

[10] Goldstein, S. (2009). Bohmian mechanics and quantum information. *Foundations of Physics*, 40, 335–355.

[11] Bouwmeester, D. and Zeilinger, A. (2010). The physics of quantum information: basic concepts. In *The Physics of Quantum Information*. Eds. Bouwmeester, D., Ekert, A., and Zeilinger, A., Springer Verlag, Berlin.

[12] Fuchs, C. A. (2002). Quantum mechanics as quantum information (and only a little more). In *Quantum Theory: Reconsideration of Foundations.* Ed. Khrennikov, A. Yu. *Series Mathematical Modelling*, 2, Växjö University Press, Växjö, Sweden, pp. 463–543.

[13] Fuchs, C. A. (2007). Delirium quantum (or, where I will take quantum mechanics if it will let me). In *Foundations of Probability and Physics-3*. Eds. Adenier, G., Fuchs, C., and Khrennikov, A. Yu., *American Institute of Physics, Conference Series Proceedings*, 889, Melville, NY, pp. 438–462.

[14] Jammer, M. (1974). *The Philosophy of Quantum Mechanics*. J. Wiley, New York.

[15] Born, M. (1927). Physical aspects of quantum mechanics. *Nature*, 119, 354–357 (reprint).

[16] Bohm, D. (1959). *Quantum Theory*. Prentice Hall.

[17] Bohm, D. (1952a). A suggested interpretation of the quantum theory in terms of hidden variables. *Physical Review*, 85, 166–179.

[18] Bohm, D. (1952b). A suggested interpretation of the quantum theory in terms of hidden variables. *Physical Review*, 85, 180–193.

[19] Jauch, J. M. and Piron, C. (1963). Can hidden variables be excluded in quantum mechanics? *Helvetica Physica Acta*, 36, 827–837.

[20] Kochen, S. and Specker, E. P. (1967). The problem of hidden variables in quantum mechanics. *Journal of Mathematics and Mechanics*, 17, 59–87.

[21] Bohm, D. and Bub, J. (1966). A refutation of the proof by Jauch and Piron that hidden variables can be excluded in quantum mechanics. *Review of Modern Physics*, 38, 470–475.

[22] Bell, J. S. (1964). *On the Problem of Hidden Variables in Quantum Mechanics*. Preprint, SLAC-PUB-44, Stanford University.

[23] Bell, J. S. (1966). On the problem of hidden variables in quantum mechanics. *Reviews of Modern Physics*, 38, 447–452.

[24] Baaquie, B. (2009). Interest rates in quantum finance: the Wilson expansion and Hamiltonian. *Physical Review E*, 80, 046119.

[25] Volovich, I. V. (2011). Randomness in classical mechanics and quantum mechanics. *Foundations of Physics*, 41, 3, 516–528.

[26] Wolfraim, D. H. (2010). *The Unity of Psyche and World*. www.psybernet.co.nz/files/The-Unity-of-Psyche-and-World.txt.

[27] Jung, C., Pauli, W., and Hull, R. (1955). *The Interpretation of Nature and the Psyche*. Routledge & Kegan Paul.
[28] Nelson, E. (2010). *Confessions of an Apostate Mathematician*. http://www.math.princeton.edu/~nelson/papers.html.
[29] Haven, E. (2010). Metaphors and interdisciplinary research. In *Proceedings of the First CHESS Interdisciplinary Conference*. Eds. Rangacharyulu, C. and Haven, E., University of Saskatchewan (Canada), Worldscientific Press, Singapore.
[30] Verhulst, F. (1999). The Validation of Metaphors. In *Validation of Simulation*. Eds. van Dijkum, C., DeTombe, D., and van Kuijk, E., SISWO Publication 403, Amsterdam, pp. 30–45.
[31] Shepard, R. N. (1987). Towards a universal law of generalization for psychological science. *Science*, 237, 4820, 1317–1323.
[32] Baaquie, B. and Martin, F. (2005). Quantum psyche: quantum field theory of the human psyche. *NeuroQuantology*, 1, 7–42.
[33] Aerts, D. and Durt, T. (1994). Quantum, classical and intermediate: an illustrative example. *Foundations of Physics*, 24, 1353–1369.
[34] Khrennikov, A. (2010). *Ubiquitous Quantum Structure: From Psychology to Finance*. Springer Verlag.
[35] Segal, W. and Segal, I. E. (1998). The Black–Scholes pricing formula in the quantum context. *Proceedings of the National Academy of Sciences of the USA*, 95, 4072–4075.
[36] Accardi, L. and Boukas, A. (2007). The quantum Black–Scholes equation. *Global Journal of Pure and Applied Mathematic*, 2, 2, 155–170.
[37] Hudson, R. L. and Parthasarathy, K. R. (1984). Quantum Ito's formula and stochastic evolutions. *Communications in Mathematical Physics*, 93, 301–323.
[38] Baaquie, B. (2005). *Quantum Finance*. Cambridge University Press.
[39] Varadarajan, V. S. (2008). The quantum dice. Talk given at U.C.L.A., Slides 1-20.
[40] Brandenburger, A. and Yanofsky, N. (2007). A classification of hidden-variable properties. *Journal of Physics A: Mathematical and Theoretical*, 41, 425302.
[41] Choustova, O. (2006). Quantum Bohmian model for financial markets. *Physica A*, 374, 304–314.
[42] Haven, E. (2005). Pilot-wave theory and financial option pricing. *International Journal of Theoretical Physics*, 44, 1957–1962.
[43] Haven, E. (2005). Analytical solutions to the backward Kolmogorov PDE via an adiabatic approximation to the Schrödinger PDE. *Journal of Mathematical Analysis and Applications*, 311, 439–444.
[44] Baaquie, B. (1987). Quantum field theory as model of consciousness. Unpublished.
[45] Deutsch, D. (1999). Quantum theory of probability and decisions. *Proceedings of the Royal Society London A*, 455, 3129–3137.
[46] Khrennikov, A. (1999). p-adic information spaces, infinitely small probabilities and anomalous phenomena. *Journal of Scientific Exploration*, 4, 13, 665–680.
[47] Khrennikov, A. (2000). Classical and quantum mechanics on p-adic trees of ideas. *BioSystems*, 56, 95–120.
[48] Brandenburger, A. and La Mura, P. (2011). *Quantum Decision Theory*. Please see: http://arxiv.org/abs/1107.023, Cornell University.
[49] Busemeyer, J., Wang, Z., and Townsend, J. T. (2006). Quantum dynamics of human decision making. *Journal of Mathematical Psychology*, 50, 3, 220–241.
[50] Busemeyer, J. and Wang, Z. (2007). Quantum information processing explanation for interactions between inferences and decisions. In *AAAI Spring Symposium on Quantum Interaction (Stanford University)*. Eds. Bruza, P., Lawless, W., van Rijsbergen, K., and Sofge, D., AAAI Press, pp. 91–97.

[51] Khrennikov, A. Yu. and Haven, E. (2009). Quantum mechanics and violations of the sure-thing principle: the use of probability interference and other concepts. *Journal of Mathematical Psychology*, 53, 378–388.
[52] Asano, M., Ohya, M., Tanaka, Y., Basieva, I., and Khrennikov, A. (2011). Quantum-like model of brain's functioning: decision making from decoherence. *Journal of Theoretical Biology*, 281, 56–64.
[53] Asano, M., Ohya, M., and Khrennikov, A. (2011). Quantum-like model for decision making process in two players game – a non-Kolmogorovian model. *Foundations of Physics*, 41, 538–548.
[54] Accardi, L., Khrennikov, A., and Ohya, M. (2009). Quantum Markov model for data from Shafir–Tversky experiments in cognitive psychology. *Open Systems and Information Dynamics*, 16, 371–385.
[55] Danilov, V. I. and Lambert-Mogiliansky, A. (2007). Non-classical expected utility theory with application to type indeterminacy. Preprint *Paris-Jourdan Sciences Economiques*.
[56] La Mura, P. (2006). Projective expected utility. Mimeo. Leipzig Graduate School of Management.
[57] Conte, E., Todarello, O., Federici, A., Vitiello, F., Lopane, M., Khrennikov, A., and Zbilut, J. (2007). Some remarks on an experiment suggesting quantum-like behavior of cognitive entities and formulation of an abstract quantum mechanical formalism to describe cognitive entity and its dynamics. *Chaos Solitons Fractals*, 31, 1076–1088.
[58] Khrennikov, A. Yu. (2004). On quantum-like probabilistic structure of mental information. *Open Systems and Information Dynamics*, 11, 267–275.
[59] Khrennikov, A. Yu. (2003). Quantum-like formalism for cognitive measurements. *Biosystems*, 70, 211–233.
[60] Bordley, R. F. and Kadane, J. (1999). Experiment dependent priors in psychology and physics. *Theory and Decision*, 47, 213–227.
[61] Eisert, J., Wilkens, M., and Lewenstein, M. (1999). Quantum games and quantum strategies. *Physical Review Letters*, 83, 3077–3080.
[62] Brandenburger, A. (2010). The relationship between quantum and classical correlation in games. *Games and Economic Behavior*, 69, 175–183.
[63] La Mura, P. (2005). Correlated equilibria of classical strategic games with quantum signals. *International Journal of Quantum Information*, 3, 183–188.
[64] Broekaert, J., Aerts, D., and D'Hooghe, B. (2006). The generalized liar paradox: a quantum model and interpretation. *Foundations of Science*, 11, 4, 399–418.
[65] Shubik, M. (1999). Quantum economics, uncertainty and the optimal grid size. *Economics Letters*, 64, 3, 277–278.
[66] Bordley, R. F. (1998). Quantum mechanical and human violations of compound probability principles: toward a generalized Heisenberg uncertainty principle. *Operations Research*, 46, 923–926.

Part II

Mathematics and physics preliminaries

Part II

Mathematical and physical preliminaries

4

Vector calculus and other mathematical preliminaries

4.1 Linear spaces

In essence vector spaces are quite simple mathematical objects. We can distinguish between so called 'real vector spaces' and the maybe less used 'complex vector spaces'. The former uses real numbers, while the latter uses complex numbers. For the novice reader, many textbooks will give a definition of such a space and they very often will go into more detail (such as on covering the notion of subspaces; product of spaces and span). Hubbard and Hubbard [1]; Kreyszig [2] and Ha [3] are good examples.

The real vector space V is characterized by i) commutativity in addition and hence for two vectors $a, b \in V$: $a + b = b + a$; ii) associativity in addition and hence for vectors $a, b, c \in V$: $(a + b) + c = a + (b + c)$; iii) the existence of a vector $0 \in V$ such that: $a + 0 = a$; iv) for each vector $a \in V$ there exists an additive inverse, $-a : (-a) + a = 0$. Moreover, for real scalars $\alpha, \beta \in \mathbb{R}$ and vectors $a, b \in V$, the following simple laws need to be satisfied: i) $1a = a$; ii) $(\alpha + \beta)\, a = \alpha a + \beta a$ (distributivity towards scalars); iii) $\alpha\, (a + b) = \alpha a + \alpha b$ (distributivity towards vectors): iv) $\alpha\, (\beta a) = (\alpha\beta)\, a$ (associativity). The complex vector space is obtained by just swapping the real scalars for complex scalars.

An important concept which is very often defined in relation to vector spaces is the concept of a basis of a vector space. The definition is quite immediate. A vector space V is n-dimensional if its basis contains n linearly independent vectors. n vectors are linearly independent when vectors $b_1, b_2, \ldots b_n \in V$ and n real scalars: $\alpha_1, \alpha_2, \ldots \alpha_n \in \mathbb{R}$: $\alpha_1 b_1 + \alpha_2 b_2 \ldots\ldots \alpha_n b_n = 0$ (with $0 \in V$) such that $\alpha_1 = \alpha_2 \ldots = \alpha_n = 0$ (with $0 \in \mathbb{R}$). The set of n linearly independent vectors is called a basis of dimension n.

4.2 References

[1] Hubbard, J. H. and Hubbard, B. B. (2002). *Vector Calculus, Linear Algebra and Differential Forms: A Unified Approach.* 2nd edition. Prentice Hall.
[2] Kreyszig, E. (1999). Advanced Engineering Mathematics. J. Wiley.
[3] Ha, D. M. (2006). Functional Analysis. Volume 1. Matrix Editions, New York.

4.3 State space: Hilbert space

We denote the fields of real and complex numbers by the symbol $\mathbb{C}$ and $\mathbb{R}$, respectively; for a complex number $z = u + iv,\ u, v \in \mathbb{R},\ \bar{z}$ denotes its conjugate: $\bar{z} = u - iv$; $|z| = \sqrt{z\bar{z}} = \sqrt{u^2 + v^2}$.

The basic mathematical structure of quantum mechanics is a complex Hilbert space, H: a linear space over complex numbers (i.e. it is possible to form linear combinations of vectors with complex coefficients) endowed with a scalar product. The latter is a form of mapping a pair of vectors $\psi_1, \psi_2 \in H$ into a complex number $\langle \psi_1, \psi_2 \rangle$. It has the following properties:

- $\langle \psi, \psi \rangle \geq 0$ (*positive – definiteness*);
- $\langle \psi, \psi \rangle = 0$ *if and only if* $\psi = 0$ (*non-degeneration*);
- $\langle \psi_1, \psi_2 \rangle = \overline{\langle \psi_2, \psi_1 \rangle}$ (*conjugate symmetry*);
- $\langle \psi, \alpha\psi_1 + \beta\psi_2 \rangle = \alpha\langle \psi, \psi_1 \rangle + \beta\langle \psi, \psi_2 \rangle,\ \alpha, \beta \in \mathbb{C}, \psi_k, \psi \in H$ (*linearity with respect to the second argument*).

As a consequence, we obtain that:

- $\langle \alpha\psi_1 + \beta\psi_2, \psi \rangle = \bar{\alpha}\langle \psi_1, \psi \rangle + \bar{\beta}\langle \psi_2, \psi \rangle$.

We remark that in the physical literature, the scalar product is linear with respect to the *second* argument, as in this book. In the mathematical literature, it is linear with respect to the *first* argument.

The theory is especially simple in the finite-dimensional case (which is typically considered in quantum information theory). Here $H = \mathbb{C}^n$ is the Cartesian product of n-copies of the set of complex numbers $\mathbb{C}$. Hence, a pure state:

$$\psi = (z_1, \ldots, z_n), \tag{4.1}$$

where:

$$\|\psi\|^2 = |z_1|^2 + \cdots + |z_n|^2 = 1. \tag{4.2}$$

We remarked above that a Hilbert space is a linear space. As per Messiah [1] (p. 164), this means that for two square integrable functions Ψ_1 and Ψ_2, the linear combination $\alpha\Psi_1 + \beta\Psi_2$, where α and β are complex numbers, are also square integrable functions. A square integrable function Ψ is a function such that $\int |\Psi|^2 \, dx$ converges. Still following Messiah [1] (pp. 164–166), the existence of linearity and the existence of a scalar product are two important characteristics of a Hilbert space. However, the completeness, which is a property of square integrable functions, is the hallmark property which really makes a space a Hilbert space. The completeness property requires Cauchy convergence of any sequence of square integrable functions towards a square integrable function (AND vice versa!). This then means that any square integrable function can be seen as the limit of a Cauchy converging sequence.

We note that definitions of the Hilbert space can be found in many books. The interested reader may want to peruse Aleksandrov *et al.* [2] (pp. 234–237). Messiah [1] (pp. 164–166), discusses the linearity and completeness of a Hilbert space. From a more function analytical point of view, Hilbert spaces can also be defined as so-called inner product spaces, but with some important conditions. See Ha [3] (p. 292) and (p. 296) for, respectively, the definition of an inner product space and the induced normed space. The relationship between a Hilbert space and an induced normed space is defined in Ha [3] (p. 300). We note that Gustafson and Sigal [4] (p. 6) also give a good overview of the definition of a Hilbert space.

4.4 References

[1] Messiah, A. (1961). *Quantum Mechanics*. Volume 1. North-Holland.

[2] Aleksandrov, A. D., Kolmogorov, A. N., and Lavrent'ev, M. A. (1984). *Mathematics: Its Content, Methods and Meaning.* Volume 3. The MIT Press, MA.

[3] Ha, D. M. (2006). *Functional Analysis*. Volume 1. Matrix Editions. New York.

[4] Gustafson, S. J. and Sigal, I. M. (2006). *Mathematical Concepts of Quantum Mechanics*. Springer Verlag, Berlin.

4.5 Operators

We have already hinted at the existence of operators in Chapter 1 (see for instance Section 1.13 on Heisenberg's symbolic calculus). In fact, a formal definition of a position and momentum operator is given in equation (1.50). For the benefit of the reader, we expand a little more on operators and related concepts in this section. An operator in its most basic form can just be seen as an operation which changes one function $f(x)$ into another function $g(x)$. See for instance Fong [1] for more on the real basics of operators or also Haken and Wolf [2].

An essential characteristic of the algebra of linear operators is the non-commutativity of the product of two operators. Linear operators $\widehat{w}$ and $\widehat{v}$ can be defined as: (i) $\widehat{w}\alpha\psi = \alpha\widehat{w}\psi$ and (ii) $\widehat{w}\,[\alpha\psi_1 + \beta\psi_2] = \alpha\widehat{w}\psi_1 + \beta\widehat{w}\psi_2$, where α and β are arbitrary complex constants.

Clearly, if $\widehat{w}$ is an operator and $\widehat{v}$ is another operator, then the expression $\widehat{v}\widehat{w}f(x)$, where $\widehat{v}$ acts on $\widehat{w}f(x)$, may not at all be equal to the expression generated by $\widehat{w}\widehat{v}f(x)$, where $\widehat{w}$ acts on $\widehat{v}f(x)$. Please see for instance Messiah [3] for more details. Bowman [4] indicates (pp. 53–54) that the "fundamental entity (in quantum physics) is the quantum state" and it is the behavior of such a state which is key. Measurements of quantum states are performed via the route of operators. Operators when applied on a quantum state will change that quantum state unless the operator is the identity operator.

According to Morrison [5] (p. 160), it were the mathematics luminaries Leibniz and Lagrange who founded the so-called operator method and Boole was one of the protagonists of the so-called formal theory of operators when applied to quantum mechanics. An operator, $\widehat{w}$, is defined relative to a function in the following straightforward way:[1] $g(x) = \widehat{w}f(x)$. Here is a simple example of an operator $\widehat{w}$: take the derivative of the function $f(x) : \widehat{w}f(x) = \frac{df(x)}{dx}$. Operators carry the distributive property towards addition and substraction. Furthermore, the product of two operators yields a third operator. Recall the definition of position and momentum operators introduced in equation (1.50). Those two operators do not commute. It is simple to show why this is the case. Consider a function $f(x)$. Show that $\widehat{q}\widehat{p}f(x) \neq \widehat{p}\widehat{q}f(x)$. One just needs to write the definition of the momentum operator, $\widehat{p} = -i\hbar\frac{\partial}{\partial q}$, and apply it to: $\widehat{p}\widehat{q}f(x) = -i\hbar\frac{\partial}{\partial q}(qf(x))$. This is clearly not the same as: $q(-i\hbar\frac{\partial}{\partial q}(f(x)))$, as one takes the derivative of a product on the right-hand side of the non-equality but no such product exists on the left-hand side of the non-equality. In fact, it also is straightforward to show that $\widehat{q}\widehat{p}f(x) - \widehat{p}\widehat{q}f(x)$ is equal to $i\hbar f(x)$ and this can be slightly rewritten as $\widehat{q}\widehat{p} - \widehat{p}\widehat{q} = i\hbar\widehat{1}$, where $\widehat{1}$ is the identity operator. This equality, which is the Heisenberg uncertainty relation, was already introduced in Chapter 1, equation (1.54).

Hermitian operators are covered below (under "Hermiticity"). Projection operators can be introduced via the concept of bras and kets (please see below under "Dirac brakets and bras and kets" and for a definition see the entry below on projection operators).

Identity operators, as mentioned above, leave unchanged a quantum state.

The "unitary operators" can be defined via the notion of an adjoint operator. Please see below again under "Dirac brakets and bras and kets."

An operator $\widehat{w}$ is bounded if $\|\widehat{w}\| = \sup_{\{\Psi \in H : \|\Psi\|=1\}} \|\widehat{w}\psi\| < \infty$. We need to stress that the basic operators of quantum mechanics such as position and momentum and the majority of Hamiltonians are not bounded. Nevertheless, bounded operators play an important role in quantum mechanics, especially in

[1] We denote operators, as in Chapter 1, with a cap.

quantum information theory. Please see also the section on projection operators (Section 4.13).

Jackson [6] is a further good source for a discussion on operators.

4.6 References

[1] Fong, P. (2005). *Elementary Quantum Mechanics*. World Scientific.
[2] Haken, H. and Wolf, H. C. (1984). *Atomic and Quantum Physics*. Springer Verlag, Berlin.
[3] Messiah, A. (1961). *Quantum Mechanics*. Volume 1. North-Holland.
[4] Bowman, G. (2008). *Essential Quantum Mechanics*. Oxford University Press.
[5] Morrison, Michael A. (1990). *Understanding Quantum Physics: A User's Manual*. Volume 1. 1st edition. Pearson Education, Upper Saddle River, NJ.
[6] Jackson, J. D. (1961). *Mathematics for Quantum Mechanics*. W. A. Benjamin.

4.7 Dirac brakets and bras and kets

We now turn to Dirac's symbolism.[2] We recall the notions of ket and bra vectors. Elements of the Hilbert space H are called ket vectors and they are denoted as $|\psi\rangle$. Elements of the dual space H^*, the space of linear continuous functionals on H, are called bra vectors. They are denoted as $\langle\phi|$. Basically, the symbol $\langle\phi|\psi\rangle$ has the meaning of the action of the functional $\langle\phi|$ to the vector $|\psi\rangle$. However, since the dual space H^* is isomorphic to H, we can always identify a bra vector with the corresponding ket vector. After such identification, the pairing $\langle\phi|\psi\rangle$ has the properties of a scalar product. Thus, for two vectors $\phi, \psi \in H$ or in the Dirac notation two ket vectors $|\phi\rangle, |\psi\rangle$:

$$\langle\phi, \psi\rangle = \langle|\phi\rangle, |\psi\rangle\rangle = \langle\phi|\psi\rangle. \tag{4.3}$$

Hence, we can rewrite properties of the scalar product by using Dirac's symbolism:

- $\langle\psi|\psi\rangle \geq 0$ (*positive – definiteness*);
- $\langle\psi|\psi\rangle = 0$ *if and only if* $\psi = 0$ (*non-degeneration*);
- $\langle\psi_1|\psi_2\rangle = \overline{\langle\psi_2|\psi_1\rangle}$ (*conjugate symmetry*);
- $\langle\psi|\alpha\psi_1 + \beta\psi_2\rangle = \alpha\langle\psi|\psi_1\rangle + \beta\langle\psi|\psi_2\rangle, \ \alpha, \beta \in \mathbb{C}, \psi_k, \psi \in H$ (*linearity with respect to the second argument*).

As a consequence, we obtain that:

- $\langle\alpha\psi_1 + \beta\psi_2|\psi\rangle = \bar{\alpha}\langle\psi_1|\psi\rangle + \bar{\beta}\langle\psi_2|\psi\rangle$.

Let us now consider how operators function within Dirac brakets. Following Morrison [1] (p. 444) (equation 10.17), a Dirac braket can be defined as $\langle\psi_1|\widehat{w}\psi_2\rangle \equiv$

[2] This is a formalism which is basic in quantum information theory.

$\int \psi_1^*(\widehat{w}\psi_2)dv$, where ψ_1^* denotes the complex conjugate[3] of ψ_1. The operator $\widehat{w}$ acts on the state function ψ_2. Note that $\langle\psi_1|\widehat{w}\psi_2\rangle$ is also written sometimes as $\langle\psi_1|\widehat{w}|\psi_2\rangle$. We can write that the operator $\widehat{w}$ acts on a state as $\widehat{w}|\psi\rangle$. Now consider again $\langle\psi_1|\widehat{w}\psi_2\rangle$, and this means the operator $\widehat{w}$ acts on the state $|\psi\rangle$, i.e. on the "ket." But there is no action on the bra, when using $\langle\psi_1|\widehat{w}\psi_2\rangle$. It is the adjoint operator which operates on the bra. Bowman [2] (p. 56) defines the adjoint operator as an operator which acts on a bra, producing a new bra which then corresponds to a ket formed by another operator. If $\widehat{w}$ is an operator, then, if $\widehat{w}^+$ is the adjoint operator, one can write $\widehat{w}|\psi\rangle = |\phi\rangle \longleftrightarrow \langle\psi|\widehat{w}^+ = \langle\phi|$. This means the action of the operator $\widehat{w}$ on the ket state $|\psi\rangle$ produces a new ket state $|\phi\rangle$. This ket state corresponds to the bra state $\langle\phi|$ obtained by the action of the operator $\widehat{w}^+$ on the bra state $\langle\psi|$. Please note that operators are defined on the Hilbert space. The state function is then an element of the Hilbert space. See for instance Gustafson and Sigal [3]. The Hilbert space is further considered in Section 5.1.

The unitary operator is defined as follows. If the inverse of an operator $\widehat{w}$, denoted as $\widehat{w}^{-1}$, is equal to the adjoint operator $\widehat{w}^+$, then $\widehat{w}$ is a unitary operator. The inverse operator is such that $\widehat{w}\widehat{w}^{-1} = \widehat{w}^{-1}\widehat{w} =$ identity operator. It is important to stress that the basic property of a unitary operator is that it preserves the scalar product (and hence the norm).

The essential component in unitary transformations is the unitary operator. Following Bransden and Joachain [4] (p. 216), "the application of a unitary operator to every wave function is called a unitary transformation." Messiah [5] (p. 292) indicates that "the product of two unitary transformations is a unitary transformation." A transformation is so-called "infinitesimal" if the operator U defining the unitary transformation can be written as (this is equation VII.94 in Messiah [5] (p. 292)) $U = I + i\varepsilon F$, where F is a Hermitian operator (Hermitian operators are covered below, under "Hermiticity"), I is the identity operator, $i \in \mathbb{C}$, and ε is an infinitesimal real quantity.

If one writes the following Dirac braket $\langle\psi_1|\widehat{w}\psi_1\rangle \equiv \int \psi_1^*(\widehat{w}\psi_1)dv$, then this yields the expectation value. One can also write $\langle\widehat{w}\psi_1|\psi_2\rangle \equiv \int(\widehat{w}^*\psi_1^*)\psi_2 dv$. Note the effect of the operator now on the state function ψ_1^*.

4.8 References

[1] Morrison, Michael, A. (1990). *Understanding Quantum Physics: A User's Manual.* Volume 1. 1st Edition. Pearson Education, Upper Saddle River, NJ.

[2] Bowman, G. (2008). *Essential Quantum Mechanics.* Oxford University Press.

[3] The notion of conjugate was first introduced in Chapter 1, Section 1.13. See also Section 4.3.

[3] Gustafson, S. J. and Sigal, I. M. (2006). *Mathematical Concepts of Quantum Mechanics*. Springer Verlag, Berlin.
[4] Bransden, B. H. and Joachain, C. J. (2000). *Quantum Mechanics*. Prentice Hall.
[5] Messiah, A. (1961). *Quantum Mechanics*. Volume 1. North-Holland.

4.9 Eigenvalues/eigenfunction

Eigenvalues were already briefly mentioned in Chapter 1. We give a little more detail here. In Chapter 5, we use the concept again when we give some detail on the time-(in)dependent Schrödinger equation. The eigenvalue equation for an operator $\widehat{w}$, with eigenfunction f_l and eigenvalue l, is given by $\widehat{w} f_l = l f_l$. For the definition of an operator, please see above. The eigenvalues of the operator can be discrete valued, and/or continuous valued. When the function f_l is replaced by a quantum mechanical wave function, ψ, the eigenvalue equation is then often re-named as an eigenstate equation, $\widehat{w}\psi = l\psi$, and ψ can be seen as an eigenstate. Haken and Wolf [1] (p. 130) indicate that "eigenvalues are identical with the observed values." To put it somewhat more precisely, postulate 4 (see also below, Section 5.5, for more of a discussion on postulates in quantum mechanics) in Bransden and Joachain [2] (p. 199) indicates that the "only result of a precise measurement of the dynamical variable... is one of the eigenvalues... of the linear operator associated with (the dynamical variable)."

4.10 References

[1] Haken, H. and Wolf, H. C. (1984). *Atomic and Quantum Physics*. Springer Verlag, Berlin.
[2] Bransden, B. H. and Joachain, C. J. (2000). *Quantum Mechanics*. Prentice Hall.

4.11 Hermiticity

One of the key postulates of quantum mechanics (postulate 3, please see Section 5.5 in Chapter 5) mentions that observables are represented by operators. However, there is an essential condition which needs fulfilling if such is the case, and this is the characteristic that all operators must satisfy Hermiticity.[4] Although the term seems quite exotic, the meaning of it is quite straightforward. An operator, $\widehat{w}$, will be called Hermitian if $\langle \widehat{w}\psi_1|\psi_2\rangle = \langle \psi_1|\widehat{w}\psi_2\rangle$. The Hamiltonian,[5] $\widehat{\mathcal{H}}$, in quantum mechanics must be Hermitian, i.e. $\langle \widehat{\mathcal{H}}\psi_1|\psi_2\rangle = \langle \psi_1|\widehat{\mathcal{H}}\psi_2\rangle$. The momentum and position operators are also Hermitian. Please see for instance Fermi [1] (p. 33) for some other examples of Hermitian operators. An important

[4] A Hermitian matrix has been first defined in Chapter 1, Section 1.13. Hermiticity is also further discussed in Section 4.11.

[5] The Hamiltonian formalism was extensively introduced in Chapter 1, Section 1.13.

property of Hermitian operators is that their eigenvalues must be real (i.e. measurable). Another important property is that the product of two Hermitian operators will be Hermitian if and only if the two operators commute. Clearly, this means that the product of the position and the momentum operators cannot be Hermitian. Hermitian operators are a subset of the space of linear operators.

Baaquie [2] [3] shows that in finance the Hamiltonian which corresponds to the famous Black–Scholes partial differential equation[6] is *not* Hermitian. Baaquie indicates that this non-Hermiticity is traceable to the necessary requirement of the martingale condition (please see Chapter 2 for a brief discussion of what a martingale is).

We finalize this section with the remark that the existence of arbitrage (please see below for the so-called non-arbitrage theorem) should then imply the existence of Hermiticity. What does this mean? What conclusions can we derive from such an observation? What does it mean that probability is not conserved in a non-arbitrage-based financial context (i.e. when there is no Hermiticity)?

4.12 References

[1] Fermi, E. (1961). *Notes on Quantum Mechanics*. University of Chicago Press.
[2] Baaquie, B. (2005). *Quantum Finance*. Cambridge University Press.
[3] Baaquie, B. (2008). Quantum mechanics and option pricing. In *Quantum interaction*. Eds. Bruza, P., Lawless W., van Rijsbergen, K., Sofge, D. A., Coecke, B., and Clark, S. *Proceedings of the Second Quantum Interaction Symposium* (Oxford University). College Publications, London, pp. 49–53.

4.13 Projection operators

We can define a projection operator via the two properties: idempotency and Hermiticity (please see entry above on Hermiticity). An operator $\widehat{w}$ is *idempotent* if $\widehat{w}^2 = \widehat{w}$. When the operator is both idempotent and Hermitian, the operator is a *projection operator*. We can express any vector ψ in the Hilbert space in terms of two orthogonal vectors ϕ and θ by means of a projection operator. Write $\psi = \phi + \theta$ and $\phi = \widehat{w}\psi$ and $\theta = (I - \widehat{w})\,\psi$, where I is the unit (or identity) operator. One can write that $\langle \phi|\theta\rangle = \langle \widehat{w}\psi|\,(I - \widehat{w})\,\psi\rangle$ and because of Hermiticity of $\widehat{w}$, one can write that $\langle \widehat{w}\psi|\,(I - \widehat{w})\,\psi\rangle = \left\langle \psi|\widehat{w} - \widehat{w}^2|\psi\right\rangle$ and this is equal to zero since, via idempotence, $\widehat{w}^2 = \widehat{w}$. Note that $I - \widehat{w}$ is also a projection operator.

[6] This equation was covered in Chapter 2.

4.14 Probability density functions

Following Ross [1] (p. 7), a random variable X will be a "function that assigns a real value to each outcome of the sample space, S (which is the set of all possible outcomes)." By the Kolmogorov axiomatics [2], the Kolmogorov probability space is a triple $P = (\Omega, F, P)$. Points ω of Ω (which is a non-empty set) are said to be elementary events. F is a so called σ-algebra and P is a probability measure.

Following Brzeźniak and Zastawniak [3] (pp. 1–2), a σ-algebra F on the non-empty set Ω is a family of subsets of Ω such that: (i) the empty set and the set Ω belong to F; (ii) if A belongs to F, then so will the complement $\Omega \backslash A$; (iii) if $A_1, A_2, \ldots$ is a sequence of sets in F, then $A_1 \cup A_2 \ldots$ also belongs to F. Remark, thus, that elements of F are events. If F is a σ-algebra on Ω, then a probability measure P will be a function $P : F \to [0, 1]$ such that (i) $P(\Omega) = 1$ and ii) if $A_1, A_2, \ldots$ are pairwise disjoint (i.e. $A_i \cap A_j = \emptyset$ $(i \neq j)$) and belong to F then $P(A_1 \cup A_2 \ldots) = P(A_1) + P(A_2) + \ldots$.

The random variable is called "continuous" if there exists a function $f(x)$ which is called the probability density function such that $\int_A f(x)\,dx$ denotes the probability of the random variable X to be in the set A. Following Hubbard and Hubbard [4] (p. 417) for a function to be a density of a probability function it must satisfy two essential conditions: (i) $f(.) \geq 0$ and (ii) $\int_{\mathbb{R}^k} f(x)\,d^k x = 1$. The one-dimensional probability density function for the so-called "normal distribution" with mean $\overline{x}$ and standard deviation σ is given by: $f(x, \sigma) = \frac{1}{\sigma\sqrt{2\pi}} e^{-(x-\mu)^2/2\sigma^2}$.

4.15 References

[1] Ross, S. (1996). *Stochastic Processes*. J. Wiley.

[2] Kolmogorov, A. N. (1933). *Grundbegriffe der Wahrscheinlichkeitsrechnung*. Springer Verlag, Berlin; English translation (1956): Kolmogorov, A. N. *Foundations of Probability Theory*. Chelsea Publishing Company, New York.

[3] Brzeźniak, Z. and Zastawniak, T. (1999). *Basic Stochastic Processes*. Springer Verlag, Berlin.

[4] Hubbard, J. H. and Hubbard, B. B. (2002). *Vector Calculus, Linear Algebra and Differential Forms: A Unified Approach*. 2nd edition. Prentice Hall.

4.16 ODEs and PDEs

The basic distinction between an ordinary differential equation (ODE) and a partial differential equation (PDE) is that the former uses ordinary derivatives, while the latter uses partial derivatives. This distinction is then based on whether a function does depend on one independent variable or depends on more than one independent variable, respectively. An example of an ODE is equation (1.4) in Chapter 1. An example of a PDE is equation (1.59) in Chapter 1. The Schrödinger equation is a partial differential equation, since, even in its simplest form, the wave function is a function of position and time variables. We also say that the Schrödinger equation is a partial differential equation of second order. The order is nothing else other than the highest derivative which appears in the equation. Hence, in the Schrödinger PDE, the derivative with highest order is the second derivative, which happens to be the second derivative towards position. Following Boyce and Di Prima [1] (p. 5), one needs to also distinguish differential equations on whether they are linear or non-linear. Consider the ODE $F(x, y, y' \dots y^{(n)}) = 0$; where y are functions of x, and y' indicates the first ordinary derivative towards x; similarly for $y^{(n)}$, which indicates the nth derivative towards x. An ODE will be linear if F is a linear function of the variables $y, y' \dots y^{(n)}$. All basic equations of quantum mechanics are linear. The Schrödinger equation is a linear PDE of second order.

The PDE or ODE will be non-linear in the alternative case; i.e. when F is a non-linear function of the variables $y, y' \dots y^{(n)}$. As an example, consider $y''' + 2\exp(x)y'' + yy' = x^4$, which is a third-order non-linear ODE, because of the presence of yy'. Kreyszig [2] (pp. 64–65) defines a linear second-order differential equation as $y'' + p(x)y' + q(x)y = r(x)$. The equation is linear in y and its derivatives, y' and y''. $p(x)$, $q(x)$, and $r(x)$ can be any given functions of x. If the differential equation cannot be written in this form, then it is non-linear. Considering the example $f(x)y'' + p(x)y' + q(x)y = r(x)$, then this a linear second-order differential equation since we can write it as[7] $y'' + \frac{p(x)}{f(x)}y' + \frac{q(x)}{f(x)}y = \frac{r(x)}{f(x)}$. Consider in the linear second-order differential equation that $r(x) = 0$, then this differential equation is homogeneous. Otherwise, it is called non-homogeneous.

4.17 References

[1] Boyce, W. E. and Di Prima, R. C. (1992). *Elementary Differential Equations and Boundary Value Problems*. J. Wiley.

[2] Kreyszig, E. (1999). *Advanced Engineering Mathematics*. J. Wiley.

4.18 Basics of stochastic mathematics, Brownian motion, non-arbitrage condition, Itô's Lemma

4.18.1 Basics of stochastic mathematics

Here are some definitions and claims which may highlight some of the basics of stochastic mathematics.

Definition 4 *(Hwei [1] (p. 209)) A random process $X(t)$ is continuous in the mean square (m.s.) if* $\lim_{\varepsilon \to 0} E\left\{[X(t+\varepsilon) - X(t)]^2\right\} = 0$.

[7] We assume $f(x) \neq 0$.

Definition 5 *(Hwei [1] (p. 172)) The random process $X(t)$ is a Wiener process when it is characterized by the following properties: 1. $X(t)$ has stationary-independent increments; 2. the increment $X(t) - X(s)$, with $t > s$, must be normally distributed; 3. $E(X(t)) = 0$; 4. $X(0) = 0$.*

Definition 6 *(Hwei [1] (p. 209)) A random process $X(t)$ is having an m.s. derivative $X'(t)$ if $l.i.m_{\varepsilon \to 0} \frac{X(t+\varepsilon)-X(t)}{\varepsilon} = X'(t)$. We denote with $l.i.m$ the limit in the mean square sense, $\lim_{\varepsilon \to 0} E\left\{\left[\frac{X(t+\varepsilon)-X(t)}{\varepsilon} - X'(t)\right]^2\right\} = 0$.*

Claim 7 *(Hwei [1] (p. 220)) If $X(t)$ is m.s. continuous, then the mean is continuous.*

Proof. Consider the $\mathrm{Var}[X(t+\varepsilon) - X(t)] = E\left\{[X(t+\varepsilon) - X(t)]^2\right\} - \{E[X(t+\varepsilon) - X(t)]\}^2 \geq 0$ and therefore $E\left\{[X(t+\varepsilon) - X(t)]^2\right\} \geq \{E[X(t+\varepsilon) - X(t)]\}^2 = [\mu_X(t+\varepsilon) - \mu_X(t)]^2$. By the assumed mean square continuity of $X(t)$ we have that $\lim_{\varepsilon \to 0} E\left\{[X(t+\varepsilon) - X(t)]^2\right\} = 0$ and hence $\lim_{\varepsilon \to 0} [\mu_X(t+\varepsilon) - \mu_X(t)] = 0$. ■

Claim 8 *(Hwei [1] (p. 220)) An m.s. continuous random process is continuous in probability.*

Proof. Use Chebychev's inequality $P\{|X(t+h) - X(t)| \geq \varepsilon\} \leq \frac{E\left[|X(t+h)-X(t)|^2\right]}{\varepsilon^2}$. If h goes to zero, then $\frac{E\left[|X(t+h)-X(t)|^2\right]}{\varepsilon^2}$ must go to zero. Therefore, $P\{|X(t+h) - X(t)| > \varepsilon\}$ must go to zero as h goes to zero. ■

Claim 9 *(Hwei [1] (p. 222)) A Wiener process has no m.s. derivative.*

Proof. We know that $X(t)$ has mean square derivative $X'(t)$ so $E(X'(t)X'(s)) = \frac{\partial^2 E(X(t)X(s))}{\partial t \partial s}$. Now for a Wiener process, we know that $E(X(t)X(s)) = \sigma^2 \min(t, s)$. Therefore, $\frac{\partial E(X(t)X(s))}{\partial s} = \sigma^2$ if $s < t$ and zero if $s > t$. Thus, we have some step function $f(t - s)$ which is equal to unity if $s < t$ and zero for $s > t$. Clearly, this step function is not continuous at $s = t$. Hence, $\frac{\partial^2 E(X(t)X(s))}{\partial t \partial s}$ will not exist at $s = t$. ■

4.18.2 Brownian motion

A discrete time version of arithmetic Brownian motion is $\delta x = a\delta t + b\epsilon(\delta t)^{1/2}$, where ϵ follows a normal density with mean zero and variance of unity. The mean of δx is $a\delta t$, and the standard deviation is: $b(\delta t)^{1/2}$. One can generalize the above discrete time version of Brownian motion by writing $\delta x = a(x,t)\delta t + b(x,t)\epsilon(\delta t)^{1/2}$. In continuous time, we can rewrite this equation as a stochastic differential equation $dx = a(x,t)dt + b(x,t)dW$, where dW is a Wiener process. An example of a geometric Brownian motion would be $dx = axdt + bxdW$.

4.18.3 Non-arbitrage theorem

The discrete state space non-arbitrage theorem (as opposed to the continuous state space non-arbitrage theorem (see Kabanov and Stricker [2]) is a crucial theorem in asset pricing theory. See for instance the paper by Harrison and Kreps [3] for the original ideas. The formulation of the non-arbitrage theorem by Etheridge [4] is followed here closely. We repeat definition 15.1 in Etheridge [4] (p. 11) but with somewhat different notation. Assume there are N tradable assets (some assets may be risky and some not) and their prices, at time t_0 are given by $\overrightarrow{p_0} = (p_0^1, p_0^2, \ldots, p_0^N)$. Assume there exists a K (where K indicates the K states of the world) dimensional state price vector $\overrightarrow{\Phi} = (\Phi_1, \Phi_2, \ldots, \Phi_K)$ which is strictly positive in all coordinates. Consider the following model

$$\begin{pmatrix} p_0^1 \\ p_0^2 \\ \vdots \\ p_0^N \end{pmatrix} = \Phi_1 \begin{pmatrix} D_{11} \\ D_{21} \\ \vdots \\ D_{N1} \end{pmatrix} + \Phi_2 \begin{pmatrix} D_{12} \\ D_{22} \\ \vdots \\ D_{N2} \end{pmatrix} + \cdots + \Phi_K \begin{pmatrix} D_{1K} \\ D_{2K} \\ \vdots \\ D_{NK} \end{pmatrix},$$

where each N-dimensional vector $\overrightarrow{D_1}, \ldots \overrightarrow{D_K}$ is the security price vector at time t_1, if the market is, respectively, in state $1, \ldots, K$. Following theorem 1.5.2 in Etheridge [4] (p. 11), "For the market model described (here) there is no arbitrage if and only if there is a state price vector." The proof of this theorem can be found in Duffie [5].

4.18.4 Itô's Lemma

Following Øksendal [6], Itô's Lemma can be expressed in integral form or in differential notation. Equation (4.1.6) (p. 44) in Øksendal [6] defines X_t as a stochastic integral: $dX_t = udt + vdB_t$, where B_t is a one-dimensional Brownian motion, u and v are respectively drift and diffusion factors, and t is time. Theorem (4.1.2) (p. 44) in Øksendal [6], says that if a $g(t, x)$ is twice continuously differentiable on the product space $[0, \infty[\times\mathbb{R}$, then $Y_t = g(t, X_t)$ is also a stochastic integral and Itô's Lemma indicates that:

$$dY_t = \frac{\partial g}{\partial t}(t, X_t)dt + \frac{\partial g}{\partial x}(t, X_t)dX_t + \frac{1}{2}\frac{\partial^2 g}{\partial x^2}(t, X_t)(dX_t)^2. \tag{4.4}$$

4.19 References

[1] Hwei, P. H. (1997). *Probability, Random Variables and Random Processes.* Schaum's Outline Series. McGraw-Hill.

[2] Kabanov, Yu. and Stricker, C. (2005). *Remarks on the True No-arbitrage Property. Séminaire de Probabilités XXXVIII. Lecture Notes in Mathematics*, 1857, 186–194. Springer Verlag, Berlin.
[3] Harrison, J. M. and Kreps, D. M. (1979). Martingales and arbitrage in multiperiod securities markets. *Journal of Economic Theory*, 20, 381–408.
[4] Etheridge, A. (2002). *A Course in Financial Calculus*. Cambridge University Press.
[5] Duffie, D. (1996). *Dynamic Asset Pricing Theory*. Princeton University Press, NJ.
[6] Øksendal, B. (1998). *Stochastic Differential Equations*. Springer Verlag, Berlin.

5

Basic elements of quantum mechanics

5.1 Mathematical formalism of quantum mechanics: brief introduction

In this chapter, we present basic notions of quantum mechanics, and we emphasize "quantum mathematics." To posit that the mathematical formalism of quantum mechanics can be used outside of quantum physics (as in sociology, economics, or finance) may be of interest. If one can steer away from merely positing and veer instead into convincing the reader that such a position is tenable, then some achievement has been booked. But the price to pay is that the formalism of quantum mechanics needs to be uncovered: how else can one judge whether what we posit is senseful? Thus, a reader can either (i) formally accept our viewpoint about the quantum-like paradigm (see Khrennikov [1]), *or* (ii) ask for a detailed analysis of the quantum foundations which would justify this paradigm. In the first case, it may well be sufficient to skip this chapter (but we should suggest to attempt to include Chapter 3). In the second case, we invite especially those readers who are concerned about the fact that the formal usage of the mathematical formalism of quantum mechanics may transfer major problems of quantum foundations to social science, economics, and finance, to read the present chapter carefully.

Quantum mechanics formally describes states of systems, observables and the dynamics of states and observables. We explained in Chapter 4 the notion of Hilbert space.

Consider complex vectors (here describing pure states of a quantum system), i.e. $\psi \in H$ (H being a Hilbert space) such that:

$$\|\psi\|^2 = \langle\psi|\psi\rangle = 1. \tag{5.1}$$

The normalization (by one) of the above vectors is of capital importance to engender the notion of the probability of a pure state. We note that Hermitian operators are used to denote observables (such as the position observable).

In an orthonormal basis $\{e_j\}$, observables are given by Hermitian matrices $A = (a_{ij})$; here:

$$\bar{a}_{ij} = a_{ji}. \tag{5.2}$$

Remark that by using the usual scalar product notation we write the matrix elements as $\langle e_i, Ae_j \rangle$, but we can also employ $\langle e_i|A|e_j\rangle$ to denote matrix elements in the notation proposed by Dirac.

The squared absolute values of the coordinates of a state vector are interpreted as probabilities (Born's interpretation). As we have remarked in Asano *et al.* [2] (p. 63): "One can say that the quantum formalism provides a geometric representation of probabilities which is based on the Euclidean distance." In the finite-dimensional case, it is especially easy (from the mathematical viewpoint) to formulate quantum postulates describing measurements of observables. Let A be a Hermitian matrix representing a quantum observable. Consider its eigenvalues α_j and eigenvectors $e_j : Ae_j = \alpha_j e_j$. It is postulated that only values α_j can be observed in measurements of the quantum observable represented by A. Quantum theory cannot predict which concrete value α_{j_0} will be obtained. It predicts only the probabilities $P_\psi(A = \alpha_j)$ to obtain values α_j for A measurements for systems prepared in the pure state ψ. For further simplification, let us consider the case of the non-degenerate spectrum[1] $\alpha_i \neq \alpha_j, i \neq j$. In this case, any pure state ψ can be expanded with respect to the orthonormal basis consisting of eigenvectors of A:

$$\psi = \sum_j c_j e_j, \tag{5.3}$$

where $\sum_j |c_j|^2 = 1$. It is postulated that:

$$P_\psi(A = \alpha_j) = |c_j|^2 = |\langle e_j|\psi\rangle|^2. \tag{5.4}$$

This is the basic rule for establishing the coupling between quantum theory and experiment. Denote by n_j the number of measurements of A in which the result $A = \alpha_j$ was obtained, and denote the total length of the series of measurements by N. Finally, set $\mu_j(N) = \frac{n_j}{N}$, the frequency of observation of the value α_j. Then by the frequency interpretation of probability:

$$|\langle e_j|\psi\rangle|^2 = \lim_{N\to\infty} \mu_j(N). \tag{5.5}$$

The rule (5.4) is a discrete version of Born's rule: the probability to find a particle at the point x of physical space is given by the square of the absolute value of the ψ-function (complex probability amplitude):

$$P_\psi(x) = |\psi(x)|^2. \tag{5.6}$$

[1] In the finite-dimensional case, the spectrum coincides with the set of eigenvalues.

By using Born's rule, it is easy to derive the quantum representation of the average of the quantum observable given by A. For any classical random variable ξ taking values ξ_j with probabilities p_j, its average is defined as:

$$\langle \xi \rangle \equiv E\xi = \sum_j \xi_j p_j. \tag{5.7}$$

Any quantum observable can be considered as a classical random variable taking values α_j with the probabilities $p_j = P_\psi(A = \alpha_j)$. Hence:

$$\langle A \rangle_\psi = \sum_j \alpha_j p_j. \tag{5.8}$$

The reader may be surprised by the above statement. However, this is really the case: any single quantum observable is described by classical probability theory. Problems with classicality arise if two or more quantum observables represented by non-commutative operators are considered.

Sometimes it is important to know, besides the possible results of quantum measurements and corresponding probabilities, the post-measurement states. It was postulated by von Neumann that if the result $A = \alpha_j$ has been obtained, then, immediately[2] after this, the system can be found in the state e_j. Thus, the state ψ has been projected onto the eigenvector e_j corresponding to the result of the measurement. We note that this opposes Born's rule, and von Neumann's projection postulate is still a subject of debate.

Now consider an observable given by a Hermitian matrix with degenerate spectrum (still in the finite-dimensional case), i.e. the dimension of the subspace L_j consisting of eigenvectors corresponding to α_j can be larger than 1. Denote by π_{L_j} the orthogonal projector onto the subspace L_j. Then it is postulated that:

$$P_\psi(A = \alpha_j) = \|\pi_{L_j}\psi\|^2. \tag{5.9}$$

If L_j has dimension 1, then (5.9) is reduced to (5.4). What can one say about the post-measurement state? In any textbook, one can find the statement that the post-measurement state is:

$$\psi_{\alpha_j} = \frac{\pi_{L_j}\psi}{\|\pi_{L_j}\psi\rangle\|}, \tag{5.10}$$

with the projection of the original state ψ on the subspace L_j which is normalized by 1 (to get again a pure state). However, von Neumann was sure that the case of a degenerate spectrum was more complicated and it cannot be treated in the same way as the case of a non-degenerate spectrum. He pointed out that the degeneration

[2] "Immediately" indeed is somewhat fuzzy . . .

of spectrum implies that in general post-measurement states are not pure states, but instead so-called mixed states [3] (see [4] for an analysis of the consequences of von Neumann's viewpoint). The latter are described not by normalized vectors, but density matrices (we shall define them below – please see equation (5.16)). Although the rule (5.10) was used at the very beginning of quantum mechanics, e.g. in the famous paper of Einstein *et al.* [5], this rule was presented in the form of a postulate only in the 1950s by Lüders. Therefore, experts in quantum foundations refer to (5.10) as Lüders' projection postulate. Sometimes it is called the "von Neumann–Lüders postulate" or simply the "projection postulate" or even the "von Neumann projection postulate."

In the infinite-dimensional case, the formulation of measurement postulates is based on more complicated mathematics. Roughly speaking, the main difficulty is the presence of generalized eigenvectors, i.e. eigenvectors which do not belong to the Hilbert space H, the space of quantum states. For example, for the position operator, Dirac's δ-function $e_{x_0}(x) = \delta(x - x_0)$ is the eigenfunction corresponding to the eigenvalue x_0. This function does not belong to the L_2-space, the space of square summable (or integrable) functions – the state space of quantum mechanics. The same happens for the momentum observable. Neither does the complex exponent $e_{p_0}(x) = e^{ip_0x/h}$ belong to the L_2-space. However, this is a generalized eigenfunction of the position operator corresponding to the eigenvalue p_0. The formalism of generalized eigenvectors was used by Paul Dirac who created the first mathematical formalism of quantum mechanics. Von Neumann [3] used another approach, which came to be known as the theory of orthogonal projection measures. We recall, as was mentioned before already, that in quantum information one often operates in a finite-dimensional Hilbert space. However, real quantum physics is infinite dimensional.

If a quantum observable is represented by a Hermitian operator whose eigenvectors $\{e_j\}$ belong to the Hilbert space H and form the basis in this space, then all aforementioned postulates can be immediately generalized to the infinite-dimensional case. In this book, we shall not delve deeper into the operator mathematics of quantum mechanics. The reader can always proceed by keeping in mind the finite-dimensional situation.

In this part of section 5.1 we reproduce (pp. 63–64)[3] the paper by Asano *et al.* [2]. The dynamics of a pure state are described by the Schrödinger differential equation:

$$i\frac{d\psi}{dt}(t) = \mathcal{H}\psi(t),\ \psi(0) = \psi_0, \tag{5.11}$$

[3] Masanari Asano and Masanori Ohya and Yishiharu Tanaka and Irina Basieva and Andrei Khrennikov (2011). Quantum like model of brain's functioning: decision making from decoherence. *Journal of Theoretical Biology*, 281, 63–64.

where the operator $\mathcal{H}$ is the generator of evolution, also called the "Hamiltonian," the operator of energy. We remark that Schrödinger dynamics are deterministic. By fixing the initial state, we can find the state of a quantum system at any instant of time. The fundamental problem of the foundations of quantum mechanics is that this state does not determine the values of observables. Measurements induce randomness.

We remark that each pure state ψ determines a Hermitian operator, the projector onto this state; $\rho \equiv |\psi\rangle\langle\psi|$ (the last symbol is simply the Dirac notation), $\rho\phi = \langle\phi|\psi\rangle\psi$. We recall the basic properties of ρ_ψ:

(a) it is positively defined, i.e. $\langle\phi|\rho|\phi\rangle \geq 0$ for any ϕ;
(b) it is Hermitian;
(c) its trace (the sum of diagonal elements) equals to one.

The Schrödinger dynamics for pure states (vectors) can be rewritten as the dynamics for corresponding operators:

$$i\frac{d\rho}{dt}(t) = [\mathcal{H}, \rho(t)], \ \rho(0) = \rho_0, \tag{5.12}$$

where $[\mathcal{H}, \rho] = \mathcal{H}\rho - \rho\mathcal{H}$ is the commutator of operators.

Consider now a statistical mixture (in the classical sense) of a few projection operators ρ_i corresponding to pure states ψ_i with weights $p_i \geq 0, \sum p_i = 1$:

$$\rho = p_1\rho_1 + \cdots + p_n\rho_n. \tag{5.13}$$

Each operator of this form satisfies conditions (a)–(c) and vice versa. Denote the class of all operators with properties (a)–(c) by the symbol $D(H)$. This is the space of states of quantum systems. Its elements (called density operators) can be interpreted as statistical mixtures of pure states. In general, a density operator can be represented in many ways. There is one special expansion corresponding to eigenvectors of ρ. The density operator corresponding to a pure state can be characterized in the following way: in the basis of eigenvectors, its matrix has only one non-zero element (equal to one), i.e. up to a permutation of eigenvectors:

$$\rho = \begin{pmatrix} 1 & 0 \\ 0 & 0 \end{pmatrix}, \tag{5.14}$$

where the blocks of zeros have the corresponding sizes. However, this takes place only in the basis of eigenvectors.

Consider,[4] for example, the two-dimensional Hilbert space H and fix some orthonormal basis in this space, $\{e_1, e_2\}$; take a pure state:

$$\psi = xe_1 + ye_2, \tag{5.15}$$

[4] This part of the text until and including equation 5.16 is not part of the Asano *et al.* [2] article.

where $|x|^2 + |y|^2 = 1$. The density matrix ρ corresponding to this pure state has the form:

$$\rho = \begin{pmatrix} |x|^2 & x\bar{y} \\ \bar{x}y & |y|^2 \end{pmatrix}. \tag{5.16}$$

The dynamics of a quantum state are described by the equation (5.12). This dynamical model can be used only in the absence of interaction of a quantum system with an environment, a bath. If such interaction is essential (so a system cannot be considered as isolated), the dynamics (5.12) have to be modified and additional terms have to be included. The basic postulate of quantum theory is that the state dynamics are linear. Therefore, the modified dynamics have the form:

$$i\frac{d\rho}{dt}(t) = [\mathcal{H}, \rho(t)] + L(\rho), \rho(0) = \rho_0, \tag{5.17}$$

where L is a linear operator. This operator has to be chosen in such a way that starting with $\rho_0 \in D(H)$, we shall obtain a trajectory $t \to \rho(t)$ in $D(H)$. The corresponding conditions for linear systems were formulated by Gorini, Kossakowski, Sudarshan, and Lindblad, see, e.g., [7] for details.

Consider a mixed state ρ and a quantum observable A. The average of A in this state is given by the formula (see e.g. von Neumann [3]):

$$\langle A \rangle_\rho = \mathrm{Tr}\rho A, \tag{5.18}$$

where Tr denotes the operator trace. This formula can be derived on the basis of Born's rule and the probabilistic interpretation of the classical statistical mixture (5.17).

5.2 References

[1] Khrennikov, A. (2009). *Contextual Approach to Quantum Formalism*. Springer Verlag, Berlin, Heidelberg, New York.

[2] Asano, M., Ohya, M., Tanaka, Y., Basieva, I., and Khrennikov, A. (2011). Quantum-like model of brain's functioning: decision making from decoherence. *Journal of Theoretical Biology*, 281, 56–64.

[3] Von Neumann, J. (1955). *Mathematical Foundations of Quantum Mechanics*. Princeton University Press, NJ.

[4] Khrennikov, A. (2009). Bell's inequality, Einstein, Podolsky, Rosen arguments and von Neumann's projection postulate. *Laser Physics*, 19, 346–356.

[5] Einstein, A., Podolsky, B., and Rosen, N. (1935). Can quantum-mechanical description of physical reality be considered complete? *Physical Review*, 47, 777.

[6] Dirac, P. A. M. (1995). *The Principles of Quantum Mechanics*. Clarendon Press, Oxford.

[7] Ingarden, R. S., Kossakowski, A., and Ohya, M. (1997). *Information Dynamics and Open Systems: Classical and Quantum Approach*. Kluwer, Dordrecht.

5.3 Double slit experiment: rationale for the existence of probability waves

Varadarajan [1] (slide 11) mentions that "the double slit experiment suggests that the logic of quantum theory is not classical, i.e. is not a Boolean algebra, because of the complementarity principle." There are many variant descriptions of the double slit experiment. Let us introduce the setting of the experiment with a description given by Varadarajan. Citing Varadarajan [1] (slide 8): "Electrons from a source, all with the same energy, pass through two slits on a screen and fall on a plate. We place a counter at a variable point x on the plate and record the arrival of the electrons. If the source is a weak one, we will be able to count the electrons one by one as they arrive." Using the same notation as in Varadarajan [1] (slide 8), denote by $A_i(x)$ the event that slit i $(= 1, 2)$ is open. Denote $A(x)$ as the event when both slits are open. From a classical point of view, one could then claim that $A(x) = A_1(x) \cup A_2(x)$ and the probability of the event $A(x)$ to occur would then be (assuming mutually exclusive events) $p(A(x)) = p(A_1(x)) + p(A_2(x))$. However, the experimental evidence, does show that this is not the case. How can that now be explained? The standard explanation revolves very often around the concept of so-called "wave–particle" duality.

Let us re-set the décor of the experiment in a slightly different way. We want to transit first from a classical mechanical situation to a quantum mechanical situation. Consider the description below (see Khrennikov [2], Khrennikov [3], Khrennikov and Haven [4]). We follow closely Haven [5] (pp. 43–45) (up to Section 5.4).

Imagine an experimenter firing a gun containing very tiny plastic pellets onto a screen, similar as in the Varadarajan description, which has two equally spaced slits. We imagine there is a detector screen, behind the screen containing the slits, upon which the pellets land. Assume there exists a detector which can count the pellets landings in the various locations on the detector screen. Denote the top and bottom slits as respectively slits 1 and 2. We then have the following three scenarios:

- slit 1 is open and slit 2 is closed,
- slit 1 is closed and slit 2 is open,
- slit 1 is open and slit 2 is open.

We imagine that the diameter of the pellets is substantially smaller than the slits' width. When carrying out the experiment, the following (expected) result occurs. In scenario 1, pellets start accumulating behind slit 1. Some pellets also land close to slit 2, and go even further as they are deflected on the edges of slit 1. In scenario 2, pellets accumulate behind slit 2 and some pellets land close to slit 1 and even further. When both slits are open we have an accumulation behind both slits and some scattering also.

The key issue now consists in remarking that if one were to convert the *pellets into electrons*, the result of the experiment would be altogether seriously different.

Essentially, with electrons, an interference pattern will form. Electrons will first behave like particles and with the onset of time, they will start behaving like waves. This in essence, exemplifies, the idea of wave–particle duality. The most striking (and frankly very much counter-intuitive) event is the fact that there is interference between an electron and itself. This result is substantially different from what we would expect if the electrons had been plastic pellets. Those results laid the basis for the formal development of quantum mechanics.

The probabilistic description of the electrons landing on the detector screen is different from the probabilistic description of plastic pellets landing on the detector screen. In fact, the difference in description is so resolutely different that a new notion was born: "probability interference." In a nutshell, the observed interference in the double slit experiment when electrons are fired almost requires that we must superpose probability distributions. Note the use of the word "almost." As is well know, such functions cannot be superposed.

Let us denote with $p_1(x)$ the probability that the electron arrives at position x when slit 1 is open. Similarly, for $p_2(x)$. We can now query what the expression would be when both slits 1 and 2 are open. Would it be $p_{12}(x) = p_1(x) + p_2(x)$? This probability formula would reflect the situation when both slits are open but *when we use plastic pellets* rather than electrons. Hence, this formulation is not reflecting the interference pattern experimenters found when they used electrons.

As we have indicated above, it is not possible to capture interference by superposing probability distributions. The mathematical machinery needed for the superposition of probability waves (or probability amplitudes) is well explained in Morrison [6] (pp. 55–56) and we follow it here closely. But before getting into the mechanics of this machinery, let us quickly remind ourselves about the notion of "conjugate" of a complex number, which we covered in Chapter 1, Section 1.13. We also discussed it briefly in Chapter 4, Section 4.3. The wave function is essential in the determination of the probability density function on the position. The pdf, denoted as $|\psi_1(x)|^2$, is obtained by writing that $|\psi_1(x)|^2$ is the product of the wave function with its complex conjugate, ψ^*. The wave function when written in terms of phase and amplitude can be expressed, for when slit 1 is open, as $|\psi_1(x)| \exp(iS_1(x))$, where $S_1(x)$ is the phase of the wave and $|\psi_1(x)|$ is the amplitude of the wave function; $|\psi_2(x)|$, for when slit 2 is open, can be expressed in the same way. The transition of going from the probability amplitude to the pdf is via the procedure of the complex conjugate. The complex conjugate of $|\psi_1(x)| \exp(iS_1(x))$ is simply $|\psi_1(x)| \exp(-iS_1(x))$. Similarly, for when slit 2 is open.

At this stage, we have not yet indicated how we arrive at probability interference. We follow now closely Morrison [6] (pp. 55–56). We need one more ingredient: superposition of the probability amplitudes. This can be defined as $\psi_{12}(x) = \psi_1(x) + \psi_2(x)$, where $\psi_{12}(x)$ is the superposed state. If we write that $p_{12}(x) \propto |\psi_1(x) + \psi_2(x)|^2$, then substituting $|\psi_i(x)| \exp(iS_i(x))$ into that expression, one obtains:

$$p_{12}(x) = |\psi_1(x)|^2 + |\psi_2(x)|^2 + 2\,|\psi_1(x)|\,|\psi_2(x)| \cos(S_1 - S_2). \tag{5.19}$$

A complex number z (see Chapter 4, Section 4.3) can be denoted as $z = x + iy$, where x is the real part and y is the imaginary part; z can also be written as $z = r\exp(i\theta)$, where $r = \sqrt{x^2 + y^2}$. r is often denoted as $|z|$. The angle $\theta = \tan^{-1}\left(\frac{y}{x}\right)$, where $y = r\sin\theta$ and $x = r\cos\theta$. In analogy with the wave function, we also say that $|z|$ is the amplitude and θ is the phase.

Equation (5.19), is the probability formula, which now includes the probability interference term $2\,|\psi_1(x)|\,|\psi_2(x)| \cos(S_1 - S_2)$. When this term is not zero, it renders the probability in a quantum context to be either sub- or super-additive.

5.4 References

[1] Varadarajan, V. S. (2008). The quantum dice. Talk given at U.C.L.A. Slides 1–20.
[2] Khrennikov, A. (1999). *Interpretations of Probability*. VSP International Science Publishers, Utrecht, Tokyo.
[3] Khrennikov, A. (2009). *Interpretations of Probability*. 2nd edition. De Gruyter, Berlin.
[4] Khrennikov, A. and Haven, E. (2009). Quantum mechanics and violations of the sure-thing principle: the use of probability interference and other concepts. *Journal of Mathematical Psychology*, 53, 5, 378–388.
[5] Haven, E. (2008). Elementary quantum mechanical principles and social science: is there a connection? *Romanian Journal of Economic Forecasting*, 9, 1, 41–58.
[6] Morrison, Michael, A. (1990). *Understanding Quantum Physics: A User's Manual*. Volume 1. 1st edition. Pearson Education, Upper Saddle River, NJ.

5.5 Quantum mechanical postulates

Aerts and Aerts [1] remark (p. 27) that there does not exist "a uniquely accepted set of axioms for quantum theory." The authors go on to show some examples of approaches used in quantum physics. They mention [1] (pp. 27–28) the Bohmian mechanics approach (see Chapter 6); the Feynman path integral[5] approach (which we have already mentioned in Chapter 1 and Chapter 3; the Birkhoff–von Neumann approach [3]; and the convexity of states approach [4]. They propose that any theory would have to contain at least three parts which link theory to experiment. They are as follows (p. 28):

[5] Holland [2] (p. 268) views the Feynman approach as a way to obtain "quantum action from the set of all classical actions."

1. "there exists a set $\sum_S$ of possible states for a system S";
2. from experiments one can infer a set of observable quantities, $\mathcal{O}_S$, and to "each observable there corresponds a set of discrete or continuous outcomes, X_A";
3. a map P exists which is as follows: "$P : \sum_S \times \mathcal{O}_S \times X_A \to [0, 1]$ and $\sum_{x_i \in X_A} P = 1$."

As Aerts and Aerts [1] (p. 28) remark, for a classical deterministic system, one defines $P : \sum_S \times \mathcal{O}_S \times X_A \to \{0, 1\}$ and $\sum_{x_i \in X_A} P = 1$.

The objective of this section does not consist of discussing an axiomatic structure of quantum mechanics. Many books exist on this topic and it is certainly not the objective of this present book to engage in this very complex topic. The title of this section in fact hints at the existence of postulates. Indeed, there are some basic postulates which are closely allied to the three parts presented above. In Chapter 1 and in other parts of this book, we mention the word "postulate" several times. For instance, in Chapter 1, we discussed (amongst others) the Bohr postulate about electrons taking on discrete energy values; the Bohr–Sommerfeld quantization rule; and the Heisenberg postulate (this chapter) that operators on position and momentum should satisfy a peculiar commutation relation (the Heisenberg uncertainty principle). We also mentioned (amongst others) the von Neumann projection postulate (or also the von Neumann–Lüders postulate). In Chapter 8, we will discuss Kolmogorov's postulate of being able to embed "context" in probability space. See also Kahneman [5] for a very interesting interpretation of "context." In Chapter 14, where we discuss the possibility of a quantum-like brain; we also will mention several postulates.

Morrison [6] distinguishes four postulates in quantum mechanics. Other authors may bring forward slightly different postulates. For instance, Bransden and Joachain [7] (pp. 194–205) propose six postulates. Omnès [8] (pp. 467–477) proposes the following rules:

Rule 1 (p. 467). "The theory of an individual isolated physical system is entirely formulated in terms of a specific Hilbert space and a specific algebra of operators, together with the mathematical notions associated with them." We remark that some of those mathematical notions were covered in Chapter 4.

Rule 2 (p. 468) "The vectors ψ in the Hilbert space evolve in time according to the Schrödinger equation."

Rule 3 (p. 468) "A physical system S can be said to consist of two non-interacting systems[6] S' and S''."

[6] Omnès does give more mathematical detail on how the Hilbert space of S relates to the two Hilbert spaces of S' and S''. We do not mention this here. Please consult, pp. 468–469 in Omnès [8].

Rule 4 (p. 476) "Every description of a physical system should be expressed in terms of properties belonging to a common consistent logic. A valid reasoning relating these properties should consist of implications holding in that logic."

Following Morrison [6], the four postulates he distinguishes are:

Postulate 1 (p. 58). "Every physically-realizable state of a system is described in quantum mechanics by a state function ψ that contains all accessible physical information about the system in that state."

This postulate has been discussed before. See also Bransden and Joachain [7] (p. 194).

Postulate 2 (p. 70). "If a system is in a quantum state represented by a wave function ψ, then $P dv = |\psi|^2 dv$ is the probability that in a position measurement at time t the particle will be detected in the infinitesimal volume element dv."

We have discussed this postulate already in Section 5.1 of this chapter. See equation (5.6).

Postulate 3 (p. 158). "In quantum mechanics, every observable is represented by an operator that is used to obtain physical information about the observable from state functions."

Postulate 4 (p. 186). "The time development of the state functions of an isolated quantum system is governed by the time-dependent Schrödinger equation . . . "

We have already mentioned this fourth postulate in Chapter 1, where we described it around equations (1.69) and (1.70).

Relative to postulate 3, we note that in the conventional axiomatics of quantum mechanics, observables are represented by Hermitian operators. However, in applications to economics and finance we can ignore this restriction and proceed in the generalized quantum mechanical framework, i.e. by operating with observables represented by non-Hermitian operators, e.g. Hamiltonians, see Chapter 13.

5.6 References

[1] Aerts, S. and Aerts, D. (2008). When can a data set be described by quantum theory? *Proceedings of the Second Quantum Interaction Symposium.* (Oxford University), pp. 27–33.

[2] Holland, P. (2000). *The Quantum Theory Of Motion: An Account of the de Broglie–Bohm Causal Interpretation of Quantum Mechanics.* Cambridge University Press.

[3] Birkhoff, G. and von Neumann, J. (1936). The logic of quantum mechanics. *Annals of Mathematics*, 37, 823–843.
[4] Mielniek, B. (1974). Generalized quantum mechanics. *Communications in Mathematical Physics*, 37, 3, 221–256.
[5] Kahneman, D. (2003). Maps of bounded rationality: psychology for behavioral economics. *American Economic Review*, 93, 1449–1475.
[6] Morrison, Michael, A. (1990). *Understanding Quantum Physics: A User's Manual.* Volume 1. 1st edition. Pearson Education, Upper Saddle River, NJ.
[7] Bransden, B. H. and Joachain, C. J. (2000). *Quantum Mechanics.* Prentice Hall.
[8] Omnès, R. (1994). *The Interpretation of Quantum Mechanics.* Princeton Series in Physics. Princeton University Press.

5.7 States and state functions

Postulate 1 above gives us a good idea about the meaning of a state. As per Morrison [1] (p. 58), note the word "accessible" in that postulate, which shows that in quantum mechanics we can never know with precision either position or momentum. This inherent imprecision is fully connected to the Heisenberg uncertainty principle which we also discuss below. Omnès [2] (p. 118) indicates that "a state of a system is well defined when one can assign a definite probability to every conceivable property."

Note also that position and time are *independent* quantities in quantum mechanics and that is because classical quantum mechanics does not allow for paths. The state function, before complex conjugation is applied, is indicative of the probability wave or the probability amplitude. Note also that only normalizable state functions can serve as representing quantum states. We also distinguish between so called "stationary" and "non-stationary" states. In the case where we have a conservative system, i.e. with a real potential which is time-independent, stationary states will exist. When the wave function is not separable, then we must obtain a non-stationary state. A stationary state is an example of a so-called eigenstate. Following Bowman [3] (p. 146), consider the Hamiltonian operator of the Schrödinger equation which is Hermitian. An eigenstate of this Hamiltonian is time invariant. For a simple overview between the difference of a classical and a quantum description of a state of a system, see Eisberg and Resnick [4], pp. 409–410.

An important distinction is to be made between so-called "pure" states and "mixed" states. Following Bransden and Joachain [5] (p. 641), when a system is described by a single wave function then it is called a pure state. In the case of a mixed state, a statistical mixture of wave functions is needed. Mixed states are represented by density operators (in the finite-dimensional case, density matrices), see Section 5.1.

5.8 References

[1] Morrison, Michael, A. (1990). *Understanding Quantum Physics: A User's Manual.* Volume 1. 1st edition. Pearson Education, Upper Saddle River, NJ.

[2] Omnès, R. (1994). *The Interpretation of Quantum Mechanics*. Princeton Series in Physics. Princeton University Press.
[3] Bowman, G. (2008). *Essential Quantum Mechanics*. Oxford University Press.
[4] Eisberg, R. and Resnick, R. (1985). *Quantum Physics of Atoms, Molecules, Solids, Nuclei and Particles*, J. Wiley.
[5] Bransden, B. and Joachain, C. (2000). *Quantum Mechanics*. Prentice Hall.

5.9 Wave packets – constructive and destructive interference

Wave packets can be expressed in the form of:

$$\psi(x,t) = \frac{1}{\sqrt{2\pi}} \int_{-\infty}^{\infty} A(k) \exp(i(kx - \omega t))dk, \tag{5.20}$$

where k is a wave number, $A(k)$ is the amplitude function, t is time, x is position, and ω is the angular frequency. This expression shows, as per Morrison [1] (p. 109) that "there is a superposition of an infinite number of plane wave functions with infinitesimally differing wave numbers k." Note that there are infinite number of regions of constructive interference which are combined with an infinite number of regions of destructive interference. See Morrison [1] (p. 108). The above expression is an example of Fourier integration. Kreyszig [2] (p. 570), defines a Fourier transform of a function f as:

$$\widehat{f}(w) = \frac{1}{\sqrt{2\pi}} \int_{-\infty}^{\infty} f(x) \exp(-iwx)dx, \tag{5.21}$$

where i is a complex number. Note that $f(x)$: (i) needs to be piecewise continuous on every finite interval and (ii) $f(x)$ is absolutely integrable on the X-axis. Kreyszig [2] (p. 570) defines the Fourier integral as:

$$f(x) = \frac{1}{\sqrt{2\pi}} \int_{-\infty}^{\infty} \widehat{f}(w) \exp(iwx)dw. \tag{5.22}$$

Remark that the wave number k can be written via the de Broglie relation as $k = \frac{p}{\hbar}$, where p is momentum.

5.10 References

[1] Morrison, Michael, A. (1990). *Understanding Quantum Physics: A User's Manual*. Volume 1. 1st edition. Pearson Education Inc., Upper Saddle River, NJ.
[2] Kreyszig, E. (1999). *Advanced Engineering Mathematics*. J. Wiley.

5.11 Heisenberg's uncertainty principle

This equality was already introduced in Chapter 1, equation (1.54), and also in the section on "Operators" in Chapter 4. The original discussion can be found in Heisenberg [1]. The relation expresses the fact that there is no full precision on measuring either position or momentum. If one measures the position with a high precision, the trade off will be that there will be more uncertainty on the measurement of momentum and vice versa. "Simultaneous eigenfunctions of (position and momentum operators) cannot (have) a physically realizable quantum state" (see Morrison [2] (p. 490)). The principle does *not* say anything about the imprecision of the measurement apparatus. Instead, the principle does indicate there exists a intrinsic uncertainty at the quantum scale. In its easy form, the relation can be expressed as $\Delta x \Delta p \geq \frac{\hbar}{2}$. For a Gaussian wave packet, the relation is said to be minimal and becomes $\Delta x \Delta p = \frac{\hbar}{2}$. The Heisenberg uncertainty principle can be proven with the use of the Schwartz inequality. See equation 6.7.13 in Holland [3] (p. 257). Finally, the Heisenberg uncertainty principle is a special case of the so-called generalized uncertainty principle.

5.12 References

[1] Heisenberg, W. (1927). Über den anschaulichen Inhalt der quantentheoretischen Kinematik und Mechanik. *Zeitschrift für Physik*, 43, 172–198.
[2] Morrison, Michael, A. (1990). *Understanding Quantum Physics: A User's Manual*. Volume 1. 1st edition. Pearson Education, Upper Saddle River, NJ.
[3] Holland, P. (2000). *The Quantum Theory of Motion: An Account of the de Broglie–Bohm Causal Interpretation of Quantum Mechanics*. Cambridge University Press.

5.13 The time-dependent and time-independent Schrödinger PDE

Equation (1.69) in Chapter 1, Section 1.20, formulated the time-dependent Schrödinger PDE for an N-particle system. This is the most general version of that PDE. The time-independent Schrödinger PDE obviously does not contain the time parameter and is substantially easier to solve. The separation of variables as a solution technique is well known, and it implies that one can write the time-dependent wave function as a product of two other functions $\Psi(x, t) = \psi(x).\alpha(t)$. This approach will require a time-independent real potential function and thus will imply the use of a conservative system.

The time-independent Schrödinger PDE for a 1-particle system can be written in shorthand as $\widehat{\mathcal{H}}\Psi = E\Psi$, where E is the particle's energy. If we recall our brief entry on the meaning of eigenvalues and eigenfunctions in Chapter 4, then the above 1-particle time-independent Schrödinger equation can be seen to very much resemble what we entered in Section 4.9, $\widehat{w}\psi = l\psi$. Clearly, $\widehat{w}$ is now the Hamiltonian operator. The wave function, Ψ, is the eigenfunction and the energy, E, is now the eigenvalue. Section 4.11 in Chapter 4 also discussed Hermiticity. We know the Hamiltonian operator is Hermitian and hence this implies that E is real. See also Morrison [1] (p. 464) for a more extensive discussion.

Messiah [2] (p. 61) points out that the choice of the wave function (i.e. the Schrödinger equation) is restricted to certain "a priori conditions" such as the (i) linearity and homogeneity of the equation, and (ii) the PDE must be "first order with respect to time." The latter condition is then linked to the fact that the wave function once known at an instant of time t will determine the function at later times.

The Schrödinger equation exhibits some interesting properties. Auletta *et al.* [3] (pp. 107–110) mention some of those properties for the cases when position variables are one- and three-dimensional. Here are some regularity properties which are interesting. The wave function Ψ, [3] (p. 107), "has to be single valued and continuous." As per Auletta, Fortunato, and Parisi [3] (p. 107) there "can not be two different probability amplitudes for the same position" and continuity is necessary since the Schrödinger equation requires differentiability. When the real potential were to be infinite in some regions, then continuity of the wave function is still needed (see (ii) in [3] (p. 108)). However, its first derivatives may not be continuous. Finally, another interesting regularity property says that whatever the dimension, "the wave function of the ground state never vanishes" [3] (p. 108).

5.14 References

[1] Morrison, Michael, A. (1990). *Understanding Quantum Physics: A User's Manual.* Volume 1. 1st edition. Pearson Education, Upper Saddle River, NJ.
[2] Messiah, A. (1961). *Quantum Mechanics.* Volume I. North-Holland.
[3] Auletta, G., Fortunato, M., and Parisi, G. (2009). *Quantum Mechanics.* Cambridge University Press.

5.15 Classical limit ideas: Ehrenfest's approach and the correspondence principle

Morrison [1] (p. 169) reports that the correspondence principle proclaims that in the so-called "classical limit" quantum mechanical laws must become Newtonian laws. Clearly, the correspondence principle must establish a relationship between an operator (on position and momentum) and the classical equivalent of the operator. The obtaining of the classical limit can sometimes be defined with the argument that one wants to see what occurs when $\hbar \to 0$ and this means really (following Morrison [1] (p. 170)) that in the classical limit "the quantum effects are immeasurably small." Holland [2] (p. 219) refers to a proposal by Berry [3] who says that the smallness of $\hbar$ "has no absolute meaning because its value depends on the system of units." The uncertainty principle can apply to macroscopic particles, but as Morrison [1] (p. 171) indicates the meaning of that uncertainty principle is the same, i.e. there is *no* limit on the precision of measurement at the macroscopic scale. Here is a precise way to express the correspondence principle: Morrison [1] (p. 172): "the classical function $Q(t)$ is the expectation value of the operator $\widehat{Q}$ with respect to the wave function $\psi(x,t)$ that represents the state." Thus, the average momentum, $\langle p\rangle = m\frac{d\langle x\rangle}{dt}$ becomes $p = m\frac{dx}{dt}$ in the classical limit. This forms the first part of Ehrenfest's theorem; similarly, for the average position. Ehrenfest [4] proposed the following relationship (this is the second part of Ehrenfest's theorem) $\frac{d\langle p\rangle(t)}{dt} = \left\langle -\frac{\partial V}{\partial x}\right\rangle$ (see also Morrison [1] (p. 520)), which indeed gives the quantum mechanical equivalent of Newton's second law. We follow closely Morrison [1] (pp. 521–522) (equations 11.105–11.112) for a sketch of the proof of this theorem. The key ingredient needed is the formulation for the time development of the expectation of a Hermitian operator $\frac{d\langle Q(t)\rangle}{dt} = \left\langle \frac{\partial \widehat{Q}}{\partial t}\right\rangle + \frac{1}{\hbar}\left\langle i\left[\widehat{\mathcal{H}}, \widehat{Q}\right]\right\rangle$, where $\widehat{Q}$ is an arbitrary Hermitian operator and $\widehat{\mathcal{H}}$ is the Hermitian Hamiltonian operator. To obtain the position equivalent in the classical limit, one replaces the arbitrary Hermitian operator $\widehat{Q}$ by a position operator. We can write $\frac{d\langle x\rangle(t)}{dt} = \frac{1}{m}\langle p\rangle(t)$, where use was made of the fact that the commutator relation for position is $\left[\widehat{\mathcal{H}}, \widehat{x}\right] = \frac{-i\hbar}{m}\widehat{p}$. To obtain the momentum equivalent in the classical limit, one follows the same strategy using now the commutator relation for momentum $\left[\widehat{\mathcal{H}}, \widehat{p}\right] = i\hbar\frac{\partial V}{\partial x}\widehat{1}$, where V is the real potential and $\widehat{1}$ is the identity operator. The result is then $\frac{d\langle p\rangle(t)}{dt} = -\left\langle\frac{\partial V}{\partial x}\right\rangle$. And from the latter one can find immediately that $m\frac{d^2\langle x\rangle(t)}{dt^2} = \langle F(x,t)\rangle$. The right-hand side can in effect be written as $-\int_{-\infty}^{\infty}\psi^*(x,t)\left[\frac{\partial V(x,t)}{\partial x}\right]\psi(x,t)dx$ (Morrison [1] (p. 522) (equation 11.112)), and note the minus sign which appears because of the fact that force is the negative derivative of the real potential towards position. In the classical limit, $\int_{-\infty}^{\infty}\psi^*(x,t)F(x,t)\psi(x,t)dx$ can be written as $F(\langle x\rangle(t))$. Thus, in the classical limit the time evolution of $\langle x\rangle$ and $\langle p\rangle$ appears as in Newtonian mechanics.

As is remarked in Holland [2] (p. 255), "the mean quantum (and classical) motion will coincide rigorously with that of a classical particle when the following condition is satisfied: $\langle F(x)\rangle = F(\langle x\rangle)$" (see Messiah [5] (p. 217), as quoted in Holland [2]). Holland [2] (p. 255) remarks that this equality is obeyed for polynomial potentials up to and including second degree.

We note that within a Bohmian mechanics context (please see the next chapter for the basics of Bohmian mechanics), Holland [2] (p. 224) makes the argument that the "quantum dynamics will coincide with, or approach, the classical dynamics when the quantum force (the negative gradient of the quantum potential) and power become negligible in comparison with the classical force and power."[7] It can be shown that if the quantum potential tends to zero and this is inserted into the

[7] Please see the next chapter for a discussion on the emergence of the quantum potential.

quantum Hamilton–Jacobi and continuity equations (which are equations (6.14) and (6.15) in the next chapter), one obtains the classical Hamilton–Jacobi and continuity equations. Thus when the quantum potential tends to zero, one can argue this as a correspondence principle (see Holland [2] (p. 226)).

There exists another well-known approach to the classical-quantum duality world and that is the Wentzel [6], Kramers [7], Brillouin [8] approach or the so-called WKB approach. Holland [2] (pp. 231–234) shows convincingly that this is not necessarily a foolproof method to argue for a classical limit. As per Holland [2] (p. 232), when the real potential "is a slowly varying function of (position) x and (the energy difference) $E - V$ is not too small,"[8] one can find a "classical wave function" (see Holland [2] (p. 232 (6.4.9)) and the quantum potential using the amplitude from that wave function can then be generated. Please note the interpretation of this wave function is not obvious. We come back to the WKB method in Chapter 13 of this book.

We finish this sub-section with some of the words out of the "caveat lector" section of the paper by de Gosson and Hiley [10]. De Gosson and Hiley cite (p. 1418 in their paper) the work of Mackey [11] (p. 106) who mentions that "quantum mechanics is not just an algorithm for attaching a self-adjoint operator to every classical Hamiltonian, because such a programme would overlook many facts: first quantum mechanics rules out a large number of conceivable Hamiltonians, and secondly there are features of quantum mechanics (such as spin) which do not manifest themselves in the classical limit."

5.16 References

[1] Morrison, Michael, A. (1990). *Understanding Quantum Physics: A User's Manual.* Volume 1. 1st edition. Pearson Education, Upper Saddle River, NJ.
[2] Holland, P. (2000). *The Quantum Theory of Motion: An Account of the de Broglie–Bohm Causal Interpretation of Quantum Mechanics.* Cambridge University Press.
[3] Berry, M. (1991). Some quantum-to-classical asymptotics. In *Chaos and Quantum Physics. Proceedings of the Les Houches Summer School (1989).* Eds. M. Giannoni, A. Voros, and J. Zinn-Justin. North-Holland, pp. 251–303.
[4] Ehrenfest, P. (1927). Bemerkung über die angenäherte Gültigkeit der Klassischen Mechanik innerhalb der Quantenmechanik. *Zeitschrift für Physik*, 45, 455–457.
[5] Messiah, A. (1961). *Quantum Mechanics.* Volume 1. North-Holland.
[6] Wentzel, G. (1926). Eine Verallgemeinerung der Quantenbedingungen für die Zwecke der Wellenmechanik. *Zeitschrift für Physik*, 38, 518–529.
[7] Kramers, H. (1926). Wellenmechanik und halbzahlige Quantisierung. *Zeitschrift für Physik*, 39, 828–840.
[8] Brillouin, L. (1926). La mécanique ondulatoire de Schrödinger: une méthode générale de résolution par approximations successives. *Comptes Rendus*, 183, 24–26.

[8] Please see also Bohm [9] for more detail.

[9] Bohm, D. (1951). *Quantum Theory*. Prentice Hall.
[10] de Gosson, M. A. and Hiley, B. (2011). Imprints of the quantum world in classical mechanics. *Foundations of Physics*, 41, 1415–1436.
[11] Mackey, G. W. (1998). The relationship between classical and quantum mechanics. *Contemporary Mathematics*, 214. American Mathematical Society.

6

Basic elements of Bohmian mechanics

6.1 Short introduction to Bohmian mechanics

Sheldon Goldstein in a *Foundations of Physics* paper [1] indicates (p. 341): "Since macroscopic objects are normally regarded as built out of microscopic constituents . . . there can be no problem of macroscopic reality per se in Bohmian mechanics." Goldstein in the same paper remarks (p. 342) that this $|\psi|^2$ has "a status very much the same as that of a thermodynamic equilibrium."

Basil Hiley the closest collaborator of David Bohm, says the following (Hiley [2] (p. 2)): "What Bohm (1952a; 1952b) did was to show how to retain a description of all the usual properties of a classical world and yet remain completely within the quantum formalism." The Bohm (1952) [3] [4] references in that quote refer to the original work of Bohm in which he sets out the basics of the Bohmian mechanics.

As we will see later in the next section of this chapter, the appearance of an "additional term" in the Hamilton–Jacobi equation, is the hallmark of Bohmian mechanics and it is very often termed the "quantum potential." Hiley [2] (p. 2) remarks that since the Bohmian momentum is a well-defined[1] function of position and time, an ensemble of trajectories can be found when the quantum potential is non-zero. A unique, classical path is found when the quantum potential is zero. From the outset, we hope the reader can savor the beauty of this simple but very powerful result. Bohmian mechanics indeed shows this very gentle transition from quantum mechanics to classical mechanics via the value of the quantum potential. A perhaps less gentle transition can be found when considering the Ehrenfest limit theorem, which is the theorem we covered in Section 5.15 of Chapter 5.

A very important characteristic of Bohmian mechanics is that paths *do* have simultaneous determined position and momentum, i.e. we can legitimately talk about the existence of paths. But as Hiley remarks [2] "even though as observers

[1] For those who want to know (but please see the sections below in this chapter), it is the gradient of the phase of the polar version of the wave function which is the momentum.

(we) do not know the value of one of these variables." The quantum potential's significance from a physics point of view seems to indicate the (Hiley [2] (p. 3)) "quantum potential had simply 'locked together' the particles described by entangled states."

6.2 References

[1] Goldstein, S. (2009). Bohmian mechanics and quantum information. *Foundations of Physics*, 40, 335–355.
[2] Hiley, B. (2008). Quantum reality unveiled through process and the implicate order. *Proceedings of the Second Quantum Interaction Symposium*. University of Oxford, pp. 1–10.
[3] Bohm, D. (1952a). A suggested interpretation of the quantum theory in terms of hidden variables. *Physical Review*, 85, 166–179.
[4] Bohm, D. (1952b). A suggested interpretation of the quantum theory in terms of hidden variables. *Physical Review*, 85, 180–193.

6.3 Mathematical formalism

The mathematical formalism involved in deriving the core arguments of the Bohmian mechanical set up is quite easy. Holland [1] (pp. 68–69 and p. 134) provides for some of the essential steps. See also Bohm and Hiley ([2] (pp. 28–29)) and Auletta, Fortunato, and Parisi [3]. For the development below (equations (6.1) until (6.16)), we follow Choustova [4] (pp. 9–12) (equations (2.3)–(2.11)).

The wave function in polar form is considered:

$$\psi(q,t) = R(q,t)e^{i\frac{S(q,t)}{h}}, \tag{6.1}$$

where $R(q,t) = |\psi(q,t)|$; $S(q,t)/h$ is the argument of the complex number $\psi(q,t)$. This polar form is put into the Schrödinger equation:

$$ih\frac{\partial\psi}{\partial t} = -\frac{h^2}{2m}\frac{\partial^2\psi}{\partial q^2} + V(q,t)\psi(q,t). \tag{6.2}$$

Begin with the left-hand side of the Schrödinger equation:

$$ih\frac{\partial\psi}{\partial t} = ih\left(\frac{\partial R}{\partial t}e^{i\frac{S}{h}} + R\frac{i}{h}\frac{\partial S}{\partial t}e^{i\frac{S}{h}}\right) \tag{6.3}$$

$$= ih\frac{\partial R}{\partial t}e^{i\frac{S}{h}} - R\frac{\partial S}{\partial t}e^{i\frac{S}{h}}. \tag{6.4}$$

Consider now the right-hand side of the Schrödinger equation. Write out first $\frac{\partial\psi}{\partial q}$:

$$\frac{\partial R}{\partial q}e^{i\frac{S}{h}} + R\frac{i}{h}\frac{\partial S}{\partial q}e^{i\frac{S}{h}}. \tag{6.5}$$

Now write out $\frac{\partial^2\psi}{\partial q^2}$:

$$\frac{\partial^2 R}{\partial q^2}e^{i\frac{S}{h}} + \frac{i}{h}\frac{\partial R}{\partial q}\frac{\partial S}{\partial q}e^{i\frac{S}{h}} + \frac{\partial R}{\partial q}\frac{i}{h}\frac{\partial S}{\partial q}e^{i\frac{S}{h}} + R\frac{i}{h}\frac{\partial^2 S}{\partial q^2}e^{i\frac{S}{h}} - \frac{R}{h^2}\left(\frac{\partial S}{\partial q}\right)^2 e^{i\frac{S}{h}}, \quad (6.6)$$

which can then be simplified to:

$$\frac{\partial^2 R}{\partial q^2}e^{i\frac{S}{h}} + \frac{2i}{h}\frac{\partial R}{\partial q}\frac{\partial S}{\partial q}e^{i\frac{S}{h}} + R\frac{i}{h}\frac{\partial^2 S}{\partial q^2}e^{i\frac{S}{h}} - \frac{R}{h^2}\left(\frac{\partial S}{\partial q}\right)^2 e^{i\frac{S}{h}}. \quad (6.7)$$

Substitute all this into the Schrödinger equation: $ih\frac{\partial\psi}{\partial t} = -\frac{h^2}{2m}\frac{\partial^2\psi}{\partial q^2} + V(q,t) \times \psi(q,t)$.

One obtains:

$$ih\frac{\partial R}{\partial t}e^{i\frac{S}{h}} - R\frac{\partial S}{\partial t}e^{i\frac{S}{h}} = \frac{-h^2}{2m}\left[\begin{array}{c}\frac{\partial^2 R}{\partial q^2}e^{i\frac{S}{h}} + \frac{2i}{h}\frac{\partial R}{\partial q}\frac{\partial S}{\partial q}e^{i\frac{S}{h}} + \\ R\frac{i}{h}\frac{\partial^2 S}{\partial q^2}e^{i\frac{S}{h}} - \frac{R}{h^2}\left(\frac{\partial S}{\partial q}\right)^2 e^{i\frac{S}{h}}\end{array}\right] + V\psi.$$

Multiply the above with $e^{-i\frac{S}{h}}$, so as to get:

$$ih\frac{\partial R}{\partial t} - R\frac{\partial S}{\partial t} = \frac{-h^2}{2m}\left[\begin{array}{c}\frac{\partial^2 R}{\partial q^2} + \frac{2i}{h}\frac{\partial R}{\partial q}\frac{\partial S}{\partial q} + \\ R\frac{i}{h}\frac{\partial^2 S}{\partial q^2} - \frac{R}{h^2}\left(\frac{\partial S}{\partial q}\right)^2\end{array}\right] + VR,$$

where $VR = V\psi e^{-i\frac{S}{h}} = VR\, e^{i\frac{S}{h}}e^{-i\frac{S}{h}}$.

The real and imaginary parts can now be separated:
First the imaginary part:

$$\frac{\partial R}{\partial t} = \frac{-h^2}{2m}\frac{1}{ih}\left[\frac{2i}{h}\frac{\partial R}{\partial q}\frac{\partial S}{\partial q} + R\frac{i}{h}\frac{\partial^2 S}{\partial q^2}\right]. \quad (6.8)$$

This can be simplified into:

$$\frac{\partial R}{\partial t} = \frac{-1}{2m}\left[2\frac{\partial R}{\partial q}\frac{\partial S}{\partial q} + R\frac{\partial^2 S}{\partial q^2}\right]. \quad (6.9)$$

For the real part:

$$-R\frac{\partial S}{\partial t} = \frac{-h^2}{2m}\left[\frac{\partial^2 R}{\partial q^2} - \frac{R}{h^2}\left(\frac{\partial S}{\partial q}\right)^2\right] + VR. \quad (6.10)$$

Multiply now (6.9) (left-hand side and right-hand side) by $2R$, so as to get:

$$2R\frac{\partial R}{\partial t} = \frac{-1}{2m}\left[2R2\frac{\partial R}{\partial q}\frac{\partial S}{\partial q} + 2RR\frac{\partial^2 S}{\partial q^2}\right]. \quad (6.11)$$

The left-hand side of (6.11) can also be written as:

$$2R\frac{\partial R}{\partial t} = \frac{\partial R^2}{\partial t}. \tag{6.12}$$

Remark also that $2R\frac{\partial R}{\partial q}\frac{\partial S}{\partial q} + RR\frac{\partial^2 S}{\partial q^2}$ (which is in the right-hand side of (6.11)) can also be written as:

$$2R\frac{\partial R}{\partial q}\frac{\partial S}{\partial q} + RR\frac{\partial^2 S}{\partial q^2} = \frac{\partial}{\partial q}\left(R^2\frac{\partial S}{\partial q}\right). \tag{6.13}$$

Thus, (6.11) can now be rewritten as:

$$\frac{\partial R^2}{\partial t} + \frac{1}{m}\frac{\partial}{\partial q}\left(R^2\frac{\partial S}{\partial q}\right) = 0. \tag{6.14}$$

Equation (6.14) is also known under the name of the "continuity equation." We can see this equation expresses the evolution of a probability distribution, since $R^2 = |\psi|^2$.

We can simplify (6.10) a little (divided by $-R$):

$$\frac{\partial S}{\partial t} + \frac{1}{2m}\left(\frac{\partial S}{\partial q}\right)^2 + \left(V - \frac{h^2}{2mR}\frac{\partial^2 R}{\partial q^2}\right) = 0. \tag{6.15}$$

Equation (6.15) is important for our purposes. If $\frac{h^2}{2m} << 1$ and let $\frac{h^2}{2mR}\frac{\partial^2 R}{\partial q^2}$ be negligibly small, then (6.15) is a Hamilton–Jacobi equation. Bohm [5] interprets the above equation by claiming that $-\frac{h^2}{2mR}\frac{\partial^2 R}{\partial q^2}$ is a so-called quantum potential.

The Newton–Bohm equation is:

$$m\frac{d^2 q(t)}{dt^2} = -\frac{\partial V(q,t)}{\partial q} - \frac{\partial Q(q,t)}{\partial q} \tag{6.16}$$

and $Q(q,t)$ depends on the wave function and the wave function evolves according to the Schrödinger equation.

The initial conditions are $q(t_0) = q_0$ and $q'(t_0) = q_0'$ (momentum).

6.4 References

[1] Holland, P. (2000). *The Quantum Theory of Motion: An Account of the de Broglie–Bohm Causal Interpretation of Quantum Mechanics.* Cambridge University Press.

[2] Bohm, D. and Hiley, B. J. (1993). *The Undivided Universe: An Ontological Interpretation of Quantum Theory.* Routledge.

[3] Auletta, G., Fortunato, M., and Parisi, G. (2009). *Quantum Mechanics.* Cambridge University Press.

[4] Choustova, O. (2007). Quantum Bohmian model for financial market. Department of Mathematics and System Engineering, International Center for Mathematical Modeling, Växjö University, Sweden.
[5] Bohm, D. (1952). A suggested interpretation of the quantum theory in terms of "hidden" variables, Parts I and II. *Physical Review*, 85, 166–193.

6.5 Non-locality

Non-locality is a topic which has been explained at multiple points in Chapter 1. An essential issue which relates intrinsically to the concept of non-locality is the fact that in a classical mechanics setting forces do "fall off with interparticle distance" (Holland [1] (p. 282)). And Holland [1] (p. 282), continues "the contribution to the total force acting on the ith particle coming from the quantum potential . . . does not necessarily fall off with distance." The most telling formulation of non-locality can probably be found back in the use of the Newton–Bohm law, which was equation (6.16) above, and which for a one body system is simply:

$$m.a = -\nabla (V + Q), \tag{6.17}$$

where m is mass, a is acceleration, V is the real potential, and Q is the quantum potential.

As per Holland [1] (p. 282), for a "many body system in the limit of large separations" one obtains

$$m_i a_i = -\left[\nabla_i Q(q_1, q_2, \ldots, q_n) + \nabla_i V_i(q_i)\right]_{q_j = q_j(t)}, \tag{6.18}$$

where i is the ith particle (this is equation 7.1.12; Holland [1] (p. 282)). Holland ([1] (p. 282)) would see three characteristics as typical of what he calls a "non-local connection":

- "dependence of each particle orbit on all the others";
- "response of the whole to localized disturbances";
- "the extension of actions to large interparticle distances."

Holland ([1] (p. 290)) reports that the "condition for non-locality is . . . non-factorizability" of the wave function. The factorizability of, say, a two body system-based wave function would be $\psi(x_1, x_2) = \psi_A(x_1)\psi_B(x_2)$ (Holland [1] (p. 287, equation (7.2.1)). Particle 1 is associated with ψ_A and satisfies its own Schrödinger equation. Similarly, particle 2 is associated with ψ_B and also satisfies its own Schrödinger equation. Holland writes [1] (p. 288) "factorizability implies physical independence." Bohm and Hiley [2] (p. 61) compare the factorization to the existence of independent pools of information.

As Holland indicates ([1] (p. 462)), it can be that "quantum mechanics is both incomplete and non-local." If the wave function is interpreted as a probability amplitude and not as a kind of a physical wave (the ensemble interpretation of wave function – Einstein, Margenau, Ballentine, see Chapter 1), then "quantum non-locality" is not surprising at all. The same property can be assigned with the Smoluchowski equation for a few Brownian particles, see again Chapter 1. Hence, the non-locality of Bohmian mechanics is so intriguing only because the wave function is assigned with an individual (may be composite) system (the pilot wave), but not an ensemble of such systems. We urge the reader also to consult again our discussion on non-locality in Chapter 1, Section 1.20, just under equation (1.78).

Now consider again equation (6.18) and one can indeed see that the non-locality is intimately connected to the existence of the quantum potential. Please also refer to Bohm and Hiley [2] (pp. 134–159), who provide for a full chapter on the topic of non-locality.

6.6 References

[1] Holland, P. (2000). *The Quantum Theory of Motion: An Account of the de Broglie–Bohm Causal Interpretation of Quantum Mechanics*. Cambridge University Press.

[2] Bohm, D. and Hiley, B. J. (1993). *The Undivided Universe: An Ontological Interpretation of Quantum Theory*. Routledge.

6.7 Criticisms of Bohmian mechanics

In Chapter 3, we already mentioned that Bohm in effect revived hidden variables. Historically speaking, we should mention that one of the very early critics of Bohmian mechanics was also a very unexpected critic: Louis de Broglie. Recall that we mentioned in Chapters 1 and 3 that de Broglie interpreted quantum waves as physical waves. Jammer [1] (p. 287) indicates that originally de Broglie rejected Bohm's work, but later on he came to accept it more. Jammer [1] (p. 289) cites Takabayasi [2], who said that "the proposed hidden variables would become physically significant only through a modification of the Schrödinger equation or law of motion, a modification which is 'artificial and improbable.' " Holland [3] (pp. 25–26) sums up very well the usual objections which are voiced against Bohmian mechanics. We mention some of them here (Holland [3] (p. 25)):

1. "you can not prove the trajectories are there";
2. "it predicts nothing new";
3. "it attempts to return to classical physics";
4. "the price to be paid is nonlocality."

Critique '1', as Holland remarks [3] (p. 25), is indeed partially true since the Heisenberg uncertainty principle holds. But indeed, as Holland also further indicates, this in itself is not sufficient proof that no trajectories can exist. Critique '2' can be replied to as follows. Holland [3] indicates (pp. 184–185), that "In the quantum theory of motion we can make more detailed statements about the behavior of individual elements in the ensemble than are contained in the distribution function, of the type possible in classical mechanics." Holland's [3] (p. 25) main argument in reply to the third objection is that there exists a "'state' of a mechanical system that lies beyond the material points" and therefore this makes that Bohmian mechanics is not just a straight return to classical physics.

A critique which is sometimes put forward against the use of Bohmian mechanics in a macroscopic system is that there would *seem* to be no action of the particle on the wave. The "one-way" direction, i.e. of the wave function guiding the particle (but no opposite influence), may have a connection to the fact that equation (6.16), $m\frac{d^2q(t)}{dt^2} = -\frac{\partial V(q,t)}{\partial q} - \frac{\partial Q(q,t)}{\partial q}$, is the particle equation of motion. However, we hasten to add that we must tread very carefully here. It is essential to remark that in fact "one can deduce the wave function from the trajectories" (Holland [4]). Please see Holland [5] and Holland [6] for recent results.

Finally, as to the last point of critique (i.e. non-locality), one can safely claim that, following Holland [3] (p. 25), non-locality is indeed "an intrinsic feature of the de Broglie–Bohm theory." However, non-locality, although may be not a desirable physical property, seems to be found in physics approaches other than Bohmian mechanics. For instance, Bohm and Hiley [7] (p. 144) mention that Aspect *et al.* [8] [9] [10] provided for an "experimental proof that, if there are hidden variables, they must be non-local."

6.8 References

[1] Jammer, M. (1974). *The Philosophy of Quantum Mechanics*. J. Wiley, New York.
[2] Takabayasi, T. (1952). On the formulation of quantum mechanics associated with classical pictures. *Progress of Theoretical Physics*, 8, 143–182.
[3] Holland, P. (2000). *The Quantum Theory of Motion: An Account of the de Broglie–Bohm Causal Interpretation of Quantum Mechanics*. Cambridge University Press.
[4] Holland, P. (2012). Private communication with one of the authors of this book.
[5] Holland, P. (2005). Computing the wavefunction from trajectories: particle and wave pictures in quantum mechanics and their relation. *Annals of Physics*, 315, 505–531.
[6] Holland, P. (2006). Quantum back-reaction and the particle law of motion. *Journal of Physics A: Mathematical and General*, 39, 559–564.
[7] Bohm, D. and Hiley, B. J. (1993). *The Undivided Universe: An Ontological Interpretation of Quantum Theory.* Routledge.

[8] Aspect, A., Grangier, P., and Roger, G. (1981). Experimental tests of realistic local theories via Bell's theorem. *Physical Review Letters*, 47, 460–463.
[9] Aspect, A., Grangier, P., and Roger, G. (1982). Experimental realization of Einstein–Podolsky–Rosen–Bohm gedankenexperiment: a new violation of Bell's inequalities. *Physical Review Letters*, 49, 2, 91–94.
[10] Aspect, A., Dalibard, J., and Roger, G. (1982). Experimental test of Bell's inequalities using time-varying analyzers. *Physical Review Letters*, 42, 1804–1807.

Part III

Quantum probabilistic effects in psychology: basic questions and answers

7
A brief overview

7.1 Decision making in social science: general overview

Decision-making models are central in economics and psychology. Finding appropriate models which can approach human decision-making behavior is, as expected, a very challenging task. Economics has for a long time embraced models which are based on a particular axiomatic skeleton. The axioms are thought to be "reasonable" approximations of general decision situations.

A central starting point in preference modeling in economics consists in proving that there exists an equivalence between the preference relation of an object x over an object y, denoted as $x \succ y$, if and only if there exists a utility function $u(.)$, which maps a set of objects into $\mathbb{R}$, such that $u(x) > u(y)$. This is not an easy task. However, for a finite set of objects X (to which x and y belong), as any good micro-economic theory textbook will show, the equivalence is quite easy to show. It is substantially more difficult to show when the set X is countable infinite or uncountable. As David Kreps [1] indicates (p. 24), for an uncountable X there may exist a preference relation $\succ$, but this does not mean that there exists a utility function $u(.)$! The lexicographic preference relation is an example.

The utility function has many practical uses in social science. Here we mention two very basic uses:

1. In economics, the utility function plays an essential role in the derivation of a demand function. The classical approach to deriving such a demand function consists in maximizing a function subject to an equality constraint. In economics, this is known as the "method of Lagrange." Ingersoll [2] (p. 8) describes the method in a nutshell: to maximize a function f subject to an (equality) constraint $g(x) = a$, the so-called Lagrangian is $\mathcal{L}(x, \lambda) \equiv f(x) - \lambda(g(x) - a))$, where λ is the Lagrangian multiplier. At this point, it needs to be mentioned that the word "Lagrangian" does have a different meaning in physics. Following Baaquie [3] (p. 78), a "Lagrangian . . . is the fundamental mathematical structure

in the path-integral formulation of quantum mechanics."[1] The use of the (economics version) of the Lagrangian can be seen in the following simple example. Any good undergraduate textbook in economics will discuss how a consumer maximizes a utility function subject to the budget he or she is constrained by. The basic formulation is then $\max u(\overrightarrow{c})$ constrained by the budget equation $\overrightarrow{p}.\overrightarrow{c} = I$, where $\overrightarrow{p}$ is a $1 \times n$ price vector and $\overrightarrow{c}$ is a $n \times 1$ vector of quantities of goods which are part of the consumption bundle of the consumer, I is the budget figure, and $u(.)$ is a utility function. The Lagrangian is immediate, $\mathcal{L} = u(\overrightarrow{c}) - \lambda(\overrightarrow{p}\,\overrightarrow{c} - I)$, where λ is the Lagrangian multiplier. One can obtain from this equation that the level of satisfaction will be maximized where the so-called "marginal rate of substitution" is equal to the slope of the budget curve. The interested reader may query whether, besides equality constrained optimization, there is room in economics for optimization procedures with inequality constraints. The answer is affirmative, and the "Kuhn–Tucker" conditions (Kuhn and Tucker [4]) are invoked in that respect. The so-called "sufficiency theorem" by Kuhn and Tucker provides for conditions to obtain a maximum/minimum. For an excellent discussion, see Chiang [5] (Chapter 21, Section 2 (p. 722); Section 4 (p. 738 and following) and especially Section 5 (p. 744)).

2. The utility function plays also an essential role as an input in risk aversion. The concavity and/or convexity of the utility function expresses risk aversion/risk loving behavior of a decision maker. Let us make the simple assumption that we derive utility from using wealth. We can then compare the expected utility of consuming wealth, denoted as $E(u(w))$, with the utility derived from expected wealth, $u(E(w))$. Let us imagine a gamble with two outcomes, for instance. The $u(E(w))$ corresponds then to the utility received from obtaining a sure outcome, while $E(u(w))$ then corresponds to the utility of gambling (that is to accept the lottery). Hence, it is intuitive that, if $u(E(w)) > E(u(w))$, the individual is risk averse. If the opposite equality occurs, $E(u(w)) > u(E(w))$, the individual is a risk lover. Equality occurs if the individual is indifferent to risk. Those relationships are also tied to the concavity, convexity, and linearity, respectively of the utility function. Please see also below for the next section where we propose other risk aversion measures.

Let us revisit the relationship between the preference over objects (whatever those are) and the corresponding relationship expressed via the mathematical inequality over levels of utility (as given by the utility function). The following

[1] Professor Baaquie flagged up this issue of the meaning of the Lagrangian whilst he was staying for his sabbatical at the University of Leicester.

sequence of "events" is important. Kreps [1] (p. 31) says it succinctly as follows:[2] "we'll make assumptions about the mathematical structure of the objects in the set X – we will want $x \in X$ somehow to represent uncertain prospects. And by imposing further conditions on $\succ$ having to do with the mathematical structure of X, we will try to specialize the form of the function u." The central question which is now being asked is this: (Kreps [1] (p. 31)): "what corresponding forms of functions u will we seek?" We are now getting back to the three grand groups of expected utility models already mentioned in Chapter 2. The workhorse of most economics models is the von Neumann–Morgenstern model [6], where the probabilities used to calculate the expected utility are "objective." When "subjective" probabilities are used, then the Savage model [7] will be very relevant. In case of a mixture of objective and subjective probability, the Anscombe–Aumann [8] model is used. The famous decision-making paradoxes (such as Allais and Ellsberg's paradox) relate to the axiomatic underpinnings of those models. The next chapter will study in detail the violation of the sure-thing principle (an important axiom in the Savage expected utility model).

The classical expected utility models in economics, we mentioned above, have now been "superseded" by augmented models. An excellent survey paper is by Machina [9].

Finally, new decision modeling movements have appeared on the international research scene. We note for instance the work of Colin Camerer at Caltech. Camerer's approach is directly inspired from brain science and it attempts to mimic human decision making in an as close way as possible. This new movement is also known under the name of "Neuroeconomics" [10], which may be seen as a subfield of a larger more general discipline known under the name of "behavioral economics."

The relationship between a preference relation and the utility function is tightly connected with probably the most fundamental "building"[3] block of economics: "economic rationality." We follow Danthine and Donaldson [11] (p. 25) to summarize the key assumptions:

1. Economic participants (or investors, or economic agents) have a complete preference relation (i.e. agents always can express a preference over objects (incidentally they can be indifferent between objects too)).
2. Using the same notation we introduced at the beginning of this chapter, a preference relation satisfies transitivity: $\forall x, y, z \in X$: if $x \succeq y$ and $y \succeq z$, then $x \succeq z$.

[2] Please note that X is the set of objects. The mapping: $X \to \mathbb{R}$, expresses the levels of utility (as given by the utility function).

[3] Or should we say "stumbling" block?

3. Preference relations are stable over time.
4. A preference relation $\succeq$ on X is continuous, that is for any sequence of pairs $\{x_n, y_n\}_{n=1}^{\infty}$ with $x_n \succeq y_n$; for all n, we have $x = \lim_{n\to\infty} x_n$ and $y = \lim_{n\to\infty} y_n$; we have $x \succeq y$. The continuity assumption says that consumer preferences cannot exhibit jumps.

As Danthine and Donaldson [11] (p. 25) remark, the above four key assumptions guarantee (please see the beginning of this chapter) that if $x \succ y$, if and only if there exists a utility function $u(.)$, which maps a set of objects into $\mathbb{R}$, such that $u(x) > u(y)$.

The above was a brief discussion of choice under certainty. When uncertainty is built into the utility spectrum another assumption route is to be taken. The choice then becomes a choice over lotteries. We follow Varian [12] (pp. 172–174). Varian makes the important distinction between two properties: (i) the existence of a utility function and (ii) the property that a utility function is (Varian [12] (p. 174)) "additively separable over the outcomes and linear in the probabilities." We do not go into detail on the necessary assumptions to guarantee the existence of a utility function (in the uncertainty environment). It is quite telling though to consider the axioms needed to guarantee the second property we mentioned above. We follow Varian [12] (p. 174):[4]

1. "lotteries with indifferent prices are indifferent";
2. "there is some best lottery... and some worst lottery . . . ";
3. "if one lottery between the best prize and the worst prize is preferred to another it must be because it gives higher probability of getting the best prize."

With those axioms, a utility function with the second property will exist. The proof can be found in Varian [12] (pp. 174–175).

The three chapters in this part of the book will analyze human decision making in a still different way. The objective here consists in trying to uncover, via the use of quantum physics concepts, whether human decision making can be better explained by such concepts.

7.2 References

[1] Kreps, D. (1988). *Notes on the Theory of Choice*. Westview Press, CO.
[2] Ingersoll, J. E. (1987). *Theory of Financial Decision Making*. Rowman & LittleField.
[3] Baaquie, B. (2004). *Quantum Finance*. Cambridge University Press.
[4] Kuhn, H. W. and Tucker, A. W. (1951). Nonlinear programming. In *Proceedings of the Second Berkeley Symposium on Mathematical Statistics and Probability*. Ed. Neyman, J. University of California Press, Berkeley, CA, pp. 481–491.

[4] We omit the continuity axiom in our discussion.

[5] Chiang, A. C. (1984). *Fundamental Methods of Mathematical Economics.* McGraw-Hill.
[6] von Neumann, J. and Morgenstern, O. (1947). *Theory of Games and Economic Behavior*. Princeton University Press.
[7] Savage, L. J. (1954). *The Foundations of Statistics*. J. Wiley, New York.
[8] Anscombe, F. and Aumann, R. (1963). A definition of subjective probability. *Annals of Mathematical Statistics*, 34, 199–205.
[9] Machina, M. (1987). Choice under uncertainty: problems solved and unsolved. *Journal of Economic Perspectives*, 1, 121–154.
[10] Camerer, C., Loewenstein, G., and Prelec, D. (2005). Neuroeconomics: how neuroscience can inform economics. *Journal of Economic Literature*, 43, 9–64.
[11] Danthine, J. P. and Donaldson, J. B. (2005). *Intermediate Financial Theory*. Academic Press Advanced Finance Series.
[12] Varian, H. (1992). *Microeconomic Analysis*. 3rd edition. Norton.

7.3 Modeling risk: some basic approaches

As is well known in the micro-economics literature, the so-called "St. Petersburg" paradox is often invoked, as per Blavatskyy [1] (p. 677) to argue against "expected value and in favour of expected utility." The key paper on this paradox is by Samuelson [2] (also cited in Blavatskyy [1]). The paradox is quite easy to explain. We follow Samuelson [2] (p. 25): "Peter offers to let Paul toss a fair coin an indefinite number of times, paying him 2 ducats[5] if heads first come up at the first toss, 4 ducats if its first comes up at the second toss ... 2^i ducats for heads first coming up at the tth toss ... and so forth *ad infinitum*. Hence, if Paul has linear utility, he could be made to pay for the 'fair price of playing the game' the value of the mathematical expectation of the gains: $E(2^i) = \frac{1}{2}2 + \frac{1}{4}4 + \cdots \frac{2^k}{2^k} + \cdots = \infty$."

It is quite straightforward to argue that if a utility functional were to be introduced which is concave (i.e. with second derivative weakly negative), the infinite expectation problem could be converted into a finite expectation (using then expected utility). The last bracketed words are important, because they rejoin exactly the point made by Blavatskyy [1] that this paradox is often invoked to defend expected *utility* (as opposed to expected value).

We remarked already in the beginning of this chapter that utility functions can be characterized by degrees of convexity and/or concavity. A linear utility function is both concave and convex.

The concavity of the expected utility function indicates risk aversion. But as a measure of risk aversion it is a little incomplete since it would be scale dependent. An appropriate measure which counters scale dependency, is the so-called "Arrow–Pratt measure of risk aversion." Textbooks devoted to the basics of micro-economic

[5] This is a currency unit.

theory will define this measure. We follow here Varian [3] (p. 178):

$$ARA(w) = -\frac{u''(w)}{u'(w)}, \tag{7.1}$$

where u'' is the second derivative of the utility function towards wealth, w; and u' denotes the first derivative of the utility function towards wealth.

Let us consider an agent with utility function $u(w)$ and another agent with utility function $v(w)$. The agent with utility function $u(w)$ is more risk averse than the agent with utility function $v(w)$ when $-\frac{u''(w)}{u'(w)} > -\frac{v''(w)}{v'(w)}$ for all levels of wealth. Using the *ARA* measure we can then also calculate whether *ARA* is increasing (*IARA*), decreasing (*DARA*), or constant (*CARA*).

As Varian [3] (p. 181) indicates ARA is a measure of local risk aversion. Two well-known measures which are appropriate for a context of so-called global risk aversion, are: (i) a measure which relates to the composition of two functions and (ii) a measure which involves a "willingness to pay approach." We mention those here and we follow again Varian ([3] (p. 181)).[6] As mentioned already above, a natural way to measure degrees of risk aversion follows from imposing degrees of concavity on the utility function. Hence, agent 1's utility function, u, is "more concave" than agent 2's utility function, v, when $u(w) = h(v(w))$ where $h(.)$ is an increasing/strictly concave function. Finally, there is another possible measure. As per Varian ([3] (p. 181)), denote[7] $a(\varepsilon)$ as the maximum amount of wealth[8] an agent would be willing to give up so as to avoid the risk, ε. Such risk is typically defined in a very basic way: a random variable with mean zero. One can write that $u(w - a(\varepsilon)) = E(u(w + \varepsilon))$. The utility for diminished wealth (the left-hand side of this simple equation) is equal to the expected utility of getting into the gamble. This indeed a very intuitive statement, and it follows that agent 1 with utility function $u(w)$ is more risk averse than agent 2 with utility function $v(w)$ when $a_u(\varepsilon) > a_v(\varepsilon)$ (and this must hold for any wealth level).

It is instructive to consider the following theorem which establishes the equivalence between the three measures we mentioned above. We adapt the proof from Varian ([3] (pp. 182–183)). Ingersoll [4] (pp. 40–41) gives also a proof.

Theorem 10 *If* $u(w)$ *and* $v(w)$ *are two differentiable, increasing and concave expected utility functions then:* $-\frac{u''(w)}{u'(w)} > -\frac{v''(w)}{v'(w)} \Leftrightarrow u(w) = h(v(w)) \Leftrightarrow a_u(\varepsilon) > a_v(\varepsilon)$, *where* $h(.)$ *and* ε *are defined as above.*

Proof. (1). $-\frac{u''(w)}{u'(w)} > -\frac{v''(w)}{v'(w)} \Rightarrow u(w) = h(v(w))$. We do not do this. (2) $u(w) = h(v(w)) \Rightarrow a_u(\varepsilon) > a_v(\varepsilon)$. $u(w - a(\varepsilon)) = E(u(w + \varepsilon))$ and by the given we can

[6] For a similar discussion, see also Ingersoll [4] (pp. 37–40).
[7] We alter somewhat notation. [8] We assume such quantity could exist.

also see that $u(w - a(\varepsilon)) = Eh(v(w + \varepsilon))$. By Jensen's inequality, which says that for a random variable x and $f(.)$ a strictly concave function of x, $Ef(x) < f(Ex)$. So we can say that $Eh(v(w + \varepsilon)) < h(Ev(w + \varepsilon))$ and we know that $h(Ev(w + \varepsilon)) = h(v(w - a_v(\varepsilon))$ and $h(v(w - a_v(\varepsilon))$ is equal to $u(w - a_v(\varepsilon))$, and therefore we can say that $u(w - a_u(\varepsilon)) < u(w - a_v(\varepsilon))$ and thus it must then be true that $a_u(\varepsilon) > a_v(\varepsilon)$. (3) That $a_u(\varepsilon) > a_v(\varepsilon) \Rightarrow -\frac{u''(w)}{u'(w)} > -\frac{v''(w)}{v'(w)}$ can also be proven. ■

A useful comparative static result is as follows. We follow again Varian ([3] (p. 184)). Consider two assets, one risky and one is risk free, and invest an amount a in the risky asset. Assume there exists an endowment of wealth of level w. Hence, $w - a$ is the amount invested in the riskless asset. Second period wealth, w^*, can be written as: $a(1 + r) + (w - a)(1 + r_f)$, where r_f is the risk free interest rate. To simplify, set[9] $r_f = 0$. A key question is then: What is the expected utility from investing a units of currency? One writes $Eu(w + ar) = v(a)$. The first derivative (towards a) is $Eu'(w + ar)r$ and the second derivative (towards a) is $Eu''(w + ar)r^2$. With an assumed existence of risk aversion, there is the requirement that the second derivative is negative. Set the first-order condition as $Eu'(w + ar)r = 0$. How does the amount of money invested, a, vary with changes in wealth, w? Denote $a(w)$ as the optimal level a function of w. Differentiating the first-order condition with respect to w yields $Eu''(w + a(w)r)r(1 + a'(w)r) = 0$, from where one finds that $a'(w) = \frac{-Eu''(w+ar)r}{Eu''(w+ar)r^2}$. In order to know what the sign behavior is of $a'(w)$, the following claim is made. We adapt the claim below from Varian [3] (p. 185).

Note: Increasing ARA means: if wealth increases, ARA increases; decreasing ARA: if wealth increases, ARA decreases.

Claim 11 *$Eu''(w + ar)r$ is positive, negative, or zero as ARA is decreasing, increasing, or constant.*

Proof. It needs to be shown that DARA (decreasing absolute risk aversion) implies that $Eu''(w + ar)r > 0$. If $r > 0$, then $ARA(w + ar) = \frac{-u''(w+ar)}{u'(w+ar)} < ARA(w)$ and this can be rewritten as $u''(w + ar) > -ARA(w)u'(w + ar)$, and since $r > 0$ we can write that $u''(w + ar)r > -ARA(w)u'(w + ar)r$. Now let us consider the case of $r < 0$. Writing that $u''(w + ar) > -ARA(w)u'(w + ar)$ and considering decreasing ARA (i.e. if wealth increases, ARA decreases) we have that $u''(w + ar) < -ARA(w)u'(w + ar)$, and since $r < 0$, we must have that $u''(w + ar)r > -ARA(w)u'(w + ar)r$. Whether $r < 0$ or $r > 0$, we get the same result. Taking expectations over $u''(w + ar)r > -ARA(w)u'(w + ar)r$ yields

[9] This is in fact not an unreasonable assumption in the current financial context where interest rates on current deposits (in some countries) carry virtually zero nominal interest rates (and possibly negative real interest rates).

$Eu''(w + ar)r > -ARA(w)Eu'(w + ar)r$ and we know that $Eu'(w + ar)r = 0$ by the first-order condition. Hence, $Eu''(w + ar)r > 0$. ■

Thus, when the investment in the risky asset, is increasing, risk aversion must be decreasing. If you reduce the investment in the risky asset, then you get increasing risk aversion.

7.4 References

[1] Blavatskyy, P. R. (2005). Back to the St. Petersburg paradox. *Management Science*, 51, 4, 677–678.
[2] Samuelson, P. (1977). St. Petersburg paradoxes: defanged, dissected, and historically described. *Journal of Economic Literature*, 15, 24–55.
[3] Varian, H. (1992). *Microeconomic Analysis*. 3rd edition. Norton.
[4] Ingersoll, J. E. (1987). *Theory of Financial Decision Making*. Rowman & LittleField.

7.5 Possible remedies to the paradox: a brief discussion

In Chapter 2, we have already mentioned the existence of paradoxes in decision making, most notably the Ellsberg paradox [1]. This paradox is discussed in detail in Chapter 8, Section 8.6. Please note the paradox is treated in condensed form but with excellent detail in Kreps [2]. We do not include the details of this paradox at this point in the book.

However, let us provide for some needed background in order to "set the tone" of the gravity of this problem for economics. To underscore this "gravity," let us consider Schoemaker [3] who at the beginning of his *Journal of Economic Literature* article noted (p. 529) that: "It is no exaggeration to consider expected utility theory the major paradigm in decision making since the Second World War." It is really in the light of this quote that one can gauge the impact a paradox such as Ellsberg's can have. After all, it affects the expected utility paradigm.

Segal [4] provides for an excellent discussion of the Ellsberg paradox, and for a proposal and summary of other approaches, to deal with that paradox within the context of the utility structure well known in economics. Please see also Section 1 in this chapter. Segal [4] remarks (p. 194) there are two important issues which relate to the Ellsberg paradox (as presented by Ellsberg himself): (i) the separation of risk and uncertainty and (ii) the fact that the so called sure-thing principle (please see again Chapter 8, Section 8.6) cannot be held up at all times as a valid axiom by actual decision makers. The second point needs emphasizing, since if indeed the axiomatic structure underlying the Savage model showed this defect, the model as such was becoming substantially weaker. As we have briefly discussed in Chapter 2, the Savage model is one of the pillar models of expected utility in economics. Hence, the economic theory community (as well as the academics in applied economics research) were quite keen to see answers to this problem. Indeed, leading economists have contributed to this important debate.

Segal also indicates [4] (p. 194) that Ellsberg never empirically verified his proposed paradox.[10] But much research subsequently did (please see Segal [4] (p. 194) for some of the relevant references).

Let us now briefly consider some of the proposals which were put forward. Segal [4] distinguishes two groups of proposals: (i) the use of so-called non-additive probabilities or (ii) the use of the additive probability approach (which is the objective of Segal's paper). In this book, we will (very) briefly review one non-additive probability approach. We use the same references Segal proposes. Schmeidler [6] (p. 572) remarks: "non-additive" probability is an objective probability and such type of "probability(ies) have been in use in physics for a long time." In this chapter, we provided for a brief discussion on the use of objective and subjective probabilities. But let us continue briefly with Schmeidler [6] (p. 572) who remarks that "objective probability is considered here as a physical concept like ... momentum." Schmeidler [6] (pp. 572–573) argues that a die can for instance be made as "perfect" as possible, i.e. by controlling for a uniform density of the material used in the manufacturing of the cube etc. By controlling for such physical aspects, the objective probability of obtaining one of the faces of the die, is indeed $1/6$. On the contrary, the subjective probability of an event, according to Schmeidler [6] (p. 573), can be seen "as the number used in calculating the expectation... of a random variable." And Schmeidler immediately adds that "This definition includes objective or physical probabilities as a special case where there is no doubt as to which number is to be used."

Let us now consider one non-additive probability approach. Fishburn, one of the most noted decision theorists of our times, proposes in his paper (Fishburn [7] (p. 1052)), six axioms, two of which are of importance to guarantee the existence of subjective probability.[11] A straightforward discussion of each of the six axioms is provided for in Fishburn [7], pp. 1052–1053. His axioms imply useful properties, and most notably one axiom implies a property ("statewise dominance") (Fishburn [7] (p. 1053)) which is a special case of Savage's sure-thing principle. Theorem 1 in Fishburn [7] (pp. 1053–1054) indicates that his set of axioms do hold if and only if there exists a unique so-called skew-symmetric bilinear functional, ρ, so that for gambles f and g, gamble g is preferred to gamble f if and only if $\rho(f, g) > 0$. As Segal [4] remarks (p. 194), the preferences in the Fishburn model are not assumed transitive, but theorem 3 in Fishburn [7] (p. 1057) allows for such transitivity, but

[10] Wu *et al.* [5] (p. 404) do mention that some data appeared in Ellsberg's doctoral thesis at Harvard.

[11] Remark that the Savage model uses subjective probability (see also Chapter 2).

as Segal [4] (p. 195) further points out, the skew-symmetric bilinear functional will not explain the Ellsberg paradox. We note that in Segal's [4] model (the additive probability approach) transitivity is satisfied.

7.6 References

[1] Ellsberg, D. (1961). Risk, ambiguity and Savage axioms. *Quarterly Journal of Economics*, 75, 643–669.
[2] Kreps, D. (1988). *Notes on the Theory of Choice*. Westview Press, CO.
[3] Schoemaker, P. J. H. (1982). The expected utility model: its variants, purposes, evidence and limitations. *Journal of Economic Literature*, 20, 529–563.
[4] Segal, U. (1987). The Ellsberg paradox and risk aversion: an anticipated utility approach. *International Economic Review*, 28, 1, 175–202.
[5] Wu, G., Zhang, J., and Gonzalez, R. (2004). Decision Under Risk. In *The Blackwell Handbook of Judgment and Decision Making*. Eds. Koehler, D. and Harvey, N. Oxford University Press.
[6] Schmeidler, D. (1989). Subjective probability and expected utility without additivity. *Econometrica*, 57, 3, 571–587.
[7] Fishburn, P. (1983). Ellsberg revisited: a new look at comparative probability. *The Annals of Statistics*, 11, 1047–1059.

7.7 The role of the law of total probability (LTP): a brief overview

One of the basic laws of classical probability theory is the law of total probability (LTP). This law is a basic theorem drawn from classical Kolmogorovian probability theory. It is a consequence of the additivity of probability and Bayes' formula for conditional probabilities. We remark that Bayes' formula is a definition in the classical Kolmogorov model. The conditional probability $\mathbf{P}(B|A)$ is defined by using the probability $\mathbf{P}(B \cap A)$ of the intersection of events A and B (their joint occurrence):

$$\mathbf{P}(B|A) = \frac{\mathbf{P}(B \cap A)}{\mathbf{P}(A)}; \ \mathbf{P}(A) > 0. \tag{7.2}$$

We recall the definition of Kolmogorov probability space. By the Kolmogorov axiomatics [1], the *probability space* is a triple:

$$\mathcal{P} = (\Omega, \mathcal{F}, \mathbf{P}). \tag{7.3}$$

Elementary events are denoted by ω (those events belong to Ω); elements of F are events, P is probability. We emphasize that Bayes' formula is not a theorem, it cannot be derived. Kolmogorov wrote that in his model, Bayes' formula is a definition of conditional probability.

We formulate the LTP in the form as it is used in decision making. Consider two dichotomous random variables $a = \pm 1$, $b = \pm 1$. The b-variable describes decisions. So, we can make the decision $b = +1$, "yes," or $b = -1$, "no." The a-variable describes possible conditions preceding the decision making. For example, $a = +1$: "the climate will change towards warming," $a = -1$; the negation. An example for the b-variable: "one buys a property near the sea," $b = -1$; the negation, $b = +1$.

LTP *The prior probability to obtain the result, e.g. $b = +1$ for the random variable b is equal to the prior expected value of the posterior probability of $b = +1$ under conditions $a = +1$ and $a = -1$:*

$$\mathbf{P}(b = j) = \mathbf{P}(a = +1)\mathbf{P}(b = j|a = +1) + \mathbf{P}(a = -1)\mathbf{P}(b = j|a = -1), \quad (7.4)$$

where $j = +1$ or $j = -1$.

The LTP gives a possibility to predict the probabilities for the b-variable on the basis of conditional probabilities and the a-probabilities. The main idea behind applications of the LTP is to split, in general, a complex condition, say C, preceding the decision making, into a family of (disjoint) conditions, in our case $C_+^a = a = +1$ and $C_-^a = a = -1$, which are less complex. One can then estimate in some way (subjectively or on the basis of available statistical data) the probabilities under these simple conditions $\mathbf{P}(b = \pm|a = \pm 1)$ and the probabilities $\mathbf{P}(a = \pm 1)$ of the realization of conditions $C_\pm^a$. On the basis of these data, the LTP provides the value of the probability $\mathbf{P}(b = j)$ for $j = \pm 1$. If, e.g., $\mathbf{P}(b = +1)$ is larger than $\mathbf{P}(b = -1)$, it is reasonable to make the decision $b = +1$, for instance say "yes."

Typically, decision making is based on two thresholds for probabilities (chosen depending on a problem): $0 \leq \varepsilon_- \leq \varepsilon_+ \leq 1$. If the probability $\mathbf{P}(b = +1) \geq \varepsilon_+$, the decision $b = +1$ should be made. If the probability $\mathbf{P}(b = +1) \leq \varepsilon_-$, i.e. $\mathbf{P}(b = -1) \geq 1 - \varepsilon_-$, the decision $b = -1$ should be made. If $\varepsilon_- < \mathbf{P}(b = +1) < \varepsilon_+$, then additional analysis should be performed.

7.8 Reference

[1] Kolmogorov, A. N. (1933). *Grundbegriffe der Wahrscheinlichkeitsrechnung*. Springer Verlag, Berlin. English translation (1956). *Foundations of the Probability Theory*. Chelsea Publishing, New York.

8

Interference effects in psychology – an introduction

8.1 Classical decision making and the Bayesian approach

In the former chapter (Chapter 7, Section 7.7), we defined the Bayes probability rule for two events A and B. The Bayesian approach has very important practical applications. The title of this section of the chapter suggests the Bayesian approach seems to have a close connection with classical decision making. This is clearly true, but it does say little on which areas of decision making this approach has high impact. In financial applications but also in the area of risk and insurance, does this theory have high levels of use. Robert [1] is an excellent source for a detailed description. We give here a slightly (but only slightly) more involved definition of the Bayesian approach. We follow here Shevchenko [2]. Denote, as in Shevchenko [2] (p. 43)[1] the density function for a random vector X of data, given a vector of parameters Y, as $F_{X|Y}$. The joint density $F_{X,Y}$ of the data and parameters is given by:

$$F_{X,Y} = \Pi_{Y|X}.F_X, \tag{8.1}$$

where $\Pi_{Y|X}$ is the density of parameters given that the data are taking a specific value (the posterior density), F_X is the marginal density of X. Bayes' theorem is very close to the above equation, and it says that the posterior density, $\Pi_{Y|X}$, can be calculated as:

$$\Pi_{Y|X} = \frac{1}{F_X} F_{X|Y} \Pi_Y, \tag{8.2}$$

where Π_Y is the density of parameters. As is indicated in Shevchenko [2] (p. 44), the density $\Pi_{Y|X}$ is characterized by combining prior knowledge (Π_Y) with information derived from the data, given thus by $F_{X|Y}$. Shevchenko [2] (pp. 119–120) also distinguishes two main approaches on how to estimate the parameters of the prior

[1] We have the adapted the notation.

distribution. In the so-called "pure Bayesian approach," the prior is subjectively specified (for instance using opinion), while in the so-called "empirical Bayesian approach," data are used for specification. Practical aspects of such specifications are given also in Shevchenko [2] (see for instance pp. 119–121). We do not expand on this here.

Schmeidler [3] makes the following very interesting statement (p. 571): "There are two . . . rules for assigning prior probabilities to events in case of uncertainty. The first says that symmetric information with respect to the occurrence of events results in equal probabilities. The second says that if the space is partitioned into k symmetric (i.e. equiprobable) events, then the probability of each event is $1/k$." Schmeidler indicates he does not agree with the second rule. Let us cite him again (Schmeidler [3] (p. 571)): "The probability attached to an uncertain event does not reflect the heuristic amount of information that led to the assignment of that probability. For example when the information on the occurrence of two events is symmetric they are assigned equal prior probabilities. If the events are complementary the probabilities will be 1/2, independently of whether the symmetric information is meager or abundant." This comment clearly puts an important qualifier on the specification of the prior (for instance in the case of the pure Bayesian approach which we mentioned above).

With this added information in mind, we are now ready to consider Bayesian decision making in a very different context: non-classical decision making.

8.2 References

[1] Robert, C. P. (2001). *The Bayesian Choice*. Springer Verlag, Berlin.
[2] Shevchenko, P. V. (2011). *Modelling Operational Risk Using Bayesian Inference*. Springer Verlag, Berlin.
[3] Schmeidler, D. (1989). Subjective probability and expected utility without additivity. *Econometrica*, 57, 3, 571–587.

8.3 Non-classical decision making: violation of the LTP (law of total probability) and the quantum Bayesian approach

The departure point for our quantum journey is the understanding that the LTP, in spite of its matching with our "natural probabilistic expectations," is really just a mathematical theorem. It was proven on the basis of a special mathematical model for probability, i.e. the *Kolmogorov probability model,* see Kolmogorov [1]. The cornerstone of Kolmogorov's approach is the postulation of a possibility to embed complex conditions (contexts) preceding the decision making into one probability space. This postulated (!) embedding provides a possibility to apply to contexts a set-theoretical algebra, known as *Boolean algebra*. Hence, operations of intersection, union, and complement can all be used. The first operation (intersection) is the subject of our main interest, since it plays the fundamental role in Bayes' formula, which we recall is a definition of conditional probability in the Kolmogorov model.[2]

Suppose now that Kolmogorov's embedding postulate has a restricted domain of application.[3] We note this is not surprising at all. All mathematical models are restricted only to special classes of natural or social phenomena. Thus, the Boolean algebra cannot be freely applied to any given collection of contexts (i.e. complexes of conditions preceding the decision making). We are especially interested in the situation where the operation of intersection cannot be used. In this case, Bayes' formula for conditional probabilities has no meaning. It then follows that the LTP cannot be derived in the absence of Bayes' formula. Hence, the natural law of classical probability, the LTP, can be violated! Is this surprising? In fact, it is not more surprising than, for instance, the appearance of *non-Euclidean geometries*, e.g. Lobachevsky's geometry in physics. In parallel to the fact that the Euclidean geometric model has its limits of application, the Kolmogorov probabilistic model has its own limits too.

In particular, the LTP is violated in quantum physics, in the *double-slit experiment.* Please see Chapter 5, Section 3 of this book for details on this experiment.

In this chapter, we consider this experiment again, but now in a way which is proper for comparing the LTP and decision making. The b-observable (see Chapter 7 (Section 7)) gives us the position of the photon on the registration screen. To remain within the objective of connecting with decision making, one can consider the problem of predicting the position of a photon's registration, that is to predict the probability that the photon hits a selected domain on the registration screen.

To make the b-variable discrete, we split the registration screen into two domains say B_+ and B_- and, if a photon produces the black dot in B_+, we set $b = +1$. We can similarly define the result $b = -1$.

The a-variable (see Chapter 7, Section 7) describes the slit which is used by a particle, say $a = +1$ the upper slit and $a = -1$ the lower slit. For simplicity, we

[2] This is a good place to make a general remark about the role of mathematical models in science. To analyze a natural or social phenomenon, one should always be aware about the mathematical details of the model which one wants to use. In the absence of such knowledge, one may risk interpreting features of the mathematical model as fundamental features of Nature or society. In applying various laws of probability, some researchers will not take note of the precise mathematical description of those laws. As an example, it is not uncommon to notice that some researchers working on applications of probability and statistics in various domains of science maybe sometimes quite ignorant about the Kolmogorov axiomatics of modern (classical) probability theory. As a showcase to this observation, this phenomenon occurs quite often in quantum physics, including quantum information theory.

[3] The main problem is that this is a hidden postulate. It does not figure in the list of Kolmogorov's axioms (see [1]). Therefore, even if one does read the book ([1]), one may risk not paying attention to this extremely important (hidden) postulate.

set $P(a = +1) = P(a = -1) = 1/2$, so the source is placed symmetrically with respect to the slits. Consider three different experimental contexts:

C : both slits are open. We can find $P(b = +1)$ and $P(b = -1)$ from the experiment as the frequencies of photons hitting the domains B_+ and B_-, respectively.

C_+^a : only one slit, labeled $a = +1$, is open. We can find $\mathbf{P}(b = j|a = +1)$, $j = \pm 1$, the frequencies of photon hitting B_+ and B_-, respectively.

C_-^a : only one slit, labeled $a = -1$, is open. We can find $\mathbf{P}(b = j|a = -1)$, $j = \pm 1$, the frequencies of photon hitting B_+ and B_-, respectively.

If we put together these frequency probabilities, which were collected in the three real experiments, then we observe that the LTP is violated, e.g. Khrennikov [2]. The classical LTP cannot be used to predict, e.g., the probability $P(b = +1)$ that the photon hits B_+ under the context C (both slits are open) or on the basis of probabilities $\mathbf{P}(b = j|a = \pm 1)$, that the photon hits B_+ under contexts $C_\pm^a$ (only one respective slit is open). We remark that here the probabilities $\mathbf{P}(b = j|a = \pm 1)$ are statistical empirical probabilities obtained in corresponding experiments. We have performed experiments under contexts $C_\pm^a$, and we have collected the corresponding data. We now want to predict the probabilities for a new experiment (under the context C), but we do not perform that experiment. We can see that the classical LTP prediction strategy does not work.

An important question is this: Can the LTP be violated outside quantum physics? Why not! What is the crucial probability issue of the previous analysis of the two slit experiment? Three experimental contexts C, $C_\pm^a$ cannot be embedded in the same space Ω, since one cannot apply to the real physical situation the Boolean algebra. It is impossible (physically) to make an "intersection" of, e.g., contexts C and C_+^a, and to create a context $C \cap C_+^a$. In philosophic terms, this story is about the *principle of complementarity.* If we specify the slit, e.g. context C_+^a, we specify particle features of a quantum system. This destroys the context C describing the wave features (interference of two waves propagating through two open slits). In quantum physics, the principle of complementarity is often coupled to *wave–particle* duality. We remark that Niels Bohr never mentioned wave–particle duality. The principle of complementarity was formulated as a general principle for the existence of incompatible (complementary) experimental contexts. From a later point of view, it is not so surprising to observe that the principle of complementarity may find applications not only in physics, but even in social science. We mention that Bohr borrowed this principle from psychology. Our presentation can be considered as the comeback of the principle of complementarity to psychology and cognitive science. However, we come well equipped with quantum mathematics.

8.4 Contextual probabilistic formalization

We now mathematically formalize the above considerations. Consider a context C (a complex of conditions: for instance physical, social, financial) and two dichotomous observables a and b. We may call a and b random variables, but we should remember that these are not random variables in the sense of the conventional Kolmogorov model, i.e. we do not suppose that they can be realized as (measurable) functions, $a, b : \Omega \to \{\pm 1\}$. These variables under context C have probabilities $\mathbf{P}(a = +1|C)$, $\mathbf{P}(a = -1|C)$, $\mathbf{P}(b = +1|C)$, $\mathbf{P}(b = -1|C)$; here, e.g., $\mathbf{P}(a = +1|C)$ is the probability that $a = +1$ under context (condition) C. We will attempt to escape fixing the interpretation of the probabilities. They can be statistical probabilities obtained via frequencies of results of measurements as well as subjective probabilities, i.e. probabilities of a priori assignments for the values of a and b.

In contrast to the LTP formulation of Chapter 7, Section 7 (see equation (7.4)), we shall proceed by carefully controlling the contextual dependence of probabilities. We emphasize that *context-conditioning* is quite different from the standard (as it pertains to the Kolmogorov model) *event-conditioning*. The reason for this is not only limited to the fact that event-conditioning is based on the Bayes' formula and Boolean algebra. A context C need not (although can in some cases) be associated with any event. For example, we can speak about different social contexts, e.g. the context of the present financial crisis. Thus, in general, a context is a complex of conditions. In some special situations, the context can be identified with an event.

An important class of such event contexts is given by *selection contexts* which correspond to conditioning upon values of some variable. Take a variable, say a, which takes two values $a = \pm 1$. Consider two contexts: C_+^a, the condition that a takes the value $a = +1$, and C_-^a, the condition that a takes the value $a = -1$.

For example, a is a question asked to a group (ensemble) of people. Here contexts $C_\pm^a$ have the ensemble representation: the C_+^a is characterized by the ensemble of people who replied "yes" and C_-^a is characterized by the ensemble of those people who replied "no." The original ensemble of people used for this experiment can be considered as the ensemble representation of some context, say C[4]. Since all contexts in this consideration can be represented by ensembles, it may induce the illusion that this experiment can be described with a unique Kolmogorov space. This is not true. It is natural to choose the C-ensemble (the original group of people) as the set Ω. This set was encountered in Chapter 7 (Section 7) as being part of the triple which defined the probability space (equation (7.3)). Then, of course,

[4] This context is determined by conditions under which the original group was created; for example, age, education, and so on.

the $C^a_\pm$-ensembles can be considered as subsets, say $A_\pm$, of Ω. However, such an embedding cannot be used for a proper description of the experiment. In general, by giving the definite answer to the a-question people change their mental states. People in $C^a_\pm$-groups (sets $A_\pm$) cannot be identified with themselves in the original C-group (set Ω).

Consider now another variable b. Under selection contexts $C^a_\pm$, we can perform the b-measurement and obtain the corresponding conditional probabilities (contextual probabilities):

$$\mathbf{P}(b = +1|C^a_+),\ \mathbf{P}(b = -1|C^a_+),\ \mathbf{P}(b = +1|C^a_-),\ \mathbf{P}(b = -1|C^a_-).$$

To make the notation closer to the standard one, we set:

$$\mathbf{P}(b = +1|C^a_+) \equiv \mathbf{P}(b = +1|a = +1),\ \mathbf{P}(b = -1|C^a_+) \equiv \mathbf{P}(b = -1|a = +1),$$
$$\mathbf{P}(b = +1|C^a_-) \equiv \mathbf{P}(b = +1|a = -1),\ \mathbf{P}(b = +1|C^a_-) \equiv \mathbf{P}(b = -1|a = -1).$$

In the above presentation, we considered measurements of the b-variables. Thus, the statistical interpretation of probabilities was in use. As was mentioned before, there are no reasons why one needs to stick to one fixed interpretation. For instance, we also can appeal to *subjective probability*. Thus, for example, $\mathbf{P}(b = +1|C^a_+)$ can be interpreted (depending on the problem) as a priori subjective probability that b takes the value $b = +1$ under the condition that a has taken the value $a = +1$.

If the LTP does not hold true (as a consequence of multi-contextuality with complementary contexts), then the left-hand side of equation (7.4) in Section 7.7 is not equal to the right-hand side. We call this difference the *interference term* by analogy with quantum mechanics in which the LTP is violated, inducing a so-called interference term. In contextual notation, the interference terms are given by:

$$\delta(b = +1|C) = \mathbf{P}(b = +1|C) - -\mathbf{P}(a = +1|C)\mathbf{P}(b = +1|a = +1)$$
$$- \mathbf{P}(a = -1|C)\mathbf{P}(b = +1|a = -1), \qquad (8.3)$$
$$\delta(b = -1|C) = \mathbf{P}(b = -1|C) - -\mathbf{P}(a = -1|C)\mathbf{P}(b = -1|a = +1)$$
$$- \mathbf{P}(a = -1|C)\mathbf{P}(b = -1|a = -1). \qquad (8.4)$$

This definition can be rewritten as the LTP which is perturbed by the interference terms:

$$\mathbf{P}(b = \pm 1|C) = \mathbf{P}(a = +1|C)\mathbf{P}(b = \pm 1|a = +1)$$
$$+ \mathbf{P}(a = -1|C)\mathbf{P}(b = \pm 1|a = -1) + \delta_\pm, \qquad (8.5)$$

where $\delta_\pm = \delta_\pm(C)$. By analogy with quantum physics, see Khrennikov [3] for details, we select the normalization:

$$\lambda_\pm = \delta_\pm / 2\sqrt{\Pi_\pm}, \tag{8.6}$$

where:

$$\Pi_\pm \equiv \Pi_\pm(C) = \mathbf{P}(a = +1|C)\mathbf{P}(b = \pm 1|a = +1)\mathbf{P}(a = -1|C) \times \mathbf{P}(b = \pm 1|a = -1) \tag{8.7}$$

Thus, the LTP with interference terms (8.5) can be written as:

$$\mathbf{P}(b = \pm|C) = \mathbf{P}(a = +1|C)\mathbf{P}(b = \pm|a = +1) + \mathbf{P}(a = -1|C)\mathbf{P}(b = \pm|a = -1) + 2\lambda_\pm\sqrt{\Pi_\pm}. \tag{8.8}$$

If the absolute values of the normalized interference term λ_q is less than 1 (for some $q = \pm 1$), we can find an angle θ_q such that:

$$\lambda_q = \cos\theta_q. \tag{8.9}$$

In the opposite case, we use the representation:

$$\lambda_q = \epsilon \cosh\theta_q, \quad \epsilon = \pm 1. \tag{8.10}$$

Here $\theta_q = \theta(q|C)$. The "probabilistic phase" θ_q depends on the context C. If (8.9) holds for both $q = \pm$, then this case yields *trigonometric interference*. If (8.10) holds for both $q = \pm$, then this case yields *hyperbolic interference*. Otherwise, we get mixed hyper-trigonometric interference.

In the trigonometric case, we have the following LTP with the interference term:

$$\mathbf{P}(b = \pm|C) = \mathbf{P}(a = +1|C)\mathbf{P}(b = \pm|a = +1) + \mathbf{P}(a = -1|C)\mathbf{P}(b = \pm|a = -1) + 2\cos\theta_\pm\sqrt{\Pi_\pm}. \tag{8.11}$$

This sort of interference between probabilities can be easily derived in the standard formalism of complex Hilbert spaces used in quantum mechanics. However, we emphasize that we did not start with this formalism. We did not start directly with the assumption that the state of the brain is described by the complex vector normalized by 1. Of course, one might proceed by starting with this assumption. But in such an approach one misses the motivation to appeal to the mathematical formalism of quantum mechanics. Why did the brain select such a representation of information? Typically to justify postulating the Hilbert space for encoding of the brain's states, people appeal to the quantum brain, i.e. the brain as the real quantum physical system. See, for instance, Roger Penrose [4] [5] and Stuart Hameroff [6] [7].

There are strong objections to claim there exists a "physical quantum brain." Therefore, we would not like to be rigidly coupled to this model which induces the interference of probabilities and violation of the LTP in a standard quantum way.

We motivate the use of the mathematical formalism of quantum physics by its natural appearance from the contextual probabilistic model for representation of information.[5] We shall see later on that our very general contextual approach induces the same LTP with an interference term just as in quantum physics. Moreover, the formalism of quantum physics based on the representation of probabilities by complex probability amplitudes, on the vector representation of probabilistic data in complex Hilbert space (Khrennikov [3]) also emerges. Thus, the brain could design such a representation by attempting to find the simplest form of the processing of contextual probabilistic data, i.e. a linear representation.

But the situation is more complicated. Besides the trigonometric interference of probabilities, well known in quantum physics, the contextual formalism induced a new kind of interference, namely hyperbolic interference:

$$\mathbf{P}(b = \pm|C) = \mathbf{P}(a = +1|C)\mathbf{P}(b = \pm|a = +1) \\ + \mathbf{P}(a = -1|C)\mathbf{P}(b = \pm|a = -1) \pm 2\cosh\theta_{\pm}\sqrt{\Pi}. \quad (8.12)$$

Such a deviation from the classical LTP has a natural explanation. We recall that the interference terms $\lambda_{\pm}$ provide the magnitudes of deviations from the classical LTP (with corresponding normalizations by products of probabilities). Trigonometric interference is the exhibition of rather small deviations (normalized by probabilities), so $|\lambda_{\pm}| \leq 1$. Hyperbolic interference is the exhibition of larger deviations, such that $|\lambda_{\pm}| \geq 1$. Of course, in our contextual model there are no reasons to assume that deviations should be bounded by 1. We mention that this may be a sign that quantum observables were selected in a "special" way, inducing only the trigonometric interference. One may expect that social systems can exhibit even stronger non-classicality than the quantum systems. It may well be that the human being is more non-classical than the electron.

It is interesting to remark that starting with the LTP with the hyperbolic interference term we can develop a formalism based on generalized Hilbert spaces over algebras of so-called *hyperbolic numbers*, which is very similar to the standard quantum formalism in complex Hilbert space. We recall that complex numbers have the form $z = x + iy$; $x, y \in \mathbb{R}$, and $i^2 = -1$. Hyperbolic numbers have the

[5] We shall see that, in fact, the canonical Dirac–von Neumann formalism based on the representation of observables by self-adjoint operators in complex Hilbert space is too restrictive to match the general contextual probabilistic model. Its generalization based on the so-called lifting of the density matrix should be used to describe probabilistic data from, for example, psychology. Please see Appendix 1 in the next chapter of this part of the book.

form $z = x + jy$; $x, y \in \mathbb{R}$, and $j^2 = +1$. In contrast to complex numbers, the system of hyperbolic numbers is not a field, but only an algebra. This means that the operations of addition, subtraction, and multiplication are well defined. However, the division is not. Nevertheless, even in the absence of division it is possible to proceed quite far (Khrennikov [3]).

8.5 Interference effects in social science: decision making based on LTP with interference terms

Our basic hypothesis is that cognitive systems (including collective social systems) can make decisions by using the LTP with non-trivial interference terms. To confirm this hypothesis, we should find experimental evidence in psychology and cognitive and social science. One should either design and perform experiments which will demonstrate the violation of the LTP or find this violation in the existing data.

We remark that the violation of the LTP supports the hypothesis of the complex vector representation of the information by cognitive systems. We remark that as in quantum physics via the interference of probabilities, the violation of the classical LTP is the only possible test for the existence of a kind of wave representation in the brain. As well as in quantum physics, in neurophysiology it is impossible (at least at this present time) to observe directly "mental complex amplitudes." We hope that our research on the violation of the LTP in the process of decision making will stimulate neurophysiologists to look for vector (wave) representations and quantum (quantum-like) information processing in the brain.

Recently Jerome Busemeyer paid attention to some famous experiments in cognitive psychology (the Shafir–Tversky [8] and Tversky–Shafir [9], experiments) which could explain violation of the LTP. Those experiments are related to the so-called "sure-thing principle" and the so-called disjunction effect (see also Rapoport [10], Hofstadter [11] [12], and Croson [13]).

8.6 Savage sure-thing principle

As we mentioned in the introduction to Chapter 7, expected utility is the tool used in traditional economics to model uncertainty. Two classical decision (expected utility) theory paradoxes are the Allais and Ellsberg paradoxes.

In this chapter, we deal extensively with the Ellsberg paradox (see the Section 8.12 in this chapter on the Tversky and Shafir experiments). The Allais paradox (which is not treated in this book) really attacks von Neumann–Morgenstern expected utility. This paradox violates the so-called substitution axiom. As we have mentioned before, the von Neumann–Morgenstern model is a model of objective probability (i.e. with probabilities which are given from outside – i.e. exogenously).

The Ellsberg paradox refers to the Savage model and in particular to a violation of the sure-thing principle, an essential axiom in this expected utility theory. Probability in the Savage model is determined by the economic agent. Hence, in this model subjective probability is used in the calculation of expected utility.

The sure-thing principle is quite easy to understand. Consider two sets of gambles. Each set contains two gambles: $G1$ and $G2$ for set 1, $G3$ and $G4$ for set 2. Imagine now that the participants in the experiment need to express a preference: (i) in a first time, do they prefer $G1$ over $G2$ (or vice versa) and (ii) in a second time, do they prefer $G3$ over $G4$ (or vice versa). We can assume that participants in the experiment are informed of the payoffs of all the gambles. We assume each gamble has three states of nature: sn_1, sn_2, and sn_3. The crux of the sure-thing principle resides now in paying close attention to the fact that the payoffs of gambles in set 1, $G1$ and $G2$ for state sn_3, are identical. Similarly, the payoffs of gambles in set 2, $G3$ and $G4$ for state sn_3, are also identical.

The sure-thing principle is now easy to understand: the preference of the participants in the experiment over $G1$ and $G2$ and $G3$ and $G4$ are not influenced by the identical outcomes given state sn_3.

In Busemeyer *et al.* [14], it is pointed out that Shafir and Tversky [8] have uncovered that participants in experiments such as the one described above, do exhibit many violations of the sure-thing principle (STP).

We can formulate the STP in another possible way (Savage [15]). Following Khrennikov [16] (p. 93), if you prefer prospect b_+, to prospect b_-, if a possible future event A happens ($a = +1$), and you still prefer prospect b_+ if the future event A does not happen ($a = -1$), then you should prefer prospect b_+, despite having no knowledge of whether or not event A will happen.

Savage's illustration refers to a person who is undecided as to whether to buy a property shortly before a presidential election. Says Savage, "Seeing that he would buy in either event, he decides that he should buy, even though he does not know which event will obtain" (Savage [15] (p. 21)). The crucial point is that the decision maker is assumed to be *rational*. Thus, the STP was used as one of the foundations of *rational decision making* and rationality in general. It plays an important role in economics in the framework of Savage's expected utility theory.

Savage's STP is a simple consequence of the LTP. Consider a Kolmogorov probability space $(\Omega, \mathcal{F}, \mathbf{P})$. Suppose that the uncertainty context C and contexts $C_\pm^a$ can be embedded in the same probability space, i.e. the situation can be described by using set-algebra (Boolean algebra). Moreover suppose that C is represented by Ω. The contexts $C_\pm^a$ are represented by subsets $A_\pm$ of Ω and $A_- = \Omega \setminus A_+$. We hope that the reader will forgive us for the introduction of new symbols $A_\pm$. We want to distinguish sharply between contexts and their set-theoretical images (if they exist). Suppose now that, for example, for $b = +1$, both conditional probabilities $\mathbf{P}(b = +1|C_+^a)$ and $\mathbf{P}(b = +1|C_-^a)$ are equal to 1. We can express contextual probabilities in the set-theoretic framework:

$$\mathbf{P}(b = +1|C) \equiv \mathbf{P}(b = +1), \ \mathbf{P}(b = \pm 1|C_\pm^a) = \mathbf{P}(b = \pm 1|A_\pm). \quad (8.13)$$

We obtain:

$$\mathbf{P}(b = +1) = \mathbf{P}(b = +1|A_+ \cup A_-) = \mathbf{P}(A_+) + \mathbf{P}(A_-) = 1. \quad (8.14)$$

Probability $\mathbf{P}(b = +1)$ can be considered as a conditional probability with respect to the disjunction of events A_+ and A_-, i.e.:

$$\mathbf{P}(b = +1|A_+ \cup A_-).$$

As we remark in Khrennikov [16] (p. 94) the "violation of (8.14) (is a nonclassical probabilistic effect)[6] induced by the creation of disjunction... the influence of disjunction (to the decision $b = +1$) cannot be reduced to separate out influences of its counterparts. In cognitive psychology this situation is called the disjunction effect."

8.7 Behavioral games: Prisoner's Dilemma

The prisoner's dilemma is a very famous example in basic game theory. Many references exist as to this game. See, for instance, Howard [17] or also Nowak *et al.* [18]. It can be formulated in the following simple way. Two individuals (denote them as 'SA' and 'SB') are suspected of committing a crime. The police, wanting to press charges must be able to ascertain who of the two individuals is the culprit. As is customary, they isolate both suspects. They can either confess or deny the crime. The situation is summarized below:

	Suspect A(SA)	PA	PA
Suspect B (SB)		Confess	Deny
PB	Confess	4 years; 4 years	1 year; 8 years
PB	Deny	8 years; 1 year	2 years; 2 years

From the table, one can see that the suspects can do well by both denying. However, the players can do better if one of them confesses (i.e. betrays the other suspect). In that case, the suspect who confesses only gets one year, while the denier gets eight years. If both of them confess, then they are worse off when compared to when both of them were denying. This is obvious. The problem in the prisoner's

[6] See the paper by Shafir and Tversky [8].

dilemma thus returns to the issue that since one suspect does not know what the other suspect will do, no best action can be taken. How should the prisoners act? Rational prisoners, i.e. prisoners who proceed on the basis of the Savage STP, should always (both) select the strategy to betray, i.e. to cooperate with police.

8.8 Violation of rationality in the experiments of Shafir and Tversky

8.8.1 Contexts involved in the Prisoner's Dilemma (PD)

This sub-sub-section reproduces[7] part of Section 7.4.1 in Khrennikov [16] (pp. 101–102).

The following mental contexts are involved in the PD:

- Context C, representing the situation when a player has no idea about the planned action of the other player, the "uncertainty context."
- Context C_+^a, representing the situation when the B-player supposes that A will stay silent ("cooperate"), and
- Context C_-^a, when B supposes that A will betray ("compete").

Such a version of the PD experiment was performed by Croson [13].

Another version of the PD experiment (which is also realistic) was performed by Shafir and Tversky [8]. In their experiments, the B-player was informed about the real actions of the A-player.[8]

We can also consider similar contexts $C_\pm^b$. We define dichotomous variables a and b corresponding to *actions* of players A and B: $a = +1$ if A chooses to cooperate and $a = -1$ if A chooses to compete; the b values are defined in the same way.

A priori the law of total probability might be violated for the PD, since player B is not able to combine contexts. If those contexts were represented by subsets of the so-called space of "elementary events," as is done in classical probability theory, then player B would be able to consider the conjunction of the contexts C and, e.g., C_+^a. Therefore, player B could operate in the context $C \wedge C_+^a$ (which would be represented by the set $C \cap C_+^a$). But the very situation of the PD is such that one cannot expect contexts C and $C_\pm^A$ to be peacefully combined. If player B obtains information about the planned action of player A (or even if he just decides that A will play in a definite way, e.g. the context C_+^a will be realized), then the context C is simply destroyed. It could not be combined with C_+^a.

We can introduce the following contextual probabilities:

[7] Andrei Khrennikov (2010). Prisoners Dilemma. *Ubiquitous Quantum Structure: From Psychology to Finance.* Springer-Verlag, Berlin, Heidelberg, pp. 101–102.

[8] It actually is a little bit more complicated. Please see the fuller description of their experiment below.

- $P(b = \pm 1 | C)$: probabilities for actions of B under the complex of mental conditions C.
- $p_{+\pm} \equiv \mathbf{P}(b = \pm 1 | C_+^a)$ and $p_{-\pm} \equiv \mathbf{P}(b = \pm 1 | C_-^a)$: probabilities for actions of B under the complexes of mental conditions C_+^a and C_-^a, respectively.
- $P(a = \pm 1 | C)$: prior probabilities that B assigns for actions of A under the complex of mental conditions C.

As we pointed out, there is no prior reason for the LTP to be true in this situation. Recently, Busemeyer *et al.* [14]; [19]–[21] have shown that the LTP can be violated by statistical data collected by Shafir and Tversky [8] many years ago.

8.9 Prisoner's dilemma-type experiment: Shafir and Tversky

This sub-section reproduces[9] part of Section 7.4.1 in Khrennikov [16] (pp. 102–103).

Shafir and Tversky [8] performed the following PD-type experiment. Participants were told that they would play a two-person PD against another participant. In fact, contrary to what they had been told, participants played against a pre-programmed strategy. All participants were told that they had been randomly assigned to a bonus group, which meant that they would occasionally be given information about the other player's already chosen move before they had to choose their own. Throughout the experiment, each participant saw three versions of each PD: one in which the other player's move was unknown, one in which the other player had cooperated, and one in which the other player had defected.

In the Shafir–Tversky [8] PD experiment: we have,

- $P(b = -1 | C) = 0.63$ and hence $P(b = +1 | C) = 0.37$;
- $p_{--} = 0.97, \ p_{-+} = 0.03; \ \ p_{+-} = 0.84, \ p_{++} = 0.16.$

As usually in probability theory, it is convenient to introduce the matrix of transition probabilities:

$$\mathbf{P}^{b|a} = \begin{pmatrix} 0.16 & 0.84 \\ 0.03 & 0.97 \end{pmatrix}.$$

We point out that this matrix is *stochastic* (as it should be). But it is clear that the *matrix obtained by Shafir and Tversky is not double stochastic!*

[9] Andrei Khrennikov (2010). Prisoners Dilemma. *Ubiquitous Quantum Structure: From Psychology to Finance.* Springer-Verlag, Berlin, Heidelberg, pp. 102–103.

8.10 Violation of double stochasticity for matrices of transition probabilities

In probability and combinatorics, a double stochastic matrix is a square matrix of non-negative real numbers, with the peculiar property that each of its columns add up to 1 and each of its rows add up to 1. When only the columns of non-negative real numbers (of the square matrix) add up to 1, then one obtains a left stochastic matrix. When only the rows of non-negative real numbers (of the square matrix) add up to 1, then one obtains a right stochastic matrix. Hence, a double stochastic matrix must be both left stochastic and right stochastic.

We emphasize that, since:

$$\mathbf{P}(b = +1|C_+^a) + \mathbf{P}(b = -1|C_+^a) = 1, \ \mathbf{P}(b = +1|C_+-^a) + \mathbf{P}(b = -1|C_-^a) = 1,$$

or in our notation:

$$p_{++} + p_{+-} = 1, \ p_{-+} + p_{--} = 1,$$

the matrix $\mathbf{P}^{b|a}$ is always a right stochastic matrix.

If $\mathbf{P}^{b|a}$ was produced by quantum observables represented by symmetric operators a and b (in the two-dimensional complex space), it should be double stochastic, since in such a case, e.g.:

$$p_{++} = |\langle +_a | +_b \rangle|^2 = |\langle +_b | +_a \rangle|^2,$$

where $|\pm_a\rangle$ and $|\pm_b\rangle$ are normalized by 1, eigenvectors of operators a and b are defined respectively:

$$a|\pm_a\rangle = \pm|\pm_a\rangle, \ b|\pm_b\rangle = \pm|\pm_b\rangle.$$

To derive double stochasticity, we apply the Parseval equality.

We point out again that *matrices of transition probabilities constructed from experimental data in cognitive psychology are not double stochastic.* We now follow Khrennikov [22] (p. 186). On the other hand, matrices of transition probabilities that should be generated by conventional quantum mechanics in the two-dimensional Hilbert space are always double stochastic.

We can present two possible explanations of this "non-double stochasticity paradox":

(a) The statistics of these experiments are neither classical nor quantum (i.e., neither the Kolmogorov measure-theoretic model nor the conventional quantum model with self-adjoint operators could describe these statistics).
(b) Observables corresponding to real and possible actions are not complete. From the viewpoint of quantum mechanics, this means that they should be represented

not in the two-dimensional (mental qubit) Hilbert space, but in a Hilbert space of a higher dimension.

8.11 Prisoner's dilemma-type experiment: Croson

This sub-section reproduces[10] part of section 7.4.1 in Khrennikov [16] (pp. 103–104).

Croson [13] performed a similar experiment with one important difference. Unlike the original Shafir–Tversky experiment, participants were now not informed that they belonged to a "bonus group." They were also not informed they would occasionally get information about actions of their co-players (in advance). Participants were proposed – in a subset of games – to guess the actions of co-players. Hence, the PD game in this version became a game with elicited rather than controlled beliefs. In such a PD experiment, Croson obtained the following data:

$P(b = -1|C) = 0.225$ and hence $P(b = +1|C) = 0.775$; so the cooperation rate was essentially higher than in the original Shafir–Tversky experiment:

$$p_{--} = 0.68, \ p_{-+} = 0.32; \ p_{+-} = 0.17, \ p_{++} = 0.83.$$

Hence, the matrix of transition probabilities is:

$$\mathbf{P}^{b|a} = \begin{pmatrix} 0.83 & 0.17 \\ 0.32 & 0.68 \end{pmatrix}.$$

We see an essential deviation from the original Shafir–Tversky experiment. It is especially interesting that both experiments were based on the same payoff matrix:

$$\begin{pmatrix} 75, 75 & 25, 85 \\ 85, 25 & 30, 30 \end{pmatrix}.$$

Croson's experiment is very important in our mental contextual model. A mental context can be changed not only by a "real change of the world," but even by the brain's self-measurement. Even by imagining something, the brain changes its state of mind, i.e. its mental context.

In Croson [13], an asymmetric version of the PD was performed. Here the payoff matrix had the form:

$$\begin{pmatrix} 85, 65 & 35, 75 \\ 95, 15 & 40, 20 \end{pmatrix}.$$

Here:

$$P(b = -1|C) = 0.375 \text{ and hence } P(b = +1|C) = 0.625,$$

[10] Ibid., pp. 103–104.

so the cooperation rate was essentially higher than in the original Shafir–Tversky experiment:

$$p_{--} = 0.65, \ p_{-+} = 0.35; \ p_{+-} = 0.47, \ p_{++} = 0.53.$$

Hence, the matrix of $b|a$-contextual probabilities is:

$$\mathbf{P}^{b|a} = \begin{pmatrix} 0.53 & 0.47 \\ 0.35 & 0.65 \end{pmatrix}.$$

8.12 Gambling experiment – 1: Tversky and Shafir

The experiment that Tversky and Shafir [9] proposed to test the disjunction effect is well described in Busemeyer [20]. We follow this description here. Experiment participants have an equal chance to win \$200 or lose \$100. The key issue consists in remarking that the gamble is played in sequence: play1 and play2. Play1 is required. Play2 is optional. The experimenters then considered three conditions. In condition (i) experiment participants were told they won play1, but in condition (ii) they were told they lost play1. In condition (iii) nothing was told to participants in the experiment as to the outcome of play1. As we have indicated before in this chapter (Section 8.6), the adding in of a sure outcome should not influence the preferences. Recall the four gambles, $G1$, $G2$, $G3$, and $G4$ we discussed in Section 8.6. Preferences of $G1$ over $G2$ and $G3$ over $G4$ should not be influenced by the identical outcomes in state sn_3. In a similar way, can we claim the same outcome with the set-up in the Tversky and Shafir experiment: if you prefer to gamble in play2, given that you know you lost in play1, and you prefer to gamble in play2, given that you know you won in play1, then it should not matter that you do not know the outcome in play1 in order to prefer to gamble in play2.

Consider still another way to look at the experiment. We follow here Khrennikov and Haven [23] (p. 380), where we show a very common form of presenting the paradox. Consider the following experiment. We have an urn with 30 red balls and 60 other balls (blue and green). We do not know the exact proportion of green and blue balls. We consider four gambles and we ask experiment participants to express a preference between gambles 1 and 2 and between gambles 3 and 4. The gamble's payoffs are as follows:

1. Gamble 1 ($G1$): you receive 1 unit of currency (uoc) if you draw a red ball.
2. Gamble 2 ($G2$): you receive 1 unit of currency (uoc) if you draw a blue ball.
3. Gamble 3 ($G3$): you receive 1 unit of currency (uoc) if you draw a red or green ball.
4. Gamble 4 ($G4$): you receive 1 unit of currency (uoc) if you draw a blue or green ball.

Most of the experiment participants (and this result has occurred in repeated experiments) will prefer $G1$ over $G2$, $G1 \succ G2$. The intuition for such preference can be explained by the fact that one knows the odds of winning in $G1$ (i.e. $1/3$ probability) but in $G2$ one is unsure about the odds. Participants in this experiment also indicated that $G4 \succ G3$. Here again, one knows the odds of winning in $G4$ are $2/3$. However, one is unsure about the odds in $G3$. Hence, the odds are ambiguous in $G2$ and $G3$.

This paradox clearly violates the sure-thing principle. This can be easily shown (see Khrennikov and Haven [24] (p. 7)) if we use the following table to summarize the payoffs (in units of currency):

	Red	Blue	Green
$G1$	1	0	0
$G2$	0	1	0
$G3$	1	0	1
$G4$	0	1	1

The constant payoffs in set 1 ($G1$ and $G2$) and in set 2 ($G3$ and $G4$) should not influence the preferences. Hence, if $G1 \succ G2$, then, using the sure-thing principle, it should be $G3 \succ G4$. As indicated already above, experiment participant will very often indicate $G3 \prec G4$.

Let us now cast the Ellsberg paradox in notation which can be used for our purposes. Here, a gambling device, e.g. roulette plays the role of A. B is a real player, his actions are $b = +1$ to play the second game, and $b = -1$ not to play the second game. The context C corresponds here to the situation when the result of the first play is unknown to B. The contexts $C^A_\pm$ correspond to the situations when B is informed of the results $a = \pm 1$ of the first play in the gamble.

Tversky–Shafir gambling experiment, the version with the same group of students.

We reproduce[11] here part of section 7.4.2. in Khrennikov [16] (p. 104).

The data given here derive from the experiment which was performed for the *same group* of students, but under different contexts C^A_+ (win context), C^A_- (lost-context), C (uncertainty context). Please note that there was a ten day pause between successive experiments. From Tversky and Shafir [9] we have:

$$P(b = +1|C) = 0.36 \text{ and hence } P(b = -1|C) = 0.64;$$

$$p_{-+} = 0.59, \ \ p_{--} = 0.41; \ \ p_{++} = 0.69, \ \ p_{+-} = 0.31.$$

[11] Andrei Khrennikov (2010). Gambling experiment. *Ubiquitous Quantum Structure: From Psychology to Finance*. Springer-Verlag, Berlin, Heidelberg, p. 104.

We get the following matrix of transition probabilities:

$$\mathbf{P}^{b|a} = \begin{pmatrix} 0.69 & 0.31 \\ 0.59 & 0.41 \end{pmatrix}.$$

This matrix of transition probabilities is not double stochastic either, compare with previously considered PD-type experiments.

8.13 Gambling experiment – 2: Tversky and Shafir

We reproduce[12] here part of section 7.4.2. in Khrennikov [16] (pp. 104–105).

In the same paper, Tversky and Shafir [9] modified the gambling experiment. Three different populations, one for each context, were involved in the b-measurement (between subject design). The data are:

$$P(b = +1|C) = 0.38 \text{ and hence } P(b = -1|C) = 0.62;$$
$$p_{-+} = 0.57,\ p_{--} = 0.43;\ \ p_{++} = 0.69,\ p_{+-} = 0.31.$$

We get the following matrix of transition probabilities:

$$P = \begin{pmatrix} 0.69 & 0.31 \\ 0.57 & 0.43 \end{pmatrix}.$$

8.14 The Hawaii vacation experiment

We reproduce[13] here Section 7.4.3 and part of Section 7.5 in Khrennikov [16] (p. 105).

Tversky and Shafir [9] considered the following psychological test demonstrating the disjunction effect. They showed that significantly more students report that they would purchase a non-refundable Hawaii vacation if they knew that they had passed or failed an important exam as opposed to the case when they did not know the outcome of the exam.

The latter context is denoted by C and the "passed" context by C_+^a and "failed" context by C_-^a.

Here the results are:

$$P(b = +1|C) = 0.32 \text{ and hence } P(b = -1|C) = 0.68;$$
$$p_{++} = 0.54,\ p_{-+} = 0.57;\ \ p_{+-} = 0.46,\ p_{--} = 0.43; \text{ and}$$
$$\mathbf{P}^{b|a} = \begin{pmatrix} 0.54 & 0.46 \\ 0.57 & 0.43 \end{pmatrix}.$$

It is again not a double stochastic matrix.

[12] Ibid. pp. 104–105.

[13] Andrei Khrennikov (2010). Exam's result and Hawaii experiment + reason Based Choice and Its Quantum Like Interpretation. *Ubiquitous Quantum Structure: From Psychology to Finance.* Springer-Verlag, Berlin, Heidelberg, p. 105.

Tversky and Shafir [9] claimed that the disjunction effect is caused by a decision process of *reason-based* choice. Participants, instead of considering the consequences of their decisions, focus on reasons to choose one thing versus another.

Let us go back to the Hawaii experiment. If the exam were passed, there would be a good reason to go to Hawaii – to celebrate. If the exam were failed, there would also be a good reason to go to Hawaii – to console oneself. However, before knowing the outcome of the exam, there is no reason to go to Hawaii. Hence, participants choose not to go. This dependence on reasons for choice leads to violation of the STP.

Bagassi and Macchi [25] show that the disjunction effect does not depend on the presence of uncertainty (pass or fail the exam), but on the introduction into the text-problem of a non-relevant goal ("paying to know").

In PD-type games, the information (real as well as obtained by imagination) on plans of the A-player induces reasons (for the B-player) that are absent in the absence of this information.

8.15 Non-classicality of statistical data: non-zero coefficients of interference

The presented experimental data demonstrate the violation of the LTP:

(1a) **Tversky–Shafir gambling experiment**: version 1 (with the same group of students) (Section 8.12). We reproduce[14] here part of Section 7.6 in Khrennikov [16] (p. 106). Here a-probabilities are equal: they were produced simply by a random generator imitating the first play of the gamble. Simple arithmetic calculations give:

$$\delta_+ = -0.28, \quad \lambda_+ = -0.44;$$
$$\delta_- = 0.28, \quad \lambda_- = 0.79.$$

The coefficients of interference are non-zero! Thus, the probabilistic data are non-classical. These coefficients are bounded by 1. Thus, a non-classical version of the LTP (with trigonometric interference) holds true. Here, the probabilistic phases are:

$$\theta_+ = 2.03, \theta_- = 0.66.$$

(1b) **Tversky–Shafir gambling experiment**: between subject design (Section 8.13). We reproduce[15] here part of Section 7.6 in Khrennikov [16]

[14] Andrei Khrennikov (2010). Coefficients of interference and quantum like representation. *Ubiquitous Quantum Structure: From Psychology to Finance*. Springer-Verlag, Berlin, Heidelberg, p. 106.

[15] Ibid.

(p. 106). Here also, the a-probabilities are equal. We find:

$$\delta_+ = -0.25, \quad \lambda_+ = -0.4, \quad \theta_+ = 1.98;$$
$$\delta_- = 0.25, \quad \lambda_- = 0.69, \quad \theta_- = 0.81.$$

Thus (as one can expect), the uncertainty context C is again non-classical; it is trigonometric.

(2) **Shafir–Tversky PD experiment**. We reproduce[16] here part of Section 7.6 in Khrennikov [16] (p. 107). In this PD experiment (Section 8.9) the B-player was given the information that the A-player had chosen to cooperate and to compete an equal number of times. Thus, here the a-probabilities are also equal. Here $\lambda_- = -0.31$ and hence the phase is $\theta_- = 1.89$. However, $\lambda_+ = 3.98$. Thus, the interference level is very high. It exceeds the possible range of the conventional trigonometric interference. This is *the case of hyperbolic interference!* Here the hyperbolic phase $\theta_+ = \text{arccosh}\,(3.98) = 2.06$.

This is the first example of hyperbolic (in fact, hyper-trigonometric) interference! It shows that students are even more non-classical than electrons and photons!

(3) **Tversky–Shafir Hawaii experiment** (Section 8.14). We reproduce [17] here part of Section 7.6 in Khrennikov [16] (p. 107). Here the a-probabilities are equal as well. We have:

$$\delta_+ = 0.17, \quad \lambda_+ = 0.3, \quad \theta_+ = 1.3;$$
$$\delta_- = -0.17, \quad \lambda_- = -0.37, \quad \theta_- = 2.$$

In summary, in all of the presented experiments, the LTP is violated. This means thus that the probabilistic behavior of players is non-classical. However, contrary to Savage, we would not call such a behavior solely irrational: it is irrational, but non-classical, behavior.

Remark (On the equality of a-probabilities) In all experiments considered above, besides Croson's experiment[18] the a-probabilities are assumed to be equal. In fact, it is a rather natural assumption for decision making in the absence of information on possible actions of the A-player to put equal probabilities for the player's possible actions.

[16] Ibid., p. 107. [17] Ibid.

[18] We were not able to get the a-probabilities for this experiment. We communicated with the author and she promised to send us data. But, finally, we did not get the data. The experiment was done a long time ago and at that time the a-probabilities were not important: Croson (as well as Tversky and Shafir) were interested only in transition probabilities, since only these probabilities were used in their discussions on irrational decision making. Hence, the a-probabilities were not included in the paper [13].

8.16 The constructive wave function approach and fit to data from the experiments of Shafir and Tversky

The Law of Total Probability (LTP) with the trigonometric interference term provides for a possibility to represent probabilities by complex probability amplitudes: i.e. on the basis of probabilistic data, one can reconstruct a probabilistic amplitude. Roughly speaking, if the cognitive systems really function on the basis of the complex vector representation of information, then such a representation can be reconstructed. This encapsulates the core of the constructive wave function approach. A detailed (quite advanced mathematical) presentation of this approach can be found (in the very general case, i.e. without coupling to cognition) in Khrennikov [3].

For the purpose of this section, let us make a short and simple presentation. We use the elementary formula for complex numbers (which can be checked by direct calculation):

$$D = A + B + 2\sqrt{AB}\cos\theta = |\sqrt{A} + e^{i\theta}\sqrt{B}|^2,$$

for real numbers $A, B > 0$, $\theta \in [0, 2\pi]$. By using the LTP with the trigonometric interference term, we can represent the probability $\mathbf{P}(b = \beta|C)$, $\beta = \pm 1$, as the square of the complex amplitude (Born's rule):

$$\mathbf{P}(b = \beta|C) = |\psi_C(\beta)|^2 . \tag{8.15}$$

Here:

$$\begin{aligned}\psi(\beta) \equiv \psi_C(\beta) = &\sqrt{\mathbf{P}(a = +1|C)\mathbf{P}(b = \beta|a = +1)} \\ &+ e^{i\theta_C(\beta)}\sqrt{\mathbf{P}(a = -1|C)\mathbf{P}(b = \beta|a = -1)}.\end{aligned} \tag{8.16}$$

The formula (8.16) gives the quantum representation algorithm. For any trigonometric context C by starting with the probabilistic data:

$$\mathbf{P}(b = \pm 1|C),\ \mathbf{P}(a = \pm 1|C),\ \mathbf{P}(b = \pm 1|a = \pm 1)$$

this algorithm produces the complex amplitude ψ_C, which we can call the "mental wave function," i.e. the quantum informational state of the brain.

Let us now reconstruct "mental wave functions" (Asano *et al.* [26]) for experiments on the disjunction effect:

(1a) **Tversky–Shafir gambling experiment,** version 1 (Section 8.12):

$$\psi(+) \approx 0.59 + e^{2.03i}\, 0.54; \quad \psi(-) \approx 0.39 + e^{0.79i}\, 0.45. \tag{8.17}$$

(1b) **Tversky–Shafir gambling experiment,** version 2, between subject design (Section 8.13):

$$\psi(+) \approx 0.59 + e^{1.98i}\, 0.53; \quad \psi(-) \approx 0.39 + e^{0.81i}\, 0.46. \tag{8.18}$$

As one can expect, the two complex vectors (8.17) and (8.18) do not differ too much.

We postpone the presentation of the Shafir–Tversky PD experiment, since it exhibited hyperbolic interference. We go directly to the Hawaii experiment:

(3) **Tversky–Shafir Hawaii experiment** (Section 8.14):

$$\psi(+) \approx 0.59 + e^{1.98i}\, 0.53; \quad \psi(-) \approx 0.39 + e^{0.81i}\, 0.46. \tag{8.19}$$

The LTP with trigonometric interference induces the representation of data by complex amplitudes and the LTP with hyperbolic interference induces the representation of data by hyperbolic amplitudes, see Khrennikov [3] for details.

In this book, we do not want to depart from the complex Hilbert space model of cognitive processing of information. We just mention that the Shafir–Tversky PD experiment provides for a sign that cognitive systems exhibit more complex information behavior than physical quantum systems. The complex Hilbert space model is too restrictive for their description. Thus, other non-classical models, including models based on Hilbert space should be explored.

We remark that even for interference coefficients which are bounded by 1 (in the case of trigonometric interference), the experiments in cognitive psychology which were presented in previous sections can be considered as a sign that the canonical (Dirac–von Neumann) quantum formalism does not work. The matrices of transition probabilities are not double stochastic. This seems to be a general feature of mental observables. Up to now, we have not found statistical data which produce double stochastic matrices. A natural generalization of the canonical (Dirac–von Neumann) quantum formalism is given by the quantum Markov chain model, developed by Accardi *et al.* [27].

We move the presentation of the quantum Markov chain description of data from the experiments in cognitive psychology (with the trigonometric interference) to Appendix 2 of the next chapter, because the model is quite complex mathematically. It is based on a quite complicated construction of quantum information theory, also known under the name of "lifting" (see Appendix 1 of the next chapter). As a numerical example, we will construct[19] in the next chapter a quantum Markov chain representation of the data from the Shafir–Tversky PD experiment (see Appendix 2 in the next chapter).

8.17 Other experiments

In this section, we propose a practical experiment to measure the level of probability interference. See for instance Khrennikov [28]. Within this section, until

[19] This is joint work of one of the authors with L. Accardi and M. Ohya [29].

Section 8.17.3, we follow entirely (with very slight modifications) Khrennikov and Haven [24] (pp. 17–24).[20] See also Haven [30] for a brief summary.

Assume there exists two mutually exclusive features A and B. Each feature has a dual outcome, "0" or "1." The outcome "0" in features A and B is denoted as respectively a_1 and b_1. Similarly, the outcome "1" in features A and B can be denoted as respectively a_2 and b_2. There exists an ensemble of experiment participants, Σ who have the same mental state. The ensemble probability is denoted as p_j^a and is defined as:

$$p_j^a = \frac{\text{number of results } a_j}{\text{total number of elements in } \Sigma}. \tag{8.20}$$

Similarly for p_j^b. A new ensemble Σ needs to be prepared to perform the measurement p_j^b.

Ensembles, Σ_i^a and $\Sigma_i^b; i = 1, 2$ need then to be prepared and they have states corresponding to the values of $A = a_j$ and $B = b_j; j = 1, 2$. The following probability is then defined:

$$p_{ij}^{a|b} = \frac{\text{number of results } a_j \text{ for the ensemble } \Sigma_i^b}{\text{total number of elements in } \Sigma_i^b}. \tag{8.21}$$

Likewise for $p_{ij}^{b|a}$. From classical probability theory, total probability is defined as:

$$p_j^a = p_1^b p_{1j}^{a|b} + p_2^b p_{2j}^{a|b}; j = 1, 2. \tag{8.22}$$

A similar definition can be made for p_j^b.

In the presence of probability interference, one obtains (see also the section on contextual probability formalization in the beginning of this chapter):

$$p_j^a = p_1^b p_{1j}^{a|b} + p_2^b p_{2j}^{a|b} + 2\sqrt{p_1^b p_2^b p_{1j}^{a|b} p_{2j}^{a|b}} \cos\theta_j; j = 1, 2, \tag{8.23}$$

where $\cos\theta_j$ is defined as:

$$\cos\theta_j = \frac{p_j^a - p_1^b p_{1j}^{a|b} + p_2^b p_{2j}^{a|b}}{2\sqrt{p_1^b p_2^b p_{1j}^{a|b} p_{2j}^{a|b}}}. \tag{8.24}$$

If the $\cos\theta_j$ is non-zero, then this would be indicative of the existence of the quantum-like behavior of cognitive systems. A similar definition can be made for p_j^b.

[20] This is Section 9.1 until Section 9.4 in that paper.

8.17.1 Description and discussion of a proposed experiment

What is the conjecture we want to test? We note that the test description which we will explain below does not contain the minute details of the experimental set up. We only provide for a fairly general discussion of how we could possibly test our conjecture.

In basic terms the experiment deals with having experiment participants recognize a list of songs from a pre-determined list of songs. The tempo of each of the songs is distorted by either lengthening or shortening the tempo. We use the definition of tempo as in Levitin and Cook [31] (p. 927): "the amount of time it takes a given note or the average number of beats occurring in a given interval of time, usually beats per minute." Kuhn [32] (p. 270) defines, beat tempo, in a very similar way as: "the rate of speed of a composition."

The time of listening exposure the test participants are allowed so as to recognize each of the songs varies. The time of exposure is lengthened when the tempo is increased. However, the time of exposure is shortened when the tempo of songs is decreased. The rationale for this time variation is discussed below.

The experiment participants are divided into several groups. We first want to discuss the "normed" group. Levitin and Cook [31] in an experiment on tempo recognition asked experiment participants to sing tunes of songs they themselves had picked from a list of available songs. They examined how well the participants could mimic the tempo of each of the chosen songs. In their experimental set up, they formed a group of experiment participants (250 students) who would fill in questionnaires which would allow, what Levitin and Cook [31] called, "norming." This questionnaire, in the words of Levitin and Cook [31] (p. 929), asked them (the test participants) "to indicate songs that 'they knew well and could hear playing in their heads'." If we were to set up such an experiment ourselves, then we would categorize experiment participants in the "normed" group, to do exactly that: indicate songs they know well. We would then also need to determine a set of songs which we can term as being "best known." However, we may want to make sure we use songs from different music categories. In a study by Collier and Collier [33], the authors examine tempo differences in different types of music (like jazz and other types). The authors find that the tempo differences are quite tied to the music genre in question. Furthermore, Attneave and Olson [34] indicate that the size of a music interval is dependent on the level of frequency of the music.

In the Levitin and Cook [31] study, the 250 students we mentioned above, were all taken from a psychology 101 course (i.e. an introductory level Psychology course). Those students were given the questionnaire and from their answers the experimenters chose 58 CDs. Our experiment proposal would not really follow the Levitin and Cook [31] study closely, since we instead would ask participants to recognize songs we played *randomly* from a list we selected from the choice of songs the "normed" group puts forward. In our experimental set up, the experiment participants have to recognize songs we play for them when *tempo and time of exposure* is varied. In the Levitin and Cook [31] study, participants are asked to sing or hum a song for a time the participant selects him/herself and the obtained tempo is then compared with the copyrighted song's tempo.

Besides the normed group, we would in our envisioned experiment, have two other groups of experiment participants. We note that none of the members of the "normed" group would participate in the experiment itself.[21] We have therefore two other groups. Group 1 whose participants are subjected to song excerpts where tempo decreases are randomly injected into each song. The time of listening exposure for each song is short. Participants in group 2 will be subjected to song excerpts where tempo increases are randomly injected into each song. The time of the listening exposure for each song is longer than in each of the songs used in group 1.

The rationale for the variation of the exposure time comes from a study by Kuhn [32],[22] where it was found that test participants were better able to detect tempo decreases rather than tempo increases. Kuhn [32] (p. 271) indicates that a study by Farnsworth [35] showed that the "listener is most likely to change affective terms with which he describes a piece of music whenever its tempo is appreciably slowed or hastened." [23] Kuhn [32] in his experiment set-up, where he solely used professional musicians, showed that the beat tempos which had been decreased were identified faster (in a statistically significant way) than beat tempos which had been increased. Kuhn [32] indicates those results also had been obtained, independently, in a study by Drake [36]. In more general terms, the work by Drake and Botte [37] addressed a fairly similar problem. They had subjects listen to two identical sequences (except for their tempo difference). One sequence had a higher tempo than the other. The subjects were asked to indicate which sequence was faster.

We note that the tempo experiments by Kuhn [32] did not make use of music. However, the Kuhn [38] study does provide for this. In this study, mention is made of the work by Geringer and Madsen [39], which actually finds, in the context of orchestral music, that tempo increases were more easily identified than tempo decreases.

The Kuhn [38] study provides for a rich context in which one can appreciate how tempo changes can be affected by extraneous variables such as melody activity and audible steady beats. The study finds that when melody activity and audible steady beat were kept constant, the experiment participants (the participants were

[21] This is probably wishful thinking. Running the experiment with a sufficiently large pool of students may well be very challenging.

[22] Also reported in the Levitin and Cook [31] paper (p. 931).

[23] Kuhn's [32] study also discusses rhythm besides beat. We omit it here.

students at an elementary school with a very strong music programme) could distinguish well between slow and fast tempo. However, experiment results in this study showed that melody activity clearly affected tempo perception, but there was much less certainty as to how tempo recognition is affected when audible beat was considered. It needs to be stressed that, according to Kuhn [38], there exists a sizable literature which documents that there is ambiguity in tempo perception. Kuhn [38] cites work by (amongst others) Madsen *et al.* [40] and Wang and Salzberg [41]. One way out, as suggested in Kuhn [38], is then maybe to only use musicians as experiment participants in groups 1 and 2 (besides the normed group participants).

Khrennikov and Haven [42] have described an experiment where we can test for the same type of variables (time and degree of deformation) by using instead a database of pictures.

In our potential experiment, two features of the song recognition would be compared:

1. time of processing of the songs,
2. the ability to recognize a song, S_0, by analyzing a deformation S of it.

We denote the time of processing of the song with the variable t. The ability to recognize the song with tempo deformations, which are either increasing or decreasing, is denoted with the variable a.

The conjecture we want to test is:

Conjecture 12 *t and a are complementary.*

Preparation

The state preparation of the experiment can be described as follows. The experimental context (state) C is given by a sequence of songs, $S_1, \ldots, S_m$. The experiment participants form a group G and each of the participants are exposed to all the songs (from the distilled list of songs taken from the normed group) excerpts they will hear. The context allows thus the experiment participants to learn the songs (some of the songs they surely will know, but some they may not know as well (or not at all)).

In the Levitin and Cook [31] study, 46 students were selected for the tempo recognition experiment. After learning has taken place, the group G is randomly divided into two equal subgroups G_1 and, G_2.

First experiment: song tempo decreases and short exposure time

A first experiment would be performed with experiment participants from group G_1. As was done in Levitin and Cook [31], test participants are to be seated in a so-called sound attenuation booth. Each of the participants is then subjected to a battery of songs (selected from our "best" songs (from the normed group)) played sequentially. Each of the songs is played for the same amount of time. In this first experiment, the tempo of each of the songs is decreased and the exposure time (to listen to the song) is kept uniformly short.[24] Experiment participants must choose from the full list of songs which the they are listening to. Songs in the full list are indicated (alphabetically) by the name of the singer or composer. Next to the authors' name is the title of the song and the record company owning the copyright of the song. We also note that experiment participants need to pick the song from the list within a uniformly prescribed time interval. This is a possible issue of contention. Kuhn [32] reports that professional musicians need less time to respond to a tempo change and they also make less mistakes than non-professional musicians. A study related to this issue is Kuhn's [43] paper which reports that, on a longitudinal basis, test participants (from all walks of life), when asked to keep a constant tempo, seem to instead increase tempo. Our study would not be able to contain only test participants who are musicians. Hence, setting the response time too short could thus create a negative bias.

The low tempo songs $S'_1, \ldots, S'_m$ are deformations of the original songs $S_1, \ldots, S_m$. The tempo change could be fine tuned in the way Madsen and Graham [44] (see also Kuhn [32]) have proposed. Those authors propose a modulation rate (of tempo) of one beat per minute change every second.

Experiment participants in group G_1 are subjected to listening to $S'_1, \ldots, S'_m$ and also to listening to a few other songs which have not been part of the training sample. We denote this set of songs as: S_{G_1}. The width of the time window, w, is a parameter of the experiment.

The task the experiment participants have to fulfill is to indicate whether they either recognize or not the songs S_{G_1} as modifications of the songs $S_1, \ldots, S_m$. We can make this experiment a little more sophisticated by using the procedure Levitt [45] used. See also Drake and Botte [37] where the Levitt [45] approach is described. Levitt [45] would decrease the tempo difference between two subsequent sequences by 1%, if the subjects gave two correct answers (to the two prior sequences). In some sense, the tempo decrease compensates for the shorter period which is given. We could also make it increasingly harder for test participants by instead increasing the tempo for every correct answer.

Let ω be an experiment participant from group G_1 performing this task. We set $t(\omega) = 1$ if ω was able to give the correct answers for $x\%$ of the songs in S_{G_1}; and $t(\omega) = 0$ in the opposite case.

We now find probabilities $P(t = 1|C)$ and $P(t = 0|C)$ through counting numbers of experiment participants who gave answers $t = 1$ and $t = 0$, respectively.

[24] The "shortness" of the exposure time is obviously a calibration issue. Please see below for some beginning discussion on this topic.

We denote respective subgroups of experiment participants by $G_1(t = 1)$ and $G_1(t = 0)$, respectively. The first subgroup consists of experiment participants who have the ability to perform song recognition (with tempo decrease) "quickly" and the second subgroup consists of experiment participants who do not possess that feature.

Second experiment: song tempo increase and long exposure time

The second experiment is performed with experiment participants from the group G_2 as well as the subgroups $G_1(t = 0)$ and $G_1(t = 1)$ of the first subgroup G_1.

The song deformations now are based on song tempo increases, which, as per the experiment of Kuhn [32], are deformations which are harder to recognize. We denote those deformations as $S_1'', \ldots, S_m''$ of the initial songs $S_1, \ldots, S_m$. As in the first experiment, "unknown" songs will be added to the group of songs. We denote the set of essentially deformed songs (with the "unknown" songs) S_{G_2}.

The task the experiment participants have to fulfill is identical to the task described in experiment 1: Can the participants recognize the songs in S_{G_2} as modifications of the initial songs $S_1, \ldots, S_m$?

The time window in experiment 2 is now longer than in experiment 1. The width of the time window, w, is a parameter of the experiment.

Let ω be an experiment participant performing this task. We set $a(\omega) = 1$ if ω was able to give the correct answers for $x\%$ of images in the series. And $a(\omega) = 0$ in the opposite case.

We now find probabilities $P(a = 1|C)$ and $P(a = 0|C)$ through counting numbers of experiment participants in the group G_2 who gave answers $a = 1$ and $a = 0$, respectively. We denote respective subgroups of experiment participants by $G_2(a = 1)$ and $G_2(a = 0)$, respectively. The first subgroup consists of experiment participants who have the ability to perform song recognition (with tempo increase) quickly and the second subgroup consists of experiment participants who do not possess that feature. Because of the tempo increase, we note that the emphasis in experiment 2 is on the carefulness of recognizing the deformation of a song.

Here again, we could use the Levitt [45] approach. We can increase (or decrease) by 1% the tempo between subsequent songs if the test participant has given two correct answers.

8.17.2 Sub- (super-) additivity

Using the above experiment, we find probabilities $P(a = \beta|t = \alpha)$; $\alpha, \beta = 0, 1$, by counting the number of people in the group $G_1(t = \alpha)$ who gave the answer $a = \beta$. After this, we calculate the coefficient λ and we find the angle θ which gives us the measure of complementarity of variables t and a. If $\lambda > 1$, we would find experimental evidence of probabilistic behavior which is neither quantum nor classical. We could also consider $\lambda(w)$ and even make λ dependent on both w and the degree of deformation. This degree of deformation could be expressed by using the tempo measure defined by Levitin and Cook [31].

8.17.3 The potential issue of the complementarity of t and a

Can the above conjecture we posed at the beginning of this section make sense from a quantum mechanical point of view? As is well known, there does not exist a quantum operator on time at all! It is not possible to find an uncertainty principle of the type $\Delta t.\Delta a = h_{\text{cogn}}/2$, where h_{cogn} is an analogue of the Planck constant. This relates to the issue of operators having discrete spectra. See Khrennikov and Haven [42]. Discussions on analogues of Planck constants in financial contexts and other non-quantum physical contexts have been provided for in Choustova [46][47], Khrennikov [48] and Haven [49].

The very objective of our experiment is just to overcome those problems of measurement and hence to be able to test for complementarity on the t and a variables.

In summary, in this chapter we have considered several experiments which indicate a clear violation of the Law of Total Probability. The various versions of the prisoner's dilemma and the Tversky–Shafir gambling experiment, as well as the Hawaii problem, all exhibit this violation. Hence, we can say that probabilistic behavior in those cases is non-classical.

8.18 References

[1] Kolmogorov, A. N. (1933). *Grundbegriffe der Wahrscheinlichkeitsrechnung*. Springer Verlag, Berlin. English translation: (1956). *Foundations of the Probability Theory*. Chelsea Publishing Company, New York.
[2] Khrennikov, A. (2009). *Interpretations of Probability*. De Gruyter, Berlin.
[3] Khrennikov, A. (2009). *Contextual Approach to Quantum Formalism*. Springer Verlag, Berlin.
[4] Penrose, R. (1989). *The Emperor's New Mind*. Oxford University Press.
[5] Penrose, R. (1994). *Shadows of the Mind*. Oxford University Press.
[6] Hameroff, S. (1994). Quantum coherence in microtubules: a neural basis for emergent consciousness? *Journal of Consciousness Studies*, 1, 91–118.
[7] Hameroff, S. (1994). Quantum computing in brain microtubules? The Penrose–Hameroff "Orch Or" model of consciousness. *Philosophical Transactions of the Royal Society*, London A 1-28.
[8] Shafir, E. and Tversky, A. (1992). Thinking through uncertainty: nonconsequential reasoning and choice. *Cognitive Psychology*, 24, 449–474.
[9] Tversky, A. and Shafir, E. (1992). The disjunction effect in choice under uncertainty. *Psychological Science*, 3, 305–309.

[10] Rapoport, A. (1988). Experiments with n-person social traps 1: prisoner's dilemma, weak prisoner's dilemma, volunteer's dilemma, and largest number. *Journal of Conflict Resolution*, 32, 457–472.
[11] Hofstadter, D. R. (1985). *Metamagical Themes: Questing for the Essence of Mind and Pattern.* Basic Books, New York.
[12] Hofstadter, D. R. (1983). The calculus of cooperation is tested through a lottery. *Scientific American*, June, 14–18.
[13] Croson, R. (1999). The disjunction effect and reasoning-based choice in games. *Organizational Behavior and Human Decision Processes*, 80, 118–133.
[14] Busemeyer, J. B., Wang, Z., and Townsend, J. T. (2006). Quantum dynamics of human decision making. *Journal of Mathematical Psychology*, 50, 220–241.
[15] Savage, L. J. (1954). *The Foundations of Statistics*. J. Wiley, New York.
[16] Khrennikov, A. (2010). *Ubiquitous Quantum Structure: From Psychology to Finance.* Springer Verlag, Berlin.
[17] Howard, J. V. (1988). Cooperation in the prisoner's dilemma. *Theory and Decision*, 24, 203–213.
[18] Nowak, M., May, R., and Sigmund, K. (1995). The arithmetics of mutual help. *Scientific American*, 272, 76–81.
[19] Busemeyer, J. B. and Wang, Z. (2007). Quantum information processing explanation for interactions between inferences and decisions. In *Quantum Interaction*. Eds. Bruza, P. D., Lawless, W., van Rijsbergen, K., Sofge, D. A., *AAAI Spring Symposium (Stanford University)*. Technical Report SS-07-08; 91-97. AAAI Press. Menlo Park.
[20] Busemeyer, J. R., Matthews, M., and Wang, Z. (2006). *A Quantum Information Processing Explanation of Disjunction Effects*. In *The 28th Annual Conference of the Cognitive Science Society and the 5th International Conference of Cognitive Science*. Eds. Sun, R. and Myake, N. Erlbaum, New Jersey, pp. 131–135.
[21] Busemeyer, J. R., Santuy, E., and Lambert-Mogiliansky, A. (2008). Comparison of Markov and quantum models of decision making. In *Quantum Interaction: Proceedings of the Second Quantum Interaction Symposium*. (Eds.) Bruza, P., Lawless, W., van Rijsbergen, K., Sofge, D. A., Coecke, B., Clark, S. Oxford University, College Publications, London.
[22] Khrennikov, A. Yu (2009). Quantum like model of cognitive decision making and information processing. *Biosystems*, 95, 179–187.
[23] Khrennikov, A. and Haven, E. (2009). Quantum mechanics and violations of the sure-thing principle: the use of probability interference and other concepts. *Journal of Mathematical Psychology*, 53, 5, 378–388.
[24] Khrennikov, A. and Haven, E. (2007). The importance of probability interference in social science: rationale and experiment. Cornell University Library, http://arxiv.org/pdf/0709.2802.pdf.
[25] Bagassi, M. and Macchi, L. (2007). The "vanishing" of the disjunction effect by sensible procrastination. *Mind and Society*, 6, 41–52.
[26] Asano, M., Khrennikov, A., and Ohya, M. (2011). Quantum-like model for decision making process in two players game: a non-Kolmogorovian model. *Foundations of Physics*, 41, 3, 538–548.
[27] Accardi, L., Ohya, M., and Watanabe, N. (1997). Dynamical entropy through quantum Markov chains. *Open Systems and Information Dynamics*, 4, 1, 71–87.
[28] Khrennikov, A. (2002). On the cognitive experiments to test quantum-like behavior of mind. *Reports from Växjö University: Mathematics, Natural Sciences and Technology*, Nr. 7.
[29] Accardi, L., Khrennikov, A., and Ohya, M. (2009). Quantum Markov model for data from Shafir–Tversky experiments in cognitive psychology. *Open Systems and Information Dynamics*, 16, 371–385.

[30] Haven, E. (2008). Bohmian mechanics in the mechanics of option pricing. In *Quantum Interaction: Proceedings of the Second Quantum Interaction Symposium.* (Eds.) Bruza, P., Lawless, W., van Rijsbergen, K., Sofge, D. A., Coecke, B., Clark, S. Oxford University, College Publications, London.
[31] Levitin, D. and Cook, P. R. (1996). Memory for musical tempo: additional evidence that auditory memory is absolute. *Perception and Psychophysics*, 58, 927–935.
[32] Kuhn, T. L. (1974). Discrimination of modulated beat tempo by professional musicians. *Journal of Research in Music Education*, 22, 270–277.
[33] Collier, G. and Collier, J. L. (1994). An exploration of the use of tempo in jazz. *Music Perception*, 11, 3, 219–242.
[34] Attneave, F. and Olson, R. K. (1971). Pitch as a medium: a new approach to psychological scaling. *The American Journal of Psychology*, 84, 147–166.
[35] Farnsworth, P. R. (1969). *The Social Psychology of Music*. Iowa State University Press, Iowa.
[36] Drake, A. H. (1968). An experimental study of selected variables in the performance of musical durational notation. *Journal of Research in Music Education*, 16, 329–338.
[37] Drake, C. and Botte, M. C. (1993). Tempo sensitivity in auditory sequences: evidence for a multiple look model. *Perception and Psychophysics*, 54, 3, 277–286.
[38] Kuhn, T. L. and Booth, G. D. (1988). The effect of melodic activity, tempo change, and audible beat on tempo perception of elementary school students. *Journal of Research in Music Education*, 36, 140–155.
[39] Geringer, J. and Madsen, C. M. (1984). Pitch and tempo discrimination in recorded orchestral music among musicians and nonmusicians. *Journal of Research in Music Education*, 32, 195–204.
[40] Madsen, C. K., Duke, R. A., and Geringer, J. M. (1986). The effect of speed alterations on tempo note selection. *Journal of Research in Music Education*, 34, 101–110.
[41] Wang, C. C. and Salzberg, R. S. (1984). Discrimination of modulated music tempo by strong students. *Journal of Research in Music Education*, 32, 123–131.
[42] Khrennikov, A. and Haven, E. (2006). Does probability interference exist in social science? In *Foundations of Probability and Physics – 4*. Eds. Adenier, G., Khrennikov, A., Fuchs, C. AIP Conference Proceedings, 899, 299–309.
[43] Kuhn, T. L. (1975). Effect of notational values, age, and example length on tempo performance accuracy. *Journal of Research in Music Education*, 23, 3, 203–210.
[44] Madsen, C. K. and Graham, R. (1970). The effect of integrated schools on performance of selected music tasks of black and white students. Music Educator's National Conference, Chicago, Illinois, USA.
[45] Levitt, H. (1971). Transformed up-down methods in psychoacoustics. *Journal of the Acoustical Society of America*, 49, 467–477.
[46] Choustova, O. (2006). Quantum Bohmian model for financial markets. *Physica A*, 374, 304–314.
[47] Choustova, O. (2007). Toward quantum-like modelling of financial processes. *Journal of Physics: Conference Series*, 70 (38pp).
[48] Khrennikov, A. (2004). *Information Dynamics in Cognitive, Psychological and Anomalous Phenomena.* Series Fundamental Theories of Physics, v. 138. Kluwer, Dordrecht.
[49] Haven, E. (2005). Pilot-wave theory and financial option pricing. *International Journal of Theoretical Physics*, 44, 11, 1957–1962.

9

A quantum-like model of decision making

9.1 Introduction

In the previous chapter, we presented experimental data from cognitive psychology which can be considered as a strong argument in favor of the complex (or more general) Hilbert space representation of information by cognitive systems and the use of the quantum information scheme in information processing. We have seen that the classical Law of Total Probability (LTP) is violated. The interference of probabilities induces a possibility to "reconstruct a mental wave function." An important lesson we can learn from this is that the standard Dirac–von Neumann approach based on self-adjoint operators is too restrictive to describe known data from cognitive psychology. More general quantum models should be used. Please see also Appendix 2, this chapter.

In this chapter, we proceed under the assumption (which we consider as sufficiently confirmed experimentally) that in order to make decisions, cognitive systems operate with "mental wave functions." We describe the scheme of such decision making[1] based on the lifting formalism, see Appendix 1, this chapter.

Since this book is oriented towards non-experts in quantum information theory, we try to avoid the use of the lifting concept as long as is tolerable. In principle, the reader can proceed without using the lifting concept. However, the formulation in terms of lifting is mathematically elegant and it provides a possibility for further generalizations of the model.

9.2 Two-player game and rational behavior

In this chapter (including Appendix 1) we reproduce[2] (with only slight modifications) the paper by Asano *et al.* [1]. We explain in more detail a two-player

[1] This is joint work of one of the authors with Asano and Ohya [1].

[2] Masanari Asano and Masanori Ohya and Andrei Khrennikov (2011). Quantum-like model for decision making process in two players game. *Foundations of Physics*, 41, 538–548.

Table 9.1. *Pay-off table of a two-player game. 0 and 1 denote the actions player A or B chooses. $C_{ij}(x, y)$ denotes the consequences of the game, and $x, y = a, b, c, d$ represent the values of the pay-off. For example, in the dilemma game, the relation of $c > a > d > b$ is assumed.*

A B	0	1
0	$C_{00}(a, a)$	$C_{01}(b, c)$
1	$C_{10}(c, b)$	$C_{11}(d, d)$

game and the rational behavior that is conventionally discussed in game theory. Two players A and B have two alternatives for actions denoted by 0 and 1. These players are under the same condition on their pay-offs as seen in the pay-off table, Table 9.1, where $C_{ij}(x, y)$ denotes the consequences of the game, and $x, y = a, b, c, d$ represent values of the pay-off.

We here explain the rational behavior of a player in the setting of the so-called "dilemma game." The dilemma game has the inequalities $c > a > d > b$ on the values of the pay-offs. In the game, a player who chooses action 1 can avoid the possibility of the minimum pay-off of d at least. Furthermore, the player can gain the maximal pay-off of c if the other player chooses 0. To choose action 1 is reasonable for a player who wants to maximize his own pay-off. Game theory considers that a rational player chooses 1 without any concern for the action of the other player. The consequence of the game is always C_{11} if both players are rational.[3]

9.3 Construction of a mental state

In the above game, in principle a player has no idea about another player's action. We assume this uncertainty creates some psychological influence on the decision making of each player, i.e. it yields irrational choices. The decision making in our model is mathematically described in the form of a quantum state called a "mental state." In this section, we discuss the definition and construction of such a mental state.

[3] Note that, in this case, the sum of pay-offs in C_{11} is less than the one in the consequence C_{00}, where both players perform the irrational choice of 0. This is an interesting property of the dilemma game.

9.3.1 A prediction state

We consider two players, named A and B. Player B only has to predict A's choice before deciding his own action. Here, we assume that player B predicts player A will choose action 0 with probability p. Action 1 is then chosen with probability $1 - p$. This does not mean that player B "judges" player A's action to be 0 (or 1). Player B should consider the possibility of A's actions simultaneously. We represent such a situation by using the quantum state $\sigma = |\phi\rangle\langle\phi|$:

$$|\phi\rangle = \alpha\,|0_A\rangle + \beta\,|1_A\rangle \in \mathcal{H} = \mathbb{C}^2, \tag{9.1}$$

where $|\alpha|^2 = p$, $|\beta|^2 = 1 - p$ and the orthogonal basis of $|0_A\rangle$ and $|1_A\rangle$ indicates the alternatives of A. We call this state a prediction state.

9.3.2 An alternative state

When player B decides his action is 0, he thinks the consequence of the game will be C_{00} or C_{10}. In a sense, he chooses a consequence described in the superposition of C_{00} and C_{10} as:

$$\alpha\,|C_{00}\rangle + \beta\,|C_{10}\rangle.$$

Now, when we put $|C_{00}\rangle = |0_A\rangle \otimes |0_B\rangle$ and $|C_{10}\rangle = |1_A\rangle \otimes |0_B\rangle$, where $|0_B\rangle$ and $|1_B\rangle$ are an orthogonal basis in $\mathcal{K} = \mathbb{C}^2$, the state vector:

$$\begin{aligned} |\Phi_0\rangle &= \alpha\,|C_{00}\rangle + \beta\,|C_{10}\rangle \\ &= |\phi\rangle \otimes |0_B\rangle. \end{aligned} \tag{9.2}$$

In a similar way, this gives another vector as:

$$\begin{aligned} |\Phi_1\rangle &= \alpha\,|C_{01}\rangle + \beta\,|C_{11}\rangle \\ &= |\phi\rangle \otimes |1_B\rangle. \end{aligned} \tag{9.3}$$

We call the states of $|\Phi_0\rangle\langle\Phi_1|$, alternative states.

9.3.3 A mental state

In the process of decision making, the player imagines the four consequences, and he decides his own action in a probabilistic way. We describe such a situation by using the following quantum state on $\mathcal{H} \otimes \mathcal{K}$;

$$\theta = |\psi\rangle\langle\psi|, \quad |\psi\rangle = x\,|\Phi_0\rangle + y\,|\Phi_1\rangle. \tag{9.4}$$

This description indicates that the player chooses 0 with probability $\lambda = |x|^2$ and 1 with probability $\mu = 1 - \lambda = |y|^2$.

9.4 A process of decision making

In this section, we discuss a process of decision making that consists of three stages: prediction, comparison, and decision.

9.4.1 Making an initial mental state

The decision making starts from the prediction of the other player's choice. Before the prediction, the player's mental state is given by:

$$\rho_0 = |\xi\rangle \langle\xi| , \ |\xi\rangle = x_0 |0_B\rangle + y_0 |1_B\rangle \in \mathcal{K}, \tag{9.5}$$

where $|x_0| = |y_0| = \frac{1}{\sqrt{2}}$. With the use of this ρ_0 and the prediction state σ (in Section 9.3.1), we define the initial mental state $\theta_I = |\psi_I\rangle \langle\psi_I|$:

$$\theta_I = \sigma \otimes \rho_0.$$

9.4.2 A comparison of consequences

The essential part of the decision making process is comparing the consequences of the game. The coefficients x, y of the mental state $|\psi\rangle = x |\Phi_0\rangle + y |\Phi_1\rangle$ become changed by the comparison. We assume its dynamics obeys the following differential equations, and approach a state of equilibrium:

$$\frac{d\lambda}{dt} = -k\lambda + \tilde{k}\mu$$
$$\frac{d\mu}{dt} = -\tilde{k}\mu + k\lambda, \tag{9.6}$$

where $\lambda = |x|^2$ and $\mu = |y|^2$. These equations are similar to a chemical equilibration in a reaction system written as:

$$\Phi_0 \underset{\tilde{k}}{\overset{k}{\rightleftharpoons}} \Phi_1, \tag{9.7}$$

where k $(\tilde{k}) \in \mathbb{R}$ means a rate of reaction from Φ_0 to Φ_1 (or from Φ_1 to Φ_0). Using k and $\tilde{k}$, we define the mental state in equilibrium:

$$\theta_E = |\psi_E\rangle \langle\psi_E| , \ |\psi_E\rangle = x_E |\phi_0\rangle + y_E |\phi_1\rangle ,$$
$$|x_E| = \sqrt{\frac{\tilde{k}}{k+\tilde{k}}}, \ |y_E| = \sqrt{\frac{k}{k+\tilde{k}}}. \tag{9.8}$$

The probabilities $\lambda_E = |x_E|^2$ and $\mu_E = |y_E|^2$ satisfy $\frac{d\lambda}{dt} = \frac{d\mu}{dt} = 0$ in the above differential equations. Here note that this equilibrium state θ_E is described as $\theta_E = \sigma \otimes \rho_E$. We call ρ_E a mental state after comparison.

The reaction rate $\tilde{k}$ or k is related with the degree of the player's tendency to choose either 0 or 1. To discuss these parameters in a simple way, we consider the case where the prediction state vector is $|\phi\rangle = |0\rangle$, i.e. $\alpha = 1$ and $\beta = 0$ in equation (9.1). In this case, the alternative state vectors in equations (9.2) and (9.3) are $|\phi_0\rangle = |C_{00}\rangle$ and $|\phi_1\rangle = |C_{01}\rangle$. Therefore, the equilibrium state $|\Psi_E\rangle$ is decided in the reaction system of:

$$C_{00} \underset{\tilde{k}}{\overset{k}{\rightleftharpoons}} C_{01}, \tag{9.9}$$

and the value of $\frac{\tilde{k}}{k}$ indicates a degree of the worth of C_{00} relative to C_{01} for the player. On the contrary, if $|\phi\rangle = |1\rangle$, k and $\tilde{k}$ decide $|\Psi_E\rangle$ in the comparison of:

$$C_{10} \underset{\tilde{k}}{\overset{k}{\rightleftharpoons}} C_{11}. \tag{9.10}$$

Here we denote k and $\tilde{k}$ of the above two comparisons with indices $k_i, \tilde{k}_i$ $(i = 1, 2)$. In general, the alternative vectors are given by:

$$|\phi_0\rangle = \alpha\,|C_{00}\rangle + \beta\,|C_{10}\rangle\,,\ \ |\phi_1\rangle = \alpha\,|C_{01}\rangle + \beta\,|C_{11}\rangle\,.$$

In this quantum mechanical representation, besides the two comparisons of $(k_1, \tilde{k}_1)$ and $(k_2, \tilde{k}_2)$, the player is allowed to make the following comparisons:

$$C_{00} \underset{\tilde{k}_3}{\overset{k_3}{\rightleftharpoons}} C_{11}\ ,\ \ C_{10} \underset{\tilde{k}_4}{\overset{k_4}{\rightleftharpoons}} C_{01}. \tag{9.11}$$

Generally, the rates of $(k, \tilde{k})$ in equation (9.7) are decided by these $(k_1, \tilde{k}_1)$, $(k_2, \tilde{k}_2)$, $(k_3, \tilde{k}_3)$, and $(k_4, \tilde{k}_4)$. If the four comparisons are made independently, we can write the rates of k and $\tilde{k}$ of equation (9.7) as a classical expectation value given by:

$$k = |\alpha|^4 k_1 + |\beta|^4 k_2 + |\alpha|^2|\beta|^2 k_3 + |\alpha|^2|\beta|^2 k_4,$$
$$\tilde{k} = |\alpha|^4 \tilde{k}_1 + |\beta|^4 \tilde{k}_2 + |\alpha|^2|\beta|^2 \tilde{k}_3 + |\alpha|^2|\beta|^2 \tilde{k}_4. \tag{9.12}$$

However, we consider that the four comparisons of equations (9.9), (9.10), and (9.11) are not made probabilistically in such classical means. In the mental state, these comparisons are made with the effect of influencing each other, in a sense simultaneously. We give k and $\tilde{k}$ in the following forms:

$$k = |s|^2, \ \tilde{k} = |\tilde{s}|^2 \ (s, \tilde{s} \in \mathbb{C}),$$
$$t = |\alpha|^2 k_1^{\frac{1}{2}} + |\beta|^2 k_2^{\frac{1}{2}} + \alpha^* \beta k_3^{\frac{1}{2}} + \alpha \beta^* k_4^{\frac{1}{2}},$$
$$\tilde{t} = |\alpha|^2 \tilde{k}_1^{\frac{1}{2}} + |\beta|^2 \tilde{k}_2^{\frac{1}{2}} + \alpha \beta^* \tilde{k}_3^{\frac{1}{2}} + \alpha^* \beta \tilde{k}_4^{\frac{1}{2}}. \tag{9.13}$$

The values of k and $\tilde{k}$ in this definition are different from that in the classical form of equation (9.12). The differences come from the effect of quantum interference. In our quantum mechanical model, those influences come from simultaneous comparisons.

Here, we introduce a quantum channel Λ^* mapping from the prediction state σ to the mental state ρ_E:

$$\Lambda^* : S(\mathcal{H}) \longrightarrow S(\mathcal{K}),$$
$$\Lambda^* \sigma = \rho_E.$$

Since the comparison of consequences is made on a compound system in $\mathcal{H} \otimes \mathcal{K}$, the channel Λ^* is defined with using a lifting $\mathcal{E}^* : S(\mathcal{H}) \longrightarrow S(\mathcal{H} \otimes \mathcal{K})$. This lifting is defined by:

$$\mathcal{E}^* \sigma = \frac{T\sigma \otimes \rho T^*}{\mathrm{tr}(T\sigma \otimes \rho T^*)},$$

where:

$$T = \begin{pmatrix} 0 & 0 & \tilde{k}_1^{\frac{1}{2}} & \tilde{k}_3^{\frac{1}{2}} \\ 0 & 0 & \tilde{k}_4^{\frac{1}{2}} & \tilde{k}_2^{\frac{1}{2}} \\ k_1^{\frac{1}{2}} & k_4^{\frac{1}{2}} & 0 & 0 \\ k_3^{\frac{1}{2}} & k_2^{\frac{1}{2}} & 0 & 0 \end{pmatrix}. \tag{9.14}$$

In general, the state $\mathcal{E}^* \sigma$ is a quantum entangled state in $\mathcal{H} \otimes \mathcal{K}$. Note that t and $\tilde{t}$ of equation (9.13) is rewritten as:

$$t = \langle \phi_1 | T | \phi_0 \rangle, \tilde{t} = \langle \phi_0 | T | \phi_1 \rangle, \tag{9.15}$$

and it is easily checked that the equilibrium state θ_E is described in the form of:

$$\theta_E = \frac{M \mathcal{E}^* \sigma M^*}{\mathrm{tr}(M \mathcal{E}^* \sigma M^*)},$$

where M is a projection defined by $M = \sigma \otimes I$. From $\theta_E = \sigma \otimes \rho_E$, we obtain:

$$\Lambda^* \sigma = \mathrm{tr}_{\mathcal{H}} \left(\frac{M \mathcal{E}^* \sigma M^*}{\mathrm{tr}(M \mathcal{E}^* \sigma M^*)} \right).$$

9.4.3 Decision of action

After the comparison, the player decides his own action. We assume the decision corresponds to a quantum measurement of the state $|0_B\rangle\langle 0_B|$ or $|1_B\rangle\langle 1_B|$ to the mental state ρ_E. Therefore, his decision is made probabilistically in general. The probabilities of decisions 0 and 1 are given by:

$$\lambda_E = \mathrm{tr}(|0_B\rangle\langle 0_B|\,\rho_E\,|0_B\rangle\langle 0_B|),$$
$$\mu_E = \mathrm{tr}(|1_B\rangle\langle 1_B|\,\rho_E\,|1_B\rangle\langle 1_B|). \tag{9.16}$$

9.5 Example: decision making in PD

In this section, we apply our model to the Prisoner's Dilemma (PD) game. We show the model makes irrational choices in this example.

In our model, the characteristics of the game are reflected in the contents of the operator T of equation (9.14). In other words, the pay-offs of the game as seen in the Table 9.1 can be factors which decide the values of k_i and $\tilde{k}_i$. We give the operator T for the example of the PD game, in the following form:

$$T = \begin{pmatrix} 0 & 0 & 0 & \tilde{k}_3^{\frac{1}{2}} \\ 0 & 0 & 0 & 0 \\ k_1^{\frac{1}{2}} & k_4^{\frac{1}{2}} & 0 & 0 \\ 0 & k_2^{\frac{1}{2}} & 0 & 0 \end{pmatrix}.$$

In this form, we assume $k_3 = 0$ and $\tilde{k}_1 = \tilde{k}_2 = \tilde{k}_4 = 0$, because the pay-off d for the consequence C_{10} is smaller than those of the other consequences and the pay-off a for C_{00} is smaller than c for C_{01} but larger than d for C_{11}. This setting of the parameters provides the probability:

$$\lambda_E = \frac{|\alpha|^2|\beta|^2\tilde{k}_3}{|\alpha|^2|\beta|^2\tilde{k}_3 + \left||\alpha|^2 k_1^{\frac{1}{2}} + |\beta|^2 k_2^{\frac{1}{2}} + \alpha\beta^* k_4^{\frac{1}{2}}\right|^2},$$

$$\mu_E = \frac{\left||\alpha|^2 k_1^{\frac{1}{2}} + |\beta|^2 k_2^{\frac{1}{2}} + \alpha\beta^* k_4^{\frac{1}{2}}\right|^2}{|\alpha|^2|\beta|^2\tilde{k}_3 + \left||\alpha|^2 k_1^{\frac{1}{2}} + |\beta|^2 k_2^{\frac{1}{2}} + \alpha\beta^* k_4^{\frac{1}{2}}\right|^2}. \tag{9.17}$$

As mentioned in Section 9.2, a rational player always chooses his action 1, i.e. $\mu_E = 1$. The player in our model becomes rational in two special cases; the case of $\tilde{k}_3 = 0$ and the case of $\alpha = 0$ or $\beta = 0$. Since the parameters of α and β decide the player's prediction for the other player's choice, the case of $\alpha = 0$ or $\beta = 0$

is unnatural to the principle of the game, which says that the player is uncertain about the other player's choice. It may correspond to a special situation where the player obtains some information and can judge another player's choice before his own choice.

Appendix 1: Channels and liftings

In quantum information theory [2], a "map" is important for describing an information transmission such as a measurement process or a signal transmission. This map is called a channel $\Lambda^* : \mathcal{S}(\mathcal{A}) \mapsto \mathcal{S}(\mathcal{B})$. Here $\mathcal{S}(\mathcal{A})$ ($\mathcal{S}(\mathcal{B})$) are state spaces of C^*-algebras $\mathcal{A}$ ($\mathcal{B}$). For example, a set of all bounded linear operators $\mathcal{B}(\mathcal{H})$ on Hilbert space $\mathcal{H}$ realizes a C^*-algebra $\mathcal{A}$. If a channel Λ^* is affine, i.e. $\Lambda^*(\sum_n \lambda_n \rho_n) = \sum_n \lambda_n \Lambda^*(\rho_n)$, $\forall \rho_n \in \mathcal{S}(\mathcal{A})$, $\forall \lambda_n \in [0, 1]$, $\sum \lambda_n = 1$, it is called a linear channel. A completely positive (CP) channel is a linear channel Λ^* such that its dual $\Lambda : \mathcal{B} \mapsto \mathcal{A}$ (i.e. $\mathrm{tr}(\Lambda^*(\rho)A) = \mathrm{tr}(\rho\Lambda(A))$ for any $A \in \mathcal{A}$) satisfies:

$$\sum_{i,j=1}^{n} A_i^* \Lambda(B_i^* B_j) A_j \geq 0,$$

for any $\{A_j\} \subset \mathcal{A}$, $\{B_j\} \subset \mathcal{B}$ and $n \in \mathbb{N}$.

Liftings are a class of channels from $\mathcal{S}(\mathcal{A})$ to $\mathcal{S}(\mathcal{A} \otimes \mathcal{B})$:

$$\mathcal{E}^* : \mathcal{S}(\mathcal{A}) \mapsto \mathcal{S}(\mathcal{A} \otimes B). \tag{9.18}$$

The following liftings are often used in physics:

1. **Linear lifting:** A linear lifting is affine and its dual is a completely positive map.
2. **Pure lifting:** A pure lifting maps pure states into pure states.
3. **Non-demolition lifting:** A lifting is a non-demolition lifting for a state $\rho \in \mathcal{S}(\mathcal{A})$ if ρ is invariant for any $A \in \mathcal{A}$ in the sense of:

   $$(\mathcal{E}^*\rho)(A \otimes 1) = \rho(A).$$

   Here, $\rho(A) \equiv \mathrm{tr}(\rho A)$, $A \in \mathcal{A}$.
4. **Compound state lifting:** A compound state lifting is a non-linear and non-demolition lifting such that for a density matrix $\rho = \sum_k \lambda_k E_k$, $E_k \in \mathcal{S}(\mathcal{A})$:

   $$\mathcal{E}^*(\rho) = \sum_k \lambda_k E_k \otimes \Lambda^* E_k.$$

5. **Transition lifting:** A transition expectation is a completely positive linear map given by $\mathcal{E} : \mathcal{A} \otimes \mathcal{B} \mapsto \mathcal{A}$, and it satisfies:

   $$\mathcal{E}(1 \otimes 1) = 1.$$

Transition expectations play a crucial role in the construction of quantum Markov chains and they appear in the framework of measurement theory. The dual of a transition expectation is an example of a lifting.

6. **Isometric lifting:** A lifting is a channel from a subsystem to a compound system, and it is useful for describing open system dynamics. Let us consider the following situation. A system interacts with another (environment) system, and a correlated state is generated in a compound system. We assume that the system is independent from another system before the interaction and initial states of the two systems are given by $\rho \in \mathcal{S}(\mathcal{A})$ and $\sigma \in \mathcal{S}(\mathcal{B})$. The compound system after the interaction is represented by $\mathcal{E}^*\rho \in \mathcal{S}(\mathcal{A} \otimes \mathcal{B})$ in general. Now let the lifting $\mathcal{E}^*$ be an **isometric lifting** defined as:

$$\mathcal{E}^*\rho = V\rho V^*,$$

where the operator $V : \mathcal{H}_\mathcal{A} \mapsto \mathcal{H}_\mathcal{A} \otimes \mathcal{H}_\mathcal{B}$ satisfies $V^*V = 1_{\mathcal{H}_\mathcal{A}}$. Then for an observable $B \in \mathcal{B}(\mathcal{B})$, the expectation value $\mathrm{tr}_\mathcal{A}\mathcal{E}^*\rho(B)$ is written as:

$$\mathrm{tr}_\mathcal{A}\mathcal{E}^*\rho(B) = \mathrm{tr}A(\rho V^*(1 \otimes B)V) = \mathrm{tr}A(\rho\mathcal{E}(1 \otimes B)) = \rho(\mathcal{E}(1 \otimes B)).$$

Here, we introduce the map $\mathcal{E}(1 \otimes B) = V^*(1 \otimes B)V$. This map is a transition expectation and its dual is the isometric $\mathcal{E}^*$. We can easily check $\mathrm{tr}_\mathcal{A}\mathcal{E}^*\rho(I) = 1$.

When the $\mathcal{E}^*$ represents a time evolution; $\mathcal{E}^*\rho = U\rho \otimes \sigma U^*$ (U is a unitary operator), the corresponding transition expectation is written by:

$$\mathcal{E}(1 \otimes B) = \mathrm{tr}_\mathcal{B}(U(1 \otimes \sigma)U^*(1 \otimes B)).$$

If B is written as $B = \sum_k \lambda_k P_k$ with projections $\{P_k\}$:

$$\mathrm{tr}_\mathcal{A}\mathcal{E}^*\rho(B) = \sum_k \lambda_k \mathrm{tr}_\mathcal{A}\mathcal{E}^*\rho(P_k) = \sum_k \lambda_k \rho(\mathcal{E}(1 \otimes P_k)) \equiv \sum_k \lambda_k \Lambda_k^*\rho(P_k).$$

In this form, values of $\Lambda_k^*\rho(P_k)$ mean probabilities $p(\lambda_k)$, and $\mathcal{E}(1 \otimes P_k)$ mean a POVM (a positive operator valued measure) Q_k that satisfies $Q_k \geq 0$, $\sum_k Q_k = I$.

Appendix 2: Quantum Markov chain description of data from experiments in cognitive psychology

In this section we construct a quantum-like representation of the experimental statistical data from the Tversky–Shafir PD experiment, please see Section 8.9 of Chapter 8. In this appendix we reproduce (pp. 378–383)[4] (with only slight

[4] Luigi Accardi and Andrei Khrennikov and Masaniori Ohya (2009). Quantum Markov model for data from Shafir-Tversky experiments in cognitive psychology. *Open Systems & Information Dynamics*, 16(4), 378–383.

modifications) the paper by Accardi *et al.* [3]. To describe these data, we have to use a more general quantum probabilistic model than the standard Dirac–von Neumann model based on Hermitian operators in a complex Hilbert space and the definition of probabilities through projections of eigenvalues of operators (Born's rule). We want to emphasize that the aforementioned data have such a complex internal structure that it is impossible to find two Hermitian operators with non-degenerate spectra which act in the Hilbert space of the dimension matching the dimension of the game, namely the two-dimensional space. This is a consequence of a simple fact: the matrix of transition probabilities is not double stochastic. We shall use so-called quantum Markov chains, a non-commutative generalization of ordinary Markov chains. Please see Accardi *et al.* [3].

This section is based on advanced mathematics. Therefore, this section is not recommended for readers who have no experience of research in quantum probability and quantum information theory. Experts from psychology, behavioral economics, finance, and decision making can learn that statistical data from games exhibiting irrational behavior can have a very complex non-commutative structure, more complex than the structure of statistical data from basic experiments on quantum foundations, e.g. the two slit experiment.

The quantum Markov model presented in this section was created by Accardi *et al.* [3].

Given a classical Markov transition matrix (p_{ij}) with state space $\{1, \ldots, d\}$ $(d \in \mathbb{N})$, a natural way to associate to it a *conditional density amplitude* is to define:

$$K = \sum_{i,j} \sqrt{p_{ij}} e_{ii} \otimes e_{jj} \in M_d \otimes M_d,$$

where M_d denotes the $d \times d$ complex matrices, (e_i) an orthonormal basis of $\mathbb{C}^d$, (e_{ij}) the associated system of matrix units $(e_{ij} e_{hk} = \delta_{j,h} e_{ik})$, and $\sqrt{p_{ij}}$ are arbitrary (usually complex) square roots of p_{ij}. Any conditional density amplitude determines a (special) linear lifting (see Appendix 1):

$$\mathcal{E}^* : \mathcal{S}(M_d) \to \mathcal{S}(M_d \otimes M_d)$$

from density matrices on M_d to density matrices on $M_d \otimes M_d$ or, dually, a *transition expectation*, i.e. a completely positive, identity preserving linear map:

$$\mathcal{E} : M_d \to M_d \otimes M_d$$

defined by:

$$\mathcal{E}(x) = \mathrm{Tr}_2(K^* x K), \qquad x \in M_d \otimes M_d, \tag{9.19}$$

where Tr_2 denotes the partial trace over the second factor $(\mathrm{Tr}_2(a \otimes b) := \mathrm{Tr}(b)a \in M_d)$ and Tr denotes the non-normalized trace on M_d. More explicitly, if $X, Y \in M_d$, one has:

$$\mathcal{E}(X \otimes Y) = \mathrm{Tr}_2(K^*(X \otimes Y)K) = \sum_{i,j}\sum_{h,k} \sqrt{p_{ij}}^* \sqrt{p_{hk}} \mathrm{Tr}(e_{ij} X e_{hk} \otimes e_{ji} Y e_{kk})$$

$$= \sum_{i,j}\sum_{h,k} \sqrt{p_{ij}}^* \sqrt{p_{hk}} e_{ii} X e_{hk} \delta_{jk} \mathrm{Tr}(Y e_{jj});$$

therefore, using the identities:

$$e_{jj} Y e_{jj} = \langle e_j, Y e_j \rangle e_{jj},$$

$$e_{ii} e_{\alpha\beta} e_{hh} = \delta_{ix} \delta_{\beta h} e_{ih},$$

$$e_{ii} X e_{hh} = \langle e_i, X_{e_h} \rangle e_{ih},$$

one obtains:

$$\mathcal{E}(X \otimes Y) = \sum_{i,j,h} \sqrt{p_{ij}}^* \sqrt{p_{hj}} \langle e_i, X e_h \rangle e_{ih} \langle e_j, Y e_j \rangle. \tag{9.20}$$

Notice that, if $X, Y \in M_d$ are diagonal in the (e_i)-basis, then this formula reduces to the usual formula for the conditional expectation in classical probability:

$$\mathcal{E}(X \otimes Y) = \sum_{i,j} p_{ij} X_{ii} e_{ii} Y_{jj} = \sum_i \left(X_{ii} \sum_j p_{ij} Y_{jj} \right) e_{ii} = X P(Y).$$

Given any density matrix in M_d:

$$\rho_0 = \sum_{\alpha',\gamma'} \rho_{\alpha',\gamma'} e_{\alpha',\gamma'},$$

the pair $\{\rho_0, \mathcal{E}\}$ uniquely determines the joint n-point correlations:

$$\varphi\left(a_0 \otimes a_1 \otimes \dots a_n\right) := \mathrm{Tr}\left(\rho_0 \mathcal{E}(a_0 \otimes \mathcal{E}(a_1 \otimes \mathcal{E}(a_2 \otimes \dots \mathcal{E}(a_n \otimes 1)\dots)\right),$$

where $a_0, a_1, a_2, \dots, a_n \in M_d$. We are interested in a time horizon with two time points: initial and final. The corresponding joint probabilities are in this case:

$$\mathrm{Tr}\rho_0 \mathcal{E}(a \otimes \mathcal{E}(b \otimes 1)).$$

Using (9.20), we obtain:

$$\mathcal{E}(b \otimes 1) = \sum_{i,j,h} \sqrt{p_{ij}}^* \sqrt{p_{hj}} . \langle e_i, b e_h \rangle e_{ih}.$$

Therefore:

$$\begin{aligned}
\mathcal{E}(a\otimes\mathcal{E}(b\otimes 1)) &= \sum_{i,j,h} \sqrt{p_{ij}}^{*}\sqrt{p_{hj}}\langle e_i, be_h\rangle \mathcal{E}(a\otimes e_{ih}) \\
&= \sum_{i,j,h} \sqrt{p_{ij}}^{*}\sqrt{p_{hj}}\langle e_i, be_h\rangle \sum_{\alpha,\beta,\gamma} \sqrt{p_{\alpha\beta}}^{*}\sqrt{p_{\gamma\beta}}\langle e_\alpha, ae_\gamma\rangle e_{\alpha\gamma}\langle e_\beta, e_{ih}e_\beta\rangle \\
&= \sum_{i,j,\alpha,\gamma} \sqrt{p_{ij}}^{*}\sqrt{p_{ij}}\langle e_i, be_i\rangle\langle e_\alpha, ae_\gamma\rangle e_{\alpha\gamma}\sqrt{p_{\alpha i}}^{*}\sqrt{p_{\gamma i}}.
\end{aligned}$$

By using the right stochasticity of the matrix of transition probabilities, $\sum_j p_{ij} = 1$ (here $p_{ij} = \mathbf{P}(b = j|a = i)$), we obtain:

$$\mathrm{Tr}\rho_0\mathcal{E}(a\otimes\mathcal{E}(b\otimes 1)) = \sum_{i,\alpha,\gamma,\alpha',\gamma'} \langle e_i, be_j\rangle\langle e_\alpha, ae_\gamma\rangle\sqrt{p_{\alpha i}}^{*}\sqrt{p_{\gamma i}}\rho_{\alpha'\gamma'}\mathrm{Tr}e_{\alpha'\gamma'}e_{\alpha\gamma}$$

and, using the identity:

$$\mathrm{Tr}e_{\alpha'\gamma'}e_{\alpha\gamma} = \delta_{\gamma'\alpha}\delta_{\alpha'\gamma}$$

and the notations:

$$\langle e_i, be_i\rangle =: b_{ii}, \ \langle e_\alpha, ae_\gamma\rangle =: a_{\alpha\gamma}$$

one finally obtains:

$$\begin{aligned}
\mathrm{Tr}\rho_0\mathcal{E}(a\otimes\mathcal{E}(b\otimes 1)) &= \sum_{i,\alpha,\gamma}\langle e_i, be_i\rangle\langle e_\alpha, ae_\gamma\rangle\sqrt{p_{\alpha i}}^{*}\sqrt{p_{\gamma i}}\rho_{\gamma\alpha} \\
&= \sum_{i,\alpha\gamma} b_{ii}a_{\alpha\gamma}\sqrt{p_{\alpha i}}^{*}\sqrt{p_{\gamma i}}\rho_{\gamma\alpha} \\
&= \sum_{\alpha,\gamma}\rho_{\gamma\alpha}a_{\alpha\gamma}\left(\sum_i b_{ii}\sqrt{p_{\alpha i}}^{*}\sqrt{p_{\alpha i}}\right) \\
&= \sum_{\alpha,\gamma}\rho_{\alpha\gamma}a_{\alpha\gamma}L(b)_{\alpha\gamma}.
\end{aligned}$$

If the observables a, b have the form:

$$a = |+_a\rangle\langle +_a| - |-_a\rangle\langle -_a, |b = |+_b\rangle\langle +_b| - |-_b\rangle\langle -_b|,$$

the marginal probabilities for a are obtained by putting $b = +1$, i.e.:

$$\begin{aligned}
\mathrm{Tr}\left(\rho_0\mathcal{E}(|+_a\rangle\langle +_a|\otimes 1)\right) &= \sum_{i,j,h}\sqrt{p_{ij}}^{*}\sqrt{p_{hj}}\langle e_i, +_a\rangle\langle +_a, e_h\rangle\rho_{hi} \\
&= \sum_{h,i}\rho_{hi}\langle e_i, +_a\rangle\langle +_a, e_h\rangle\sum_j\sqrt{p_{ij}}^{*}\sqrt{p_{hj}},
\end{aligned}$$

where we used the identity:

$$\mathrm{Tr}\rho e_{ih} = \sum_{\alpha,\gamma} \rho_{\alpha,\gamma} \mathrm{Tr} e_{\alpha\gamma} e_{ih} = \sum_{\alpha,\gamma} \rho_{\alpha\gamma} \delta_{\alpha i} \delta_{\alpha h} = \rho_{hi}.$$

Now we specialize to the case $d = 2$, so that ρ_0 has the form:

$$\rho_{11} = p, \ \rho_{22} = 1 - p, \ \rho_{12} = \overline{\rho_{21}} = z,$$

for some $p \in [0, 1]$ and $z \in \mathbb{C}$. This gives the equation:

$$p|\langle e_1, +_a\rangle|^2 + (1-p)|\langle e_2, +_a\rangle|^2 + 4Re\left(z\langle e_1, +_a\rangle\langle +_a, e_2\rangle\right) Re\left(\sqrt{p_{12}}^* \sqrt{p_{21}}\right)$$
$$= P(a = +1|C).$$

A similar calculation for the marginal probabilities for b gives the equation:

$$\mathrm{Tr}\rho_0 \mathcal{E}(1 \otimes \mathcal{E}(|+_b\rangle\langle +_b|)) = \sum_{\alpha} \rho_{\alpha\alpha} \sum_i |\langle a_1, e_i\rangle|^2 p_{\alpha i} = \mathbf{P}(b = +1|C).$$

Expanding the eigenvectors of b and a in the e-basis, we obtain, for some $p_a, p_b \in [0, 1]$:

$$\sum_i pp_{1i}|\langle +_b, e_i\rangle|^2 + \sum_i (1-p)p_{2i}|\langle +_b, e_i\rangle|^2 = P(b = +1 = |C)$$

$$|+_a\rangle =: \sqrt{p_a} e_1 + \sqrt{1 - p_a} e_2,$$
$$|+_b\rangle =: \sqrt{p_b} e_1 + \sqrt{1 - p_b} e_2,$$

and therefore the above equations become:

$$pp_a + (1-p)(1-p_a) + 4Re\left(z\sqrt{p_a}\sqrt{1-p_a}^*\right) Re\left(\sqrt{p_{12}}^* \sqrt{p_{12}}\right)$$
$$= P(a = +1|C),$$
$$pp_{11}p_b + pp_{12}(1-p_b) + (1-p)p_{12}p_b + (1-p)p_{22}(1-p_b)$$
$$= P(b = +1|C).$$

Looking for a solution with $z = 0$ leads to the simplified system:

$$pp_a - (1-p)p_a + (1-p) = (2p-1)p_a + (1-p) = P(a = +1|C),$$
$$p(p_{11} - p_{12})p_b + pp_{12} + (1-p)p_b(p_{12} - p_{22}) + (1-p)p_{22} = P(b = +1|C).$$

Recalling that the empirical values, found by Tversky and Shafir for the PD-type experiment (please see Section 8.9 in Chapter 8), are:

$$P(a = +1|C) = \frac{1}{2}, \ P(b = +1|C) = 0.63,$$

the first equation becomes equivalent to:

$$(2p-1)p_a+(1-p)=\frac{1}{2},$$

which gives, independently of p:

$$p_a=\left(p-\frac{1}{2}\right)\frac{1}{2p-1}=\frac{1}{2}.$$

This result is not really surprising, since here the A-player is a computer which produces wins and losses with equal probabilities. It is natural for the B-player (who is really a player) to simply incorporate these probabilities in the quantum-like representation of the A-strategy. This representation is given by vectors $|\pm_a\rangle$. Thus, the simplest representation:

$$|+_a\rangle=(e_1+e_2)/\sqrt{2},$$

the second vector can be chosen as any normalized vector which is orthogonal to:

$$|-_a\rangle=(e_1-e_2)/\sqrt{2}.$$

Of course, mathematically the situation is essentially more general. The vector representation of the state of A can involve complex phases:

$$|+_a\rangle=(e_1+e^{i\theta_A}e_2)/\sqrt{2},\ |-_a\rangle=(e_1-e^{i\theta_A}e_2)/\sqrt{2}.$$

In a more complicated gamble, such that the A player is not a computer, the B player may really use the additional degree of freedom, namely the phase to represent the A-strategy. However, in the present gambling experiment it seems that B can be fine with purely real eigenvectors.

The second equation becomes equivalent to:

$$\begin{aligned}
&p_b(p(p_{11}-p_{12})+(1-p)(p_{21}-p_{22}))+pp_{12}+(1-p)p_{22}=0.63\\
&\Leftrightarrow p_b(p(0.16-0.84)+(1-p)(0.03-0.97))+p0.84+(1-p)0.97=0.63\\
&\Leftrightarrow -p_bp0.68-p_b(1-p)0.94+p0.84+(1-p)0.97=0.63\\
&\Leftrightarrow p(0.84-p_b0.68)+(1-p)(0.97-p_b0.94)=0.63\\
&\Leftrightarrow p=\frac{0.63-(0.97-p_b0.94)}{0.84-p_b0.68}.
\end{aligned}$$

Thus any $p_b \in (0.1)$ such that:

$$0 < \frac{0.63 - (0.97 - p_b 0.94)}{0.84 - p_b 0.68} < 1$$

will give a unique p.

The two inequalities are equivalent to:

$$0 < 0.63 - 0.97 + p_b 0.94$$
$$0.63 - 0.97 + p_b 0.94 < 0.84 - p_b 0.68.$$

The first is equivalent to:

$$0.34 < p_b 0.94 \Leftrightarrow \frac{34}{94} < p_b.$$

The second is equivalent to:

$$p_b(0.94 + 0.68) < 0.84 + 0.34 \Leftrightarrow p_A 1.62 < 1.28 \Leftrightarrow p_b < \frac{118}{162}.$$

Since:

$$\frac{34}{94} \approx 0.36 < \frac{118}{162} \approx 0.73$$

any p_b in the (non-empty) interval $[\frac{34}{94}, \frac{118}{162}]$ will give a solution.

If $p_b = \frac{34}{94}$, then $p = 0$. Thus, in the density matrix $\rho_{11} = 0$ and $\rho_{22} = 1$, non-diagonal elements are equal to zero. The pure state $\rho = e_2 \otimes e_2$ is a "pure state of mind." In the same way, in the case $p_b = \frac{118}{162}$ we get $\rho = e_1 \otimes e_1$. We point out the crucial difference of the quantum Markov model from the conventional Dirac–von Neumann model based on Born's rule, where:

$$\mathbf{P}(b = +1|C) = \mathbf{P}_\rho(b = +1) = \mathrm{Tr}\rho(|+_b\rangle\langle +_b|)$$

and:

$$\mathbf{P}(a = +1|C) = \mathbf{P}_\rho(a = +1) = \mathrm{Tr}\rho(|+_a\rangle\langle +_a|).$$

In the brain of the B-player, vectors $|+_a\rangle$ and $|+_b\rangle$ are created. One can speculate that they provide a kind of "internal mental representation" of the context C. However, this internal representation does not determine completely the real probabilities of answers. The latter arise in the process of interaction of the internal mental representation with previous experience, which is encoded in probabilities p_{ij}, and, finally, in the lifting operator $\mathcal{E}^*$, which is determined by the operator K. Of course, it is just a first reflection to interpret moving from the conventional quantum model to the quantum Markov model: interference of the internal mental representation of context C with representations of contexts corresponding to fixed strategies, say $C_\pm^b = (b = \pm 1)$ and $C_\pm^a = (a = \pm 1)$. The latter strategies are based on previous experience. We hope that psychologists may suggest more justified motivations of the shift from the conventional quantum model to the quantum Markov model. We would not be surprised if possible psychological interpretations would be different from our interpretation.

9.6 References

[1] Asano, M., Ohya, M., and Khrennikov, A. (2011). Quantum-like model for decision making process in two players game. *Foundations of Physics*, 41, 3, 538–548.
[2] Ingarden, R. S., Kossakowski, A., and Ohya, M. (1997). *Information Dynamics and Open Systems: Classical and Quantum Approach.* Kluwer, Dordrecht, Holland.
[3] Accardi, L., Khrennikov, A., and Ohya, M. (2009). Quantum Markov model for data from Shafir-Tversky experiments in cognitive psychology. *Open Systems and Information Dynamics*, 16, 371–385.

Part IV

Other quantum probabilistic effects in economics, finance, and brain sciences

10

Financial/economic theory in crisis

In this chapter and Chapters 11, 12, and 13, we will discuss quantum-like models with financial applications in mind. The very last chapter of this book will analyze the sources of quantum-like processing in the brain.

The concept of non-arbitrage, as we have indicated before (see Section 4.18.3) is of key importance in financial models. We reiterate its importance in this chapter. We also briefly mention a popular finance model. In the coming chapters, the wave function will sometimes have an information interpretation. Therefore we also discuss the informational aspects of the classical financial theory (see Section 10.1). This chapter is meant to introduce the reader to some of the key achievements in financial theory. By doing so, we hope to convince the reader, in later chapters, that some of those very achievements can possibly be reformulated with the help of the wave function approach. Section 10.3 in this chapter will give a flavor of how Fourier integration can possibly be used to model George Soros' concept of reflexivity.

10.1 Relevance of the concepts of efficiency and non-arbitrage: a brief discussion

With the 2008 financial crisis, the concept of efficiency may have been seriously questioned. Indeed the financial market in September 2008 did take a tremendous hit and the ensuing price movements would not be consistent with a so-called efficient market.

For now, let us concentrate on what is efficiency. Ma [1] (p. 4) defines efficient portfolios as: "portfolios that are efficient in the sense that no other portfolios are available to reduce further the portfolio risk." Here is an interesting corollary out of Ma [1] (Corollary 1.3 (p. 41)): "Suppose the market is efficient with respect to $\mathcal{F}$. For all $\mathcal{G} \subseteq \mathcal{F}$ and for any arbitrary trading strategies ϕ based on information structure $\mathcal{G}$, the value of the investment must coincide with the market price of the trading strategy." This is an important statement. The set of information $\mathcal{G}$ is a subset of the set of information $\mathcal{F}$. This means that $\mathcal{G}$ contains less information than $\mathcal{F}$.

As Ma [1] (Chapter 1)) indicates, the efficiency definitions are due originally to Fama [2]. Any textbook in basic finance, will define the following types of efficiency:

- *strong form efficiency:* prices reflect all available information;
- *semi-strong form efficiency:* prices reflect all publicly available information;
- *weak form efficiency:* prices reflect all historic prices of the traded securities.

Strong form efficiency is elegantly formulated in Ma [1] (p. 52) as follows: "In particular, no individual investor can make excess returns with his possessed information set $\mathcal{F}_i$ that is no more informative than the pooling information set $\bigcap_{i\in\mathcal{I}}\mathcal{F}_i$."

At this point, we would like to make two possibly important points. In what follows, we would like to discuss (a) information manipulation, and (b) give an example of where private information may not give any profit advantage.

Let us tackle (a). Ma [1] (p. 43)) indicates that the manipulation of information can take on the forms of, what he calls: (i) "information leakage" and (ii) "uncertainty creation." These are two very interesting notions which do emulate very much realistic behavior. Ma [1] (p. 43) defines information leakage as: "situations where agents wish to reveal truthfully their private possessed information to others." Uncertainty creation is then defined as per Ma [1] (p. 43) as agents wishing to create "uncertainty in the rest of the economy." The latter type of information creation is straightforward: create information which is dubious (or even false) so it can serve your investment strategy.

Now let us tackle (b). We reproduce[1] here Haven's [3] section 1[2] of that paper. In Tirole [4] a market is considered with I $(i = 1, \ldots, I)$ risk averse or risk neutral traders. The traders exchange an asset on a price p and the (random) value of that asset is $\widetilde{p}$. The trader buys x units of the asset and if $\widetilde{p}$ is revealed, his gain will be given by:

$$G^i = (\widetilde{p} - p)x^i. \tag{10.1}$$

A set $E \subset \mathbb{R}$ is then defined containing all the potential values of $\widetilde{p}$. Each trader has a private signal s^i, which belongs to the set of signals S^i. The cross product of all

[1] Emmanuel Haven (2007). Private information and the "information function." *Theory and Decision*, 64, 2–3, 2008, pp. 200–201.

[2] This is the section "Private information and the information function" with as subsection "Private information as defined by Tirole."

those sets S^i forms a set S. The vector of all signals is denoted by $s = (\ldots, s^i, \ldots)$. The set Ω is then defined as: $\Omega \equiv E \times S$. It is then assumed that the traders have a common prior ν on Ω. Each trader is assumed to take out a quantity x^i so as to maximize his conditional expected gain. The rational expectations equilibrium is defined to be a forecast function Φ, which in the words of Tirole [4] (p. 1166) "associates with each set of signals s a price $p = \Phi(s)$ and the set of trades $x^i(p, s^i, S(p))$ for each agent i, relative to information s^i and $s \in S(p) \equiv \Phi^{-1}(p)$, such that $x^i(p, s^i, S(p))$" will maximize the expected conditional gain, which we denote as in the excellent paper by Scheinkman and Xiong [5], as:

$$\mathbb{E}_\Phi(G^i | p, s^i) \equiv \int G d\Gamma^i_{\Phi, p, s^i}, \tag{10.2}$$

where $\Gamma_{\cdot}$ is a conditional distribution. The conditional distribution is induced by the forecast function, the observed price and the signal s^i. This explains the notation Scheinkman and Xiong [5] use on this conditional distribution. Furthermore, the market will clear for each $s \in S$. The no-trade theorem then says that in a rational expectations equilibrium (see Scheinkman and Xiong [5], p. 227): $\mathbb{E}_\Phi(G^i | p, s^i) = 0$. This will thus be true if all agents are rational and have the same prior ν. Thus private information has no impact whatsoever on the conditional expected gains if the agents are behaving under the conditions of what Scheinkman and Xiong [5] call the standard rational expectations model. Thus, when those conditions are violated the conditional expected gain could be non-negative.

This section of the book also has as goal to re-iterate the concept of non-arbitrage. In Chapter 4, Section 4.18.3 we provided for a formulation of the non-arbitrage theorem following Etheridge [6]. It is not difficult to show that if one of the prices in the price vector (at time[3] $t = 0$) is set to unity, and the security prices of that one asset under the various states of nature (at time $t = 1$) are set equal to unity with the addition of a return equal to the risk free rate of interest, then one can define a set of probabilities (risk neutral probabilities) by which we can discount a risky asset at the risk free rate. Neftci [7] shows this very nicely in his book (see pp. 19–20). This is a very useful result in finance, since risky assets would require the use of preferences for risk which are notoriously difficult to estimate.[4] The use of the risk free rate does not ask for such preferences, and, in effect, there is wide agreement by market participants on the level of a risk free rate of interest in an economy (for instance it could be the prime rate set by the central bank of a country). As long as we do not worry about the *meaning* of the probability, then such discounting is warranted.

[3] When we formulated the non-arbitrage theorem in Chapter 4, Section 4.18.3, we denoted $t = 0$, as t_0 and $t = 1$, as t_1.

[4] But see below for our brief discussion of the CAPM model.

Beginning undergraduate students studying for finance degrees will all know about the so-called "CAPM" model, or also the "Capital Asset Pricing Model," which in effect does give an estimate of the return of an asset (whether risky or risk free). Following Luenberger [8] (p. 173), this model was developed mainly by Sharpe [9], Lintner [10], and Mossin [11]. The CAPM model has been extended in many ways.

The model is in fact quite easy to grasp. One needs essentially two main ingredients: (i) the so-called β of an asset and (ii) the concept of a market portfolio. The market portfolio can be thought of as proxied by for instance an index portfolio such as the Dow Jones 30. One can define the so-called Beta for a security i, β_i, as $\beta_i = \frac{cov(r_i, r_m)}{\sigma_m^2}$, where r_i is the return of asset i and r_m is the return of the market portfolio and $cov(r_i, r_m)$ is the covariance of the return of asset i and the return of the market portfolio m. The variance of the return of the market portfolio is denoted as σ_m^2. If asset i is the risk free asset (with return the risk free rate of return, r_f), then $cov(r_f, r_m) = 0$ and therefore $\beta_f = \frac{cov(r_f, r_m)}{\sigma_m^2} = 0$. Remark that if i is the market portfolio (with return r_m), then $cov(r_m, r_m) = \sigma_m^2$ and hence $\beta_m = \frac{\sigma_m^2}{\sigma_m^2} = 1$. In a non-expected form, the so-called CAPM can be written as $r_i = r_f + \beta_i(r_m - r_f) + \epsilon_i$, where ϵ_i is the error term.

The CAPM can also be written as a pricing formula. As per Luenberger [8] (p. 187), one obtains then $P = \frac{E(Q)}{r_f + \beta_i(E(r_m) - r_f) + 1}$, where P is the price (now) of an asset with expected selling price Q. The appropriate discount rate is thus $r_f + \beta_i(E(r_m) - r_f)$.

10.2 References

[1] Ma, C. (2010). *Advanced Asset Pricing Theory*. Imperial College Press.
[2] Fama, E. (1970). Efficient capital markets: a review of theory and empirical work. *Journal of Finance*, 25, 2, 383–417.
[3] Haven, E. (2008). Private information and the "information function." *Theory and Decision*, 64, 2–3, 193–228.
[4] Tirole, J. (1982). On the possibility of speculation under rational expectations. *Econometrica*, 50, 1163–1181.
[5] Scheinkman, J. and Xiong, W. (2004). *Heterogeneous Beliefs, Speculation and Trading in Financial Markets. Paris–Princeton Lectures on Mathematical Finance. Lecture Notes in Mathematics 1847*, Springer Verlag, Berlin.
[6] Etheridge, A. (2002). *A Course in Financial Calculus*. Cambridge University Press.
[7] Neftci, S. (2000). *An Introduction to the Mathematics of Financial Derivatives*. Academic Press.
[8] Luenberger, D. (1998). *Investment Science*. Oxford University Press.
[9] Sharpe, W. (1964). Capital asset prices: a theory of market equilibrium under conditions of risk. *Journal of Finance*, 19, 425–442.

[10] Lintner, J. (1965). The valuation of risk assets and the selection of risky investment in stock portfolios and capital budgets. *Review of Economics and Statistics*, 47, 13–37.
[11] Mossin, J. (1966). Equilibrium in a capital asset market. *Econometrica*, 34, 4, 768–783.

10.3 George Soros' interpretation of the crisis and the use of classical quantum physics in finance

Any beginning undergraduate student in economics will know about one of the basic principles of economics, the so-called "invisible hand." The principle is deceptively simple and roughly says that each economic agent, even though he or she acts in his or her own self-interest, will, in so doing, also help the other. Paul Samuelson, the Nobel prize winner in economics mentions in his well-known textbook (with Nordhaus) [1] that the "invisible hand" is a doctrine. As Samuelson notes in this book, the "invisible hand" is a concept which can be sourced back from Adam Smith's [2] *The Wealth of Nations*. Quoting Samuelson [1] (p. 46), "the invisible hand doctrine is a concept for explaining why the outcome of a market mechanism looks so orderly . . . We know that the market sometimes lets us down, that there are 'market failures'." Samuelson mentions that externalities are one case of such a market failure.

Nozick [3] remarks, the "invisible hand" approach can be explained via the theory of rational choice behavior amongst agents and he also indicates that this does not necessarily imply that one needs to invoke the principle of expected utility maximization. Nozick refers to his book [4], where variants of expected utility are used instead.

The economic literature has provided interesting explanations of the workings of the invisible hand. One quite condensed way of explaining its mechanics, in an albeit quite abstract setting, is by Schelling [5]. We do not expand on it here. We note that, already in 1978, it was remarked by Ullmann-Margalit [6] (p. 278) that the invisible hand approach could be viewed as "the counterpart, within the social domain, of the biological-evolutionary explanations within the domain of living organisms."

From an econophysics (statistical mechanics) point of view, the "invisible hand" could be seen as a "device" by which a complex system can be stabilized to an equilibrium state. However, we must emphasize that one needs to be very careful with the extension of physics-based theories to social systems.

It could be claimed that in the command economies of the past, the rational choice of individuals was assumed by the state. In effect, command economies "modeled" an aggregate version of the rational choice of individuals, while in free market economies the rational choice was squarely lying at the individual level. The "invisible hand" really operates at an individual level rather than at an aggregate level. Clearly, the debate on the (optimal) weight of state regulation in economic activity is an interminable one, and this book has no pretence at all to be a forum for such a discussion.

A question however, which now has come to the foreground, is whether the absence (or presence) of financial regulation could not in fact (in part) explain the events of the 2008 financial crisis. If we interpret this crisis as having been caused by the emergence of a degree of ignorance about how to price specific financial assets, then we may want to indeed connect the doctrine of the invisible hand to that crisis. One could for instance argue that the idea of efficiency, which we covered in Section 10.1 of this chapter, is connected to the invisible hand. In effect, assume very little relevant information is incorporated in prices. The very presence of "price-less assets" would disallow the proper functioning of the rational choice of individuals. After all, how could one exercise a rational choice on goods, when their prices would not be reflective of the worth of those goods? As a consequence, we would observe very random price behavior and decision making. In effect, no economic decision maker would be able to decide whether to (rationally) buy or sell. Such an economic state of affairs would indeed invalidate the proper functioning of the invisible hand principle altogether. We remark that the 2008 crisis did exhibit instances where assets seemed to be very much mispriced in some cases, merely because of the consequence of the absence of knowledge on how to price the asset in the first place. We may wonder whether it was precisely the absence of state intervention which made the existence of such "price-less" assets possible.

As one can appreciate, such a very difficult problem cannot be solved in the space of a book (and certainly not this one!). The interested reader may want to visit the editorials by the Nobel prize winner (economics), Paul Krugman in the *New York Times* to get an idea of how difficult the debate is on arguing the pros and cons of state intervention in an economy (in this case the US economy), which to some extent seems to be at a critical cross-road of its existence.

In social systems, as any economist well knows, it is impossible to perform a pure experimental test (such as in physics) by selecting a specific experimental context. The absence of such a setting does indeed make it extremely difficult to ascertain the degree of potency of having the "invisible hand" function in a society with for instance a mildly interventionist government.

One could argue that the focus needed for the "managing" of the "invisible hand" principle in a financial context is in effect *very* different from when the same principle is to be looked after in, say, an industrial production setting. Clearly, the theory of industrial organization (for instance see the work of Jean Tirole [7]), which devotes a lot of intellectual effort to modeling the economics of industrial production and organization, is different in its mathematical expression, from a theory of financial management, which can account for, say, extreme events. The stress would need to be put on the word "extreme." The attacks against financial models are very often oblivious about mentioning this important qualifier. It needs to be emphasized that a branch of applied probability has developed very sophisticated models on how to measure extreme events. Whether those models have been sufficiently incorporated into financial models remains to some degree an open question. However, to claim that no such theories exist (mathematical or financial mathematical) is an incorrect statement. Please see, for instance, the work by Embrechts [8] and Delbaen [9].

George Soros, the multi-billionaire investor, has for many years proposed an alternative theory to efficiency. The key concept which emerges from Soros's theory is the principle of "reflexivity." See for instance [10]. The connection between reflexivity and the wave function has been discussed in Khrennikov [11] [12] [13] and Choustova [14] [15] [16].

Soros [17] argues that, in his theory of reflexivity, the economic agent's thinking, has two functions: (i) "to understand reality, that is the cognitive function" and (ii) "to make an impact on the situation . . . (the) manipulative, function." Clearly, the second role, we believe, has already been alluded to when we discussed efficiency in Section 10.1 of this chapter. Recall the idea of so-called "information manipulation" we discussed above. We mentioned that such manipulation can take on the form of information leakage and (ii) uncertainty creation. We recall that Ma [18] (p. 43) defines information leakage as: "situations where agents wish to reveal truthfully their private possessed information to . . . others." Uncertainty creation is then defined as per Ma [18] (p. 43) as agents wishing to create "uncertainty in the rest of the economy."

Soros [17] proposes that if both the cognitive function (understanding the market) and the manipulative function "work independently of each other they produce determinate results." However, if both functions, in the words of Soros [17], do "operate simultaneously (then) they interfere with each other." It is this interference that Soros calls "reflexivity." Soros [17] indicates that "reflexivity introduces an element of unquantifiable uncertainty into both the participants' understanding and the actual course of events." The interested reader may sense the notion of an uncertainty principle in some macroscopic environment. We already mentioned such principle in Chapter 3, Section 3.2. But we also note the word "interference" which from a simple analogy point of view makes us possibly think about probability interference. This concept was covered in Chapter 5.

We may wonder whether we can connect this approach with the meaning of a wave function in a Bohmian mechanical environment? Chapters 11, 12, and 13 will discuss more the economic role of the Bohmian wave function. In Chapter 13 we will introduce a so-called "information risk parameter." The Wiener process there will be mimicking the source of information risk. We can also argue that the constructive and destructive interference, which are essential in the build up of the wave function via the Fourier integration (see Chapter 5, Section 5.9), could be an echo of the simultaneous interfering of what Soros calls the manipulative and cognitive functions.

We can make the following argument. We can consider cognitive and manipulative functions as variables of the model under consideration. If these observables interfere (and experimentally this fact is expressed in the violation of the formula of total probability), then we can apply the quantum-like approach to modeling Soros' theory of reflexivity, and describe the state of the market by the quantum mechanical wave function. In simple quantum mechanical terms, if normalized plane wave functions do interfere, via constructive and destructive interference (i.e. cognitive and manipulative functions), then they build up a quantum mechanical wave function. This resultant wave function, we will propose, can be used as a carrier of a degree of erroneous information. The rationale for doing so, as we will see in Chapter 13, may possibly be traced back to the idea Bohm and Hiley [19] proposed of "active information." This active information is narrowly connected to the existence of the wave function. Furthermore, the use of Fourier series and integrals in this context, also naturally leads to the uncertainty principle of momentum and position. Hence, the idea of reflexivity in effect would then be closely connected to the Fourier integration apparatus (and thus the uncertainty principle). Furthermore, recalling the definition of the wave function in Chapter 5, Section 5.9, we can see that the role of the amplitude function of the wave number is of central importance in giving shape to the emerging wave function. The *absence* of reflexivity in this proposed framework would then clearly imply the *absence* of the Fourier integration apparatus of the wave function construction. The wave number can be best interpreted in the light of the famous de Broglie relation $p = \hbar k$: p is momentum and $\hbar$ is the rationalized Planck constant.

Let us quote Soros on his idea on how interference produces effects in the financial markets. Says Soros [17]: "This two-way connection works as a feedback loop. The feedback is either positive or negative. Positive feedback reinforces both the prevailing trend and the prevailing bias – and leads to a mispricing of financial assets. Negative feedback corrects the bias." With our proposed connection of bringing reflexivity into the realm of wave function construction, we could identify the existence of a negative feedback with the appearance of an amplitude function which is very wide, producing via Fourier integration, a very tight wave function. If that wave function is dependent on the position parameter, which is price, then the density function derived from that wave function will be tight, and the *probability* of mispricing would be very much limited. Positive feedback effects would provoke the opposite effect: the amplitude function on the wave number would be very tight and, via Fourier integration, this would create a very wide wave function. If again, this wave function is dependent on price, the *probability* of mispricing would indeed be much less limited. Remark also the case where reflexivity is absent, i.e. when the manipulative and cognitive functions do not interfere with each other, no such wave function construction exists at all. In other words, the Fourier integration apparatus (and the ancillary uncertainty principle) in that case completely disappears. In fact this absence of reflexivity could possibly be connected to the existence of no-arbitrage, where it can be shown that the absence of arbitrage leads to a break down of the Fourier integration machinery we propose here. The existence of arbitrage would then allow for the existence of an amplitude function and the appearance of the Fourier integration machinery.

We finish this section with an observation on feedback loops. Feedback loops are well known in classical dynamics (and neural networks). Hence, the reader may wonder why the authors of this book advertise the usage of quantum formalism to describe such market dynamics. As was pointed out, one of the reasons is the presence of interference effects which cannot be described by classical dynamics. However, for Soros the crucial consequence of the reflexivity theory is that, contrary to models which are based on the formalism of classical statistical mechanics, his theory does not imply that a stabilization to equilibrium must occur. Following Soros, as we mentioned above, the financial market can be unstable. Such dynamical behavior can be well explained by the fact that any solution of the Schrödinger equation (which is different from the stationary solution) will fluctuate forever without approaching the equilibrium state. We remark that stationarity of solutions means that there are no such dynamics at all. In the modeling of the market, the only possibility for the realization of such a stationary state is the state of stagnation.

10.4 References

[1] Samuelson, P. and Nordhaus, W. D. (1985). *Economics*. McGraw-Hill.

[2] Smith, A. (1776). *An Inquiry into the Nature and Causes of the Wealth of Nations*. 5th edition. Methuen and Co. Ltd (publication date 1904).

[3] Nozick, R. (1994). Invisible-hand explanations. *American Economic Review*, 84, 2, 314–318.

[4] Nozick, R. (1993). *The Nature of Rationality*. Princeton University Press, NJ.

[5] Schelling, T. C. (1969). Models of segregation. *American Economic Review*, 59, 488–493.

[6] Ullmann-Margalit, E. (1978). Invisible-hand explanations. *Synthese*, 39, 2, 263–291.

[7] Tirole, J. (1992). *Theory of Industrial Organization*. MIT Press.
[8] Embrechts, P., Lindskog, F., and McNeil, A. (2003). Modelling dependence with copulas and applications to risk management. In *Handbook of Heavy Tailed Distributions in Finance*. Ed. Rachev, S. T. *Handbooks in Finance* (Book 1), Elsevier.
[9] Artzner, P., Delbaen, F., Eber, J.M., and Heath, D. (1999). Coherent measures of risk. *Mathematical Finance*, 9, 3, 203–228.
[10] Soros, G. (1987). *The Alchemy of Finance. Reading the Mind of the Market*. J. Wiley, New York.
[11] Khrennikov, A. (2010). *Ubiquitous Quantum Structure: From Psychology to Finance*. Springer Verlag, Berlin.
[12] Khrennikov, A. (2004). *Information Dynamics in Cognitive, Psychological and Anomalous Phenomena*. Series *Fundamental Theories of Physics*. v. 138. Kluwer, Dordrecht.
[13] Khrennikov, A. Yu. (1999). Classical and quantum mechanics on information spaces with applications to cognitive, psychological, social and anomalous phenomena. *Foundations of Physics*, 29, 1065–1098.
[14] Choustova, O. (2004). Bohmian mechanics for financial processes. *Journal of Modern Optics*, 51, 6/7, 1111.
[15] Choustova, O. (2006). Quantum Bohmian model for financial market. *Physica A*, 374, 304–314.
[16] Choustova, O. (2007). Quantum modeling of nonlinear dynamics of prices of shares: Bohmian approach. *Theoretical and Mathematical Physics*, 152, 7, 1213–1222.
[17] Soros, G. (2011). *Why I agree with (some of) Friedrich Hayek*. Interviews and Speeches. GeorgeSoros.com: the official website of George Soros (April 28, 2011).
[18] Ma, C. (2010). *Advanced Asset Pricing Theory*. Imperial College Press.
[19] Bohm, D. and Hiley, B. (1998). Non-locality and locality in the stochastic interpretation of quantum mechanics. *Physics Reports*, 172, 93–122.

10.5 The need for an information modeling device in economics and finance

The former section gave us a flavor on how wave functions could be used to model Soros' idea of reflexivity. The interference of wave functions can be tightly connected to the existence of reflexivity. The positive and negative feedback loops may be connected to respectively tight and wide amplitude functions. The same amplitude function could also be proposed to be connected to the prior distribution (via Bayes rule). Bayes rule was briefly discussed in Chapter 8, Section 8.1.

The concept of information has a long history in the sciences. Information theory has proposed many useful measures, such as (of course) entropy and "information gain" formulas, which aid in precisely assessing the quantity of information in systems. However, it remains very challenging to devise measures which can determine the content of information.

Haken [1] (p. 55), gives an interesting example on how one can maximize information. He considers symbols like dashes or dots. The probability of finding such a dash or a dot, in a total population (N) of N_1 and N_2 dashes and dots, he defines simply to be $p_j = \frac{N_j}{N}$; $j = 1, 2$. The measure for the information per symbol is then (Haken [1] (p. 55)): $i = -K \sum_j p_j \ln p_j$, where K is a constant and $j = 1, 2$. This measure can then be maximized subject to the fact probabilities add up to unity. Haken [1] (p. 57) also shows that by rewriting $i = -K \sum_j p_j \ln p_j$ as: $i = -\sum_j p_j f_j$ (with $f_j = -K \ln p_j$), f_j can be interpreted (Haken [1] (p. 57)) "as the information content of the symbol (dash or dot for instance)."

Haken shows that "information" and "information gain" can be precisely measured in, for instance, self-organizing systems. It is unclear whether economic systems can in fact be seen as self-organizing systems.

10.6 Reference

[1] Haken, H. (1988). *Information and Self-Organization: A Macroscopic Approach to Complex Systems*. Springer Verlag, Berlin.

11

Bohmian mechanics in finance and economics

11.1 The pilot wave function and its uses outside of quantum mechanics

Let us recall our discussion on Bohmian mechanics in Chapter 6. The idea of using Bohmian mechanics outside of quantum mechanics is not new anymore. Khrennikov [1] provides for an overview on how this model can contribute in precise terms to areas such as economics and finance.[1] Historically, the work by Bohm and Hiley [2] and also Hiley and Pylkkänen [3] brought forward the idea that the pilot wave function could be seen as a wave function containing information. In Khrennikov [4], Choustova [5], and Haven [6], the idea of using pilot wave theory to finance was investigated. Khrennikov [1] (p. 160) remarks that "the force induced by the pilot wave field does not depend on the amplitude of the wave." He cites the argument made by Bohm and Hiley [2] that because of this property the pilot wave is an information wave.

The various interpretations (with applications) of the wave function will be covered in detail in Chapter 13.

Here are some of the key features which can be of use in an financial/economics setting. When the wave function is not factorized, then a change in the price of a stock i will affect the prices of stocks, j, with $j \neq i$. See Khrennikov [1] (p. 161). The Bohmian theory, as we remarked already in Chapters 1 and 6, is non-local. The amplitudes of wave functions are inconsequential on the force emanating from the pilot wave (recall the existence of a quantum potential which is an essential part of pilot wave theory – see Chapter 6).

An interesting discussion can be engaged on the notion of what we would like to call a "pricing rule." We will briefly refer back to this notion in the next chapter. Recall the definition of the quantum potential from Chapter 6, which we could derive from equation (6.15), $Q(q) = -\frac{h^2}{2mR}\frac{\partial^2 R}{\partial q^2}$. In Khrennikov [1], several

[1] Chapter 9 in that book (pp. 151–170) provides for a discussion.

examples are given on how such a quantum potential can be used in an economics context. We reiterate two examples given there. The first example (Khrennikov [1] (pp. 162–163) (Example 9.1)) concerns the simple case of a constant amplitude, R. The force derived from the quantum potential, $\frac{-\partial Q}{\partial q}$, is then, in this case, zero. Thus, the size of the amplitude in this case is totally inconsequential. Consider a second example (see Khrennikov [1] (p. 163) (Example 10.3)) with amplitude $R(q) = c(q^4 + b)$, c, $b > 0$. The force, $\frac{-\partial Q}{\partial q}$, yields $\frac{bq-q^5}{(q^4+b)^2}$ and setting $d = \sqrt[4]{b}$, then, as the price moves from $q = 0$ to $q = d$, there is no negative force in increasing the price from the level 0 to the level d. Beyond level d, the situation is reversed.

The forces portrayed through those two simple examples show clearly that an argument can be made that such quantities could be interpreted as some "pricing rule." Such rules could indeed be very sophisticated, depending on the functional form of the quantum potential.

11.2 References

[1] Khrennikov, A. (2009). *Ubiquitous Quantum Structure: From Psychology to Finance*. Springer Verlag, Berlin.
[2] Bohm, D. and Hiley, B. (1993). *The Undivided Universe: An Ontological Interpretation of Quantum Mechanics*. Routledge.
[3] Hiley, B. and Pylkkänen, P. (1997). Active information and cognitive science – a reply to Kieseppä. In *Brain, Mind and Physics*. Eds. Pylkkänen, P., Pyllkö, P., and Hautamäki, A. IOS Press, Amsterdam, pp. 123–145.
[4] Khrennikov, A. (2004). *Information Dynamics in Cognitive, Psychological and Anomalous Phenomena. Series Fundamental Theories of Physics*, v. 138. Kluwer, Dordrecht.
[5] Choustova, O. (2009). Application of Bohmian mechanics to the dynamics of prices of shares: stochastic model of Bohm–Vigier from properties to price trajectories. *International Journal of Theoretical Physics*, 47, 252–260.
[6] Haven, E. (2006). Pilot-wave theory and financial option pricing. *International Journal of Theoretical Physics*, 44, 11, 1957–1962.

12

The Bohm–Vigier model and path simulation

12.1 The Bohm–Vigier model in finance

Bohm and Hiley [1] (p. 194) (equation (9.29)) indicate that according to Bohm and Vigier the velocity of a particle could be given by[1] $v = \frac{\nabla S}{m} + \xi$, where ∇S is the gradient of the phase of the wave function towards position, m is mass, and ξ is a random contribution with mean zero. Bohm and Hiley's interpretation (p. 195) of the term $\frac{\nabla S}{m}$ is seen as the "average velocity of the particle." Bohm and Hiley [1] (p. 195) argue that if the average velocity is so defined and the wave function is expressed via the polar form $\psi = R.e^{iS}$ (this was equation (6.1) in Chapter 6), "then Schrödinger's equation implies that":

$$\frac{d\overline{v}}{dt} = -\nabla(V + Q), \tag{12.1}$$

where the gradient (towards position) is taken on both the real and quantum potentials.

Now recall the portfolio we considered in the Black–Scholes option pricing model. In Chapter 2, equation (2.10), it was indicated that the return per unit of time of the portfolio should be equal (in case of no arbitrage) to the risk free rate of interest $\frac{d\Pi}{\Pi}\frac{1}{dt} = r$, where Π is the Black–Scholes portfolio (a collection of a stock and an option). We can make the argument that the above Bohm–Vigier model could be used in this option pricing context in the following way. As in Haven [2] (p. 334), we write in analogy with $v = \frac{\nabla S}{m} + \xi$ that $\frac{d\Pi}{dt} = \nabla S(\Pi) + \omega$, and the information function is now defined as $\psi(\Pi) = R(\Pi)\exp(iS(\Pi))$ and $S(\Pi) = \frac{r\Pi^2}{2}$ and $\omega = x\frac{\nabla S(\Pi)}{r}$, where x is a time-independent arbitrage return.

We may of course wonder why one would want to write the above? In Chapter 3, we hinted at the necessary existence of non-arbitrage in the derivation of the Black–Scholes model. In effect, the left-hand side and right-hand

[1] We omit subscripts i.

side of $\frac{d\Pi}{\Pi}\frac{1}{dt} = r$ can only be equal in the absence of arbitrage. Hence, writing $\frac{d\Pi}{dt} = \nabla S(\Pi) + \omega$, with the phase and the stochastic term as defined above, indicates allowance is made for the existence of arbitrage. Thus, in that context one can now propose $\frac{d\Pi}{\Pi}\frac{1}{dt} > r$ or $\frac{d\Pi}{\Pi}\frac{1}{dt} < r$. This leads us into an already established theory, developed mainly by Ilinski [3], Otto [4], Fedotov, Panayides [5], and Panayides [6]. See also Haven [7].

The Ilinski [3] and Otto [4] approaches specify that the return on the Black–Scholes portfolio, Π, is: $d\Pi = r\Pi dt + x\Pi dt$. See, for instance, equation (43) in Ilinski [3] (p. 237). The "x" can follow a specific random process like the Ornstein–Uhlenbeck process, and this process is defined as in Ilinski [3] (p. 240) (equation 49):

$$\frac{dx}{dt} = -\lambda x(t) + \eta(t), \tag{12.2}$$

where $\eta(t)$ is white noise and λ is a relaxation parameter.

12.2 References

[1] Bohm, D. and Hiley, B. (1993). *The Undivided Universe: An Ontological Interpretation of Quantum Mechanics*. Routledge.

[2] Haven, E. (2005). Bohmian mechanics in a macroscopic quantum system. In *Quantum Theory: Reconsideration of Foundations* – 3. Eds. Adenier, G., Khrennikov, A., and Nieuwenhuizen, T. *American Institute of Physics Conference Proceedings*, New York.

[3] Ilinski, K. (2001). *Physics of Finance: Gauge Modelling in Non-Equilibrium Pricing*. J. Wiley.

[4] Otto, M. (1999). Stochastic relaxational dynamics applied to finance: towards non-equilibrium option pricing theory. *European Physical Journal B*, 14, 383–394.

[5] Fedotov, S. and Panayides, S. (2005). Stochastic arbitrage returns and its implications for option pricing. *Physica A*, 345, 207–217.

[6] Panayides, S. (2005). Derivative pricing and hedging for incomplete markets: stochastic arbitrage and adaptive procedure for stochastic volatility. Ph.D. Thesis. The School of Mathematics, University of Manchester.

[7] Haven, E. (2008). Elementary quantum mechanical principles and social science: is there a connection. *Romanian Journal of Economic Forecasting*, 9, 1, 41–58.

12.3 The Newton–Bohm equation: path simulation

The Newton–Bohm equation was described in equation (6.16), see Chapter 6, Section 6.3. We recall this PDE as being $m.a = m\frac{d^2q(t)}{dt^2} = -\frac{\partial V(q,t)}{\partial q} - \frac{\partial Q(q,t)}{\partial q}$, where V is the real potential function and Q is the quantum potential function. The hallmark of Bohmian mechanics is the quantum potential, which depends on the wave function's amplitude. The wave function does evolve according to the Schrödinger partial differential equation.

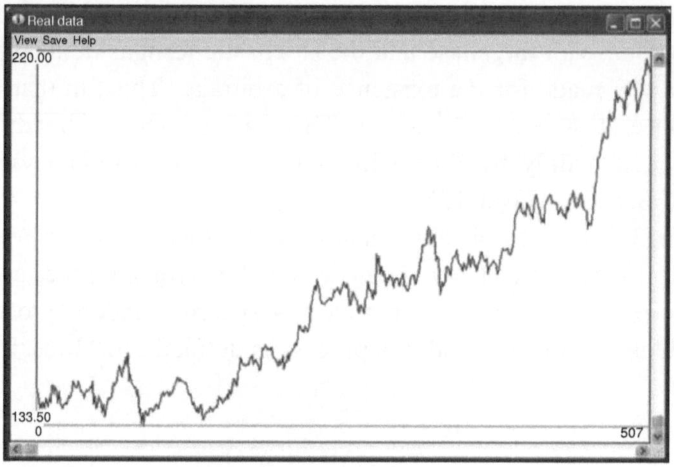

Figure 12.1. Real trajectory I, generated on the basis of price data (source: Khrennikov and Kotovich [1])

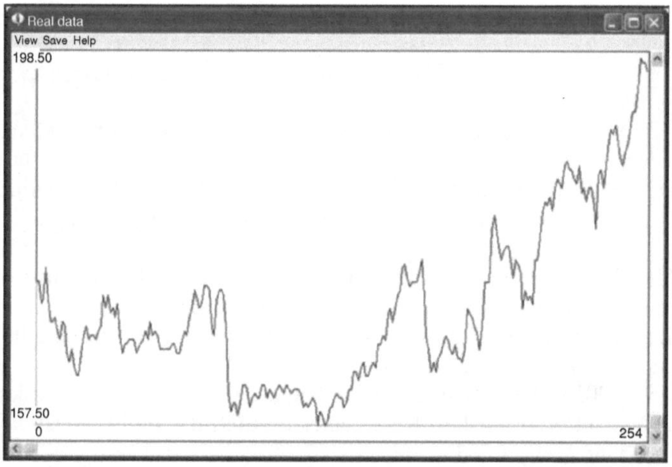

Figure 12.2. Real trajectory II, generated on the basis of price data (source: Khrennikov and Kotovich [1])

An important issue which needs to be tackled is to inquire whether real price trajectories can be simulated by the Bohmian model. The simulations below (Figures 12.1–12.6) are using as real potential a free particle in a box. The Schrödinger PDE is solved (using the above real potential function) and then the quantum potential is found. Simulation can then be started. An important issue which remains open is to know what initial condition wave function one needs to choose for the Schrödinger equation. The simulated Newton–Bohm paths do very much depend on the choice of this initial condition.

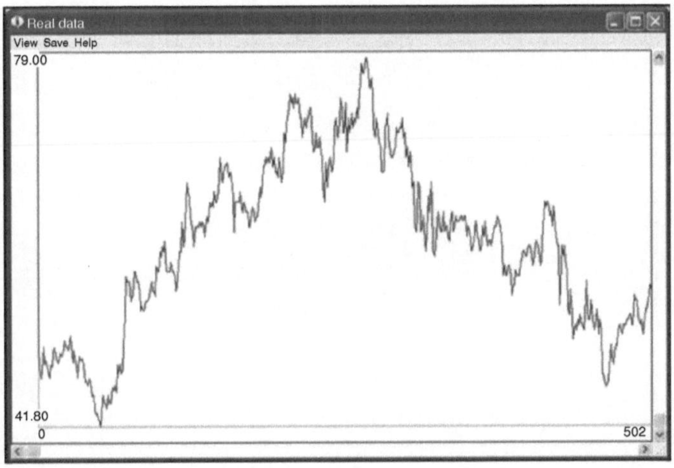

Figure 12.3. Real trajectory III, generated on the basis of price data (source: Khrennikov and Kotovich [1])

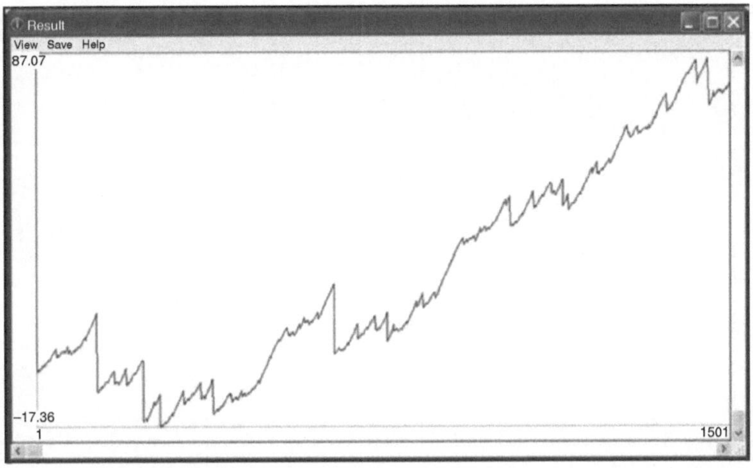

Figure 12.4. A Newton–Bohm trajectory (source: Khrennikov and Kotovich [1])

The figures above show path simulations based on real data (Figures 12.1–12.3). The Newton–Bohm paths are shown in Figures 12.4–12.6.

The Y axis in Figures 12.4, 12.5, and 12.6 represent values which are the logarithm of the values on the Y axis from Figures 12.1, 12.2, and 12.3. As indicated already above, the results of the simulation are highly conditional on the initial condition – wave function. A problem which presents itself here is how to interpret the dynamics of expectations (of say market participants), which are encapsulated at the level of the wave function. This is a problem we can also find

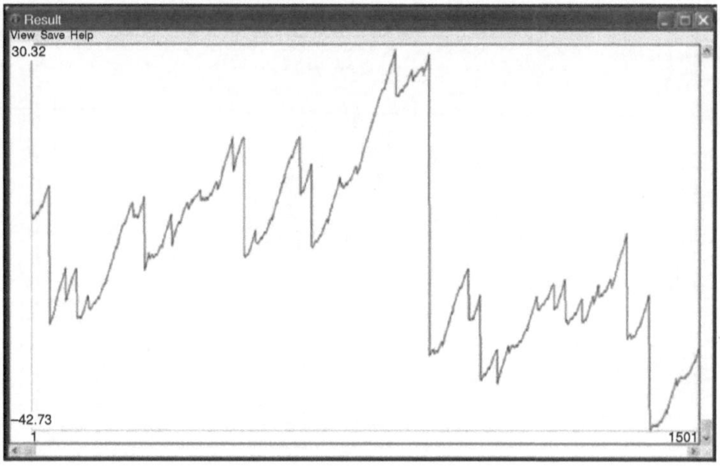

Figure 12.5. Another Newton–Bohm trajectory (source: Khrennikov and Kotovich [1])

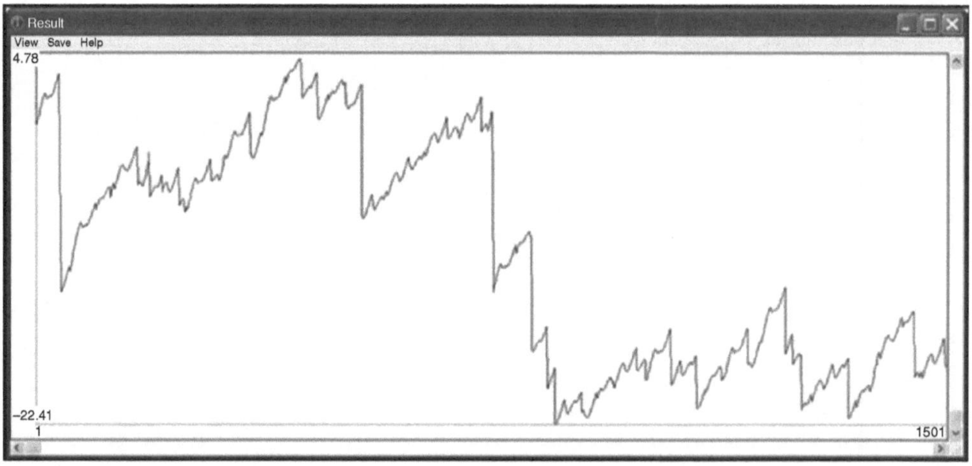

Figure 12.6. Another Newton–Bohm trajectory (source: Khrennikov and Kotovich [1])

back in economics. Do we know how to encapsulate the mood of a representative agent in a utility function? If we were to try to mimic price paths with a stochastic approach, should we not need to worry about the expected rate of interest to use (which reflects preferences for risk (of the representative agent))? We may wonder whether a pricing rule (tied to the quantum potential) cannot be guessed via certain specific historical price moments. As an example, the behavioral finance literature has come forward with several pricing effects. The pricing rule is in effect nothing else than the force derived from the quantum potential. This is in effect the last term in equation (6.16) (see Chapter 6). We must therefore pay particular attention to the amplitude function of the wave function. One very important feature of the model would then be to encapsulate in the choice of the amplitude function the possibility that a noticeable event in the market may be coming forward. One could possibly measure such upcoming event via the change in level of the LIBOR rate (London Interbank Offer Rate).

12.4 Reference

[1] Khrennikov, A. Yu. and Kotovich, N. (2010). Private communication. December 30, 2010.

13
Other applications to economic/financial theory

This chapter attempts to delve deeper into the question on how quantum mechanical techniques can be brought closer into the realm of economics and finance.

13.1 The (non-)Hermiticity of finance-based operators?

Hermiticity of operators was discussed in Chapter 4 of the book. We again take up this very important concept in the context of financial asset pricing. It is a classical result from quantum mechanics that the existence of Hermiticity of the Hamiltonian operator is intimately linked with the notion of conservation of probability. The existence of Hermiticity is also known to be closely linked to the concept of spatial localization. Please see below.

Baaquie [1] makes the important argument that the Black–Scholes Hamiltonian is non-Hermitian and this condition provides for the need to satisfy the martingale condition. Please recall that the martingale property was covered in Chapter 2, Section 2.9. It is also important to mention that Luigi Accardi has indicated[1] that it is white noise which may be the cause of non-Hermiticity in a finance context.

One can argue that within an economics/finance context, the equivalent of the state function, using Baaquie [1] [2], can be the option price function. A similar interpretation can also be found in the paper by Li and Zhang [3] (see also Haven [4]). Please note that the work of Li and Zhang is covered in Section 13.12 of this chapter. We would like to note that by interpreting the wave function as the call/put price, we depart from the conventional probabilistic interpretation of the wave function (the Born interpretation), and we move instead towards Schrödinger's original interpretation of the wave function as a real physical field. Schrödinger considered the wave function as the density of the electron charge (see Khrennikov [5]). In

[1] One of the authors recalls that Professor Accardi made this remark at the Fifth Foundations of Probability and Physics conference, which was held at Linnaeus University (Sweden), August 25–27, 2008.

particular, in his approach the wave function was not normalized. Moreover, at the beginning he considered only real valued wave functions. Hence, the analogy with the call/put price interpretation of the wave function in finance is evident. Recently Schrödinger's interpretation of the wave function was justified in the framework of so-called prequantum classical statistical field theory (see Khrennikov [6]). The main departure from Schrödinger's approach (as well as from Khrennikov [6]) are the non-unitary dynamics of such a wave function of the field type.

The underlying asset of the option pricing function (please see also Chapter 2, Section 2.7.1 for the definition of an option) could be the equivalent of the independent variable entering the state function. Since asset prices can not be negative, there is a restriction on the space where we can find (financial) particles.

However, can we necessarily argue for spatial localization? The concept of "probability conservation" tells us that the probability of finding a particle anywhere in space is time independent. To show that probability conservation is intimately related to the Hermiticity of the Hamiltonian operator, one actually needs to activate the condition of spatial localization. This condition is simple and says that the wave function $\psi(x,t) \to 0$ when $|x| \to \infty$. Clearly, if we consider the finance equivalent of $\psi(x,t)$, i.e. the option pricing function where the underlying is an asset price, then we know asset prices never go to infinity. But assume for theoretical sake that they do. Consider the martingale property $E(S_{t+1}|F_t) = S_t$. This is intimately linked to the existence of non-arbitrage, since we can write under this condition that $\frac{1}{1+r}E^P(S_{t+1}) = S_t$, i.e. the risky asset can be discounted at the risk free rate of interest, r, under the risk neutral probability measure, P.

Björk [7] (p. 9) indicates that a probability measure P is called a martingale measure if the condition $S_0 = \frac{E^P(S_1)}{1+r}$ is satisfied. Björk [7] (p. 10) also shows that there does not exist arbitrage if and only if there exists a martingale measure P.

Let us for easiness of purpose say that $r = 0$, and hence we can then write $E^P(S_{t+1}) = S_t$. When there is arbitrage, the best we can write is $E^P(S_{t+1}) = S_t(1 + r + \textit{premium})$, and, therefore, $E^P(S_{t+1}) > S_t$ in the case where the premium is positive, and $E^P(S_{t+1}) < S_t$ in the case where the premium is negative.[2] Assume now that the strike price of a call option is finite, which is a very reasonable assumption. If S_t were to go to infinity, and assuming there is a positive premium, then $E^P(S_{t+1})$ would certainly go to infinity and therefore the equivalent condition in finance terms of $\psi(x,t) \to 0$ when $|x| \to \infty$ would certainly not be met, since the call price would even "be more"[3] infinite since $S_t \to S_{t+1} \to \infty$. Thus the call price $C(S_{t+1}, t) \nrightarrow 0$. If the premium is negative, we would have possibly the reverse: S_t may tend to infinity, but, since $S_{t+1} < S_t$, we could argue that S_{t+1} is "now less"[4] tending to infinity. But if it is less tending to infinity, it may be

[2] Remark that the premium is asset specific, i.e. the level of the premium refers to the specific asset in question.
[3] We agree this is indeed very imprecise language.
[4] Again, the language we use is imprecise.

finite. Hence, although the call price $C(S_t, t) \to \infty$, one could possibly argue that $C(S_{t+1}, t) < \infty$ since S_{t+1} can be finite. If the stock price were to indeed coincide with the finite strike price, then $C(S_{t+1}, t) = 0$. Remark that no option contract exists with infinite strike prices.

In the proposition below, we use the field interpretation of the wave function, i.e. as a classical (real valued) field.[5] Here is the argument in a more formal format.

Proposition 13 *Assume the state function $\psi(x, t)$ becomes now $C(A, t)$, where A is the price of an asset and C is the price of a call option. A necessary and sufficient condition for the Black–Scholes option price Hamiltonian operator to be Hermitian is that an appropriate super-martingale on A exists.*

Proof. The non-arbitrage condition yields that: $\frac{E^P(A(t+1))}{1+r} = A(t)$, where r is the risk free rate (a constant) and P is the risk neutral probability measure. Hence, we can write this expression as a martingale (see Chapter 2, Section 2.9 for a definition):[6] $E^P(A(t+1)) = A(t)$. Assume for easiness of purpose that $r = 0$. Assume a martingale does not exist, i.e. we obtain $E^P(A(t+1)) = A(t)(1 + \text{risk premium})$. Hence, we can only write the expectation as either a so-called super-martingale $E^P(A(t+1)) < A(t)$ or a so-called sub-martingale $E^P(A(t+1)) > A(t)$. If $A(t)$ tends to the infinite, then, in the case of a sub-martingale, $A(t+1)$ will be even larger. Therefore, from a call option point of view, the intrinsic value yielded by $A(t+1)$ will be even farther removed from the strike price (and thus farther removed from zero). Under a martingale, if $A(t)$ is tending to infinity with a non-zero call price, then $A(t+1)$ will equally yield a non-zero call price. A necessary (but not sufficient) condition for the call price to tend to zero is to impose a super-martingale on $A(t)$. If $A(t)$ tends to infinity, then under the super-martingale, $A(t+1) < \infty$ will be finite. However, we need to further impose that $A(t+1)$ must tend to the finite strike price. In that case, only the call price will tend to zero. Hence, if there exists a super-martingale and for an appropriate choice of $A(t+1)$, one can obtain spatial localization. Assuming that such spatial localization exists, using the usual state functions[7] and following (Morrison [9] (p. 202)), one can write $\frac{d}{dt}p([-\infty, \infty]) = -\frac{i}{\hbar}\int_{-\infty}^{\infty} \left\{\psi^*(x,t)\widehat{\mathcal{H}}\psi(x,t) - \psi(x,t)\left[\widehat{\mathcal{H}}\psi(x,t)\right]^*\right\} dx$ and since the Hamiltonian operator can be split into a real potential energy operator (which

[5] For a nice introduction on the topic of the quantization of fields, please see Bethe [8].

[6] We have taken off, for easiness of notation, the conditionality of the expectation on the information set at time t.

[7] Adjustments would have to be made if the conversion from state function to option price is made, like remapping the fact that the product $-\frac{i}{\hbar}$ will not appear in the option pricing setting. However, this does not alter the argument of the proof.

is the real potential) and a kinetic energy operator, one can write: $\frac{d}{dt}p([-\infty,\infty]) = -\frac{i}{\hbar}\int\limits_{-\infty}^{\infty}\left\{\psi^*(x,t)\widehat{T}\psi(x,t) - \psi(x,t)\left[\widehat{T}\psi(x,t)\right]^*\right\}dx$, where $\widehat{T}$ is the kinetic energy operator which is equal to $\frac{-\hbar^2}{2m}\frac{\partial^2}{\partial x^2}$. The above expression $\frac{d}{dt}p([-\infty,\infty])$ can then be shown to be equal to (Morrison [9] (p. 203)): $-\frac{i\hbar}{2m}\left[\psi^*(x,t)\frac{\partial}{\partial x}\psi(x,t) - \psi(x,t)\frac{\partial}{\partial x}\psi(x,t)^*\right]_{\infty}^{-\infty}$ and this expression is zero because spatial localization now exists. Hence, $\frac{i}{\hbar}\int\limits_{-\infty}^{\infty}\left\{\psi^*(x,t)\widehat{\mathcal{H}}\psi(x,t) - \psi(x,t)\left[\widehat{\mathcal{H}}\psi(x,t)\right]^*\right\}dx = 0$ and therefore we can recuperate the definition for the Hermiticity of a Hamiltonian operator: $\int\limits_{-\infty}^{\infty}\psi^*(x,t)\widehat{\mathcal{H}}\psi(x,t)dx = \int\limits_{-\infty}^{\infty}\psi(x,t)\left[\widehat{\mathcal{H}}\psi(x,t)\right]^*dx.$ ■

We remark the very specific conditions needed to argue for the existence of Hermiticity of the Black–Scholes Hamiltonian operator. Here are the conditions again:

1. We assume that an equivalent analogue of the state function is the option call function (please see above Proposition 13).
2. We require a super-martingale on the underlying asset.
3. Because of 2, we must assume the existence of arbitrage on the underlying asset.
4. The underlying asset's price must tend to the strike price.

Conditions 2 and 3 are conditions which tend to confirm that Hermiticity and the existence of arbitrage may have to co-exist. Condition 4, clearly imposes a constraint, which financially is possible. Asset prices can definitely converge to a strike price in reality. However, there are many instances where that is not the case.

Proposition 14 *Assume the state function $\psi(x,t)$ becomes now $P(A,t)$, where A is the price of an asset and P is the price of a put option, then the Black–Scholes option price Hamiltonian operator is Hermitian.*

Proof. The proof follows almost entirely the proof of Proposition 13 above. When $A \to \infty$, then $P(A,t) = 0$ by definition of the intrinsic value of the put option and the fact that the strike price is always finite. Using the usual state functions,[8] there exists spatial localization. Following (Morrison [9] (p. 202)) one can write:

[8] Adjustments would have to be made if the conversion from state function to option price is made, like re-mapping the fact that the product $-\frac{i}{\hbar}$ will not appear in the option pricing setting. However, this does not alter the argument of the proof.

$\frac{d}{dt}p([-\infty,\infty]) = -\frac{i}{\hbar}\int\limits_{-\infty}^{\infty}\{\psi^*(x,t)\widehat{\mathcal{H}}\psi(x,t) - \psi(x,t)\left[\widehat{\mathcal{H}}\psi(x,t)\right]^*\}dx$ and since the Hamiltonian operator can be split into a real potential energy operator (which is the real potential) and the kinetic energy operator, one can write: $\frac{d}{dt}p([-\infty,\infty]) = -\frac{i}{\hbar}\int\limits_{-\infty}^{\infty}\{\psi^*(x,t)\widehat{T}\psi(x,t) - \psi(x,t)\left[\widehat{T}\psi(x,t)\right]^*\}dx$, where $\widehat{T}$ is the kinetic energy operator which is equal to: $\frac{-\hbar^2}{2m}\frac{\partial^2}{\partial x^2}$. The above expression $\frac{d}{dt}p([-\infty,\infty])$ can then be shown to be equal to (Morrison [9] (p. 203)): $-\frac{i\hbar}{2m}[\psi^*(x,t)\frac{\partial}{\partial x}\psi(x,t) - \psi(x,t)\frac{\partial}{\partial x}\psi(x,t)^*]_{\infty}^{-\infty}$ and this expression is zero because spatial localization exists. Hence, $\frac{i}{\hbar}\int\limits_{-\infty}^{\infty}\{\psi^*(x,t)\widehat{\mathcal{H}}\psi(x,t) - \psi(x,t)\left[\widehat{\mathcal{H}}\psi(x,t)\right]^*\}dx = 0$ and therefore we can recuperate the definition for the Hermiticity of a Hamiltonian operator: $\int\limits_{-\infty}^{\infty}\psi^*(x,t)\widehat{\mathcal{H}}\psi(x,t)dx = \int_{-\infty}^{\infty}\psi(x,t)\left[\widehat{\mathcal{H}}\psi(x,t)\right]^* dx.$ ■

If we compare proposition 14 with proposition 13, we can immediately observe a high degree of asymmetry. Here are the conditions:

1. We assume that an equivalent analogue of the state function is the option put function (please see above Proposition 13).
2. The existence of arbitrage or non-arbitrage is irrelevant.

Thus in the case of the put, the existence of Hermiticity is un-connected with the existence of arbitrage or non-arbitrage. Hence, spatial localization is always verified in that case.

13.2 References

[1] Baaquie, B. (2008). Quantum mechanics and option pricing. In *Quantum Interaction.* Eds. Bruza, P., Lawless, W., van Rijsbergen, K., Sofge, D. A., Coecke, B., and Clark, S. *Proceedings of the Second Quantum Interaction Symposium* (Oxford University), College Publications, pp. 49–53.

[2] Baaquie, B. (2005). *Quantum Finance*. Cambridge University Press.

[3] Li, Y. and Zhang, J. E. (2004). Option pricing with Weyl–Titschmarsh theory. *Quantitative Finance*, 4, 457–464.

[4] Haven, E. (2005). Analytical solutions to the backward Kolmogorov PDE via an adiabatic approximation to the Schrödinger PDE. *Journal of Mathematical Analysis and Applications*, 311, 439–444.

[5] Khrennikov, A. (2009). *Contextual Approach to Quantum Formalism*. Springer Verlag, New York.

[6] Khrennikov, A. (2010). Quantum correlations and dynamics from classical random fields valued in complex Hilbert spaces. *Journal of Mathematical Physics*, 51, art-number 082106.
[7] Björk, T. (1998). *Arbitrage Theory in Continuous Time*. Oxford University Press.
[8] Bethe, H. (1964). *Intermediate Quantum Mechanics*. W. A. Benjamin Inc.
[9] Morrison, Michael, A. (1990). *Understanding Quantum Physics: A User's Manual*. Volume 1. 1st edition. Pearson Education, Upper Saddle River, NJ.

13.3 Implications of the non-Hermiticity of a Black–Scholes Hamiltonian operator on the use of the classical limit arguments

The non-Hermiticity of the Black–Scholes Hamiltonian operator will also have an immediate implication on the classical limit arguments of Chapter 5, Section 5.15. Here is a straightforward proposition.

Proposition 15 *If the Black–Scholes Hamiltonian operator is not Hermitian, then the Ehrenfest theorem cannot hold and the classical limit can not be obtained.*

Proof. The proof of this proposition is quite straightforward. From Ehrenfest's theorem, one finds that $\frac{d\langle\widehat{p}\rangle(t)}{dt} = -\left\langle\frac{\partial V}{\partial x}\right\rangle$ and $\frac{d\langle\widehat{x}\rangle(t)}{dt} = \frac{\langle\widehat{p}\rangle}{m}$ (see Holland [1], p. 111). The key result to prove Ehrenfest's theorem, i.e. the time evolution of the expectation value of an arbitrary operator, Q, which is: $\frac{d}{dt}\langle Q(t)\rangle = \left\langle\frac{\partial\widehat{Q}}{\partial t}\right\rangle + \frac{1}{\hbar}\left\langle i\left[\widehat{\mathcal{H}}, \widehat{Q}\right]\right\rangle$ does not hold since it can be shown that $\langle\widehat{\mathcal{H}}\psi|\widehat{Q}\psi\rangle$ cannot be written as: $\langle\psi|\widehat{\mathcal{H}}\widehat{Q}\psi\rangle$. In that case, following Morrison [2] (p. 516) $\frac{d}{dt}\langle Q(t)\rangle - \left\langle\frac{\partial\widehat{Q}}{\partial t}\right\rangle = \frac{i}{\hbar}\left[\langle\widehat{\mathcal{H}}\psi|\widehat{Q}\psi\rangle - \langle\psi|\widehat{Q}\widehat{\mathcal{H}}\psi\rangle\right]$ and since $\langle\widehat{\mathcal{H}}\psi|\widehat{Q}\psi\rangle \neq \langle\psi|\widehat{\mathcal{H}}\widehat{Q}\psi\rangle$, one cannot rewrite $\frac{i}{\hbar}\left[\langle\widehat{\mathcal{H}}\psi|\widehat{Q}\psi\rangle - \langle\psi|\widehat{Q}\widehat{\mathcal{H}}\psi\rangle\right]$ as $\frac{i}{\hbar}\left[\langle\psi|\widehat{\mathcal{H}}\widehat{Q}\psi\rangle - \langle\psi|\widehat{Q}\widehat{\mathcal{H}}\psi\rangle\right]$. We note that $\frac{i}{\hbar}\left[\langle\psi|\widehat{\mathcal{H}}\widehat{Q}\psi\rangle - \langle\psi|\widehat{Q}\widehat{\mathcal{H}}\psi\rangle\right] = \frac{1}{\hbar}\left\langle i\left[\widehat{\mathcal{H}}, \widehat{Q}\right]\right\rangle$. To obtain the second result of the Ehrenfest theorem: $\frac{d\langle\widehat{x}\rangle(t)}{dt} = \frac{\langle\widehat{p}\rangle}{m}$, one would need to use (Morrison [2] (p. 517)): $\frac{d}{dt}\langle Q(t)\rangle = \left\langle\frac{\partial\widehat{Q}}{\partial t}\right\rangle + \frac{1}{\hbar}\left\langle i\left[\widehat{\mathcal{H}}, \widehat{Q}\right]\right\rangle$, and substitute for $\widehat{Q}$ the position operator. However, $\frac{d}{dt}\langle Q(t)\rangle = \left\langle\frac{\partial\widehat{Q}}{\partial t}\right\rangle + \frac{1}{\hbar}\left\langle i\left[\widehat{\mathcal{H}}, \widehat{Q}\right]\right\rangle$ cannot be used because of the absence of Hermiticity. A similar argument can be made for the first result of the Ehrenfest theorem: $\frac{d\langle\widehat{p}\rangle(t)}{dt} = -\left\langle\frac{\partial V}{\partial x}\right\rangle$. ∎

Corollary 16 *If the state function $\psi(x,t)$ now becomes $P(A,t)$, where A is the price of an asset and P is the price of a put option, then the Ehrenfest theorem does hold and the classical limit can be obtained.*

Proof. Immediate. See Propositions 14 and 15. ∎

Corollary 17 *If the state function $\psi(x,t)$ now becomes $C(A,t)$, where A is the price of an asset and C is the price of a call option and if there exists an appropriate* *super-martingale on A, then the Ehrenfest theorem does hold and the classical limit can be obtained.*

Proof. Immediate. See Propositions 13 and 15. ∎

13.4 References

[1] Holland, P. (2000). *The Quantum Theory of Motion: An Account of the de Broglie–Bohm Causal Interpretation of Quantum Mechanics.* Cambridge University Press.
[2] Morrison, Michael, A. (1990). *Understanding Quantum Physics: A User's Manual.* Volume 1. 1st edition. Pearson Education, Upper Saddle River, NJ.

13.5 Implications of the non-Hermiticity of a Black–Scholes Hamiltonian operator on the stochastic equivalent of Hamilton–Jacobi equations

When we considered the mathematical formalism of the Bohmian mechanical set up in Chapter 6, Section 6.3, we obtained equation (6.14), which was termed there the "continuity equation." This leads to the next proposition:

Proposition 18 *If the Black–Scholes Hamiltonian operator is not Hermitian, then the Bohmian continuity equation may not hold.*

Proof. This is immediate. The continuity equation was: $\frac{\partial R^2}{\partial t} + \frac{1}{m}\frac{\partial}{\partial q}\left(R^2 \frac{\partial S}{\partial q}\right) = 0$ and when conservation of probability is not obtained, then clearly $\frac{\partial R^2}{\partial t} \neq 0$. ∎

Clearly, using Propositions 13 and 14 can reverse this result.

In summary, when options would only be puts, the classical limit and the Bohmian continuity equation would all be verified. When call options were to be used, those results only obtain under the conditions of proposition 13.

13.6 Interpretations of the wave function: a brief discussion

In Chapter 10, Section 10.3, we proposed some possible uses of the wave function within the setting of Soros' idea of reflexivity. The idea of connecting this interesting concept to the use of one of the most elementary tools of quantum mechanics, i.e. Fourier integration (and series), was quite straightforward. We also considered briefly the so-called "active information" interpretation of the Bohmian mechanics-based wave function in Chapter 10.

We can consider seven possible (economics based/finance based) interpretations of the wave function.

1. the use of a basic quantum mechanical (like) wave function which represents the expectations of traders in a financial market (Chapter 10, Section 10.3);

2. the use of a basic quantum mechanical (like) wave function which is the resultant of wave packet construction: the Soros interpretation of reflexivity (Chapter 10, Section 10.3);
3. the use of a basic quantum mechanical (like) wave function as an input in the Radon–Nikodym derivative (this chapter, Section 13.14);
4. the use of a basic quantum mechanical (like) wave function as an analogue of the put and call functions (this chapter, Section 13.1); and as an analogue of the value of an asset (this chapter, Section 13.8);
5. the Bohmian mechanics-based pilot wave function as a carrier of so-called "active information": pilot waves as carriers of erroneous information (this chapter, Section 13.18);
6. the Bohmian mechanics-based pilot wave function as a carrier of so-called "active information": pilot wave in a drift dependent option (this chapter, Section 13.20);
7. the Bohmian mechanics-based pilot wave function as a carrier of so called "active information": pilot wave in the non-arbitrage theorem (this chapter, Section 13.7).

It is important to remark that the above interpretations can in fact be embedded within a more general possible framework. Depending on the role we decide the wave function should take in the macroscopic environment, we implicitly make an assumption as to what formal role is ascribed to this very wave function. It is clear that the array of formal models one can choose from is constraining if we want to define a wave function role (in a macroscopic environment) which has a quantum mechanical characteristic. In other words, if we want to step resolutely outside of quantum physics, we could consider wave functions as they are defined in, for instance, classical physics wave dynamics.

A question we may ask is whether there exist other possible interpretations of the wave function within an economics/finance-based setting? In other words, compared to, for instance, the analogue interpretation of the wave function as either a call or put option pricing function, does there exist another type of analogue? One important issue to note is the following. As Baaquie [1] remarked (p. 51), the wave function in quantum mechanics is not observable, while the put and call option pricing functions are observable. The observability of the option price is, as we know, artificially rendered; i.e. such an observation is made by man. The observability does not derive from a natural phenomenon. Clearly, the same argument can be invoked for asset prices, which are not options. The social science literature has been enriched with important research, which attempts to document how in fact the option market was generated via the development of the Black–Scholes formula mentioned in Chapter, Section 2.7. In other words, the phenomenon of an option pricing market may have been generated by the uncovering of the Black–Scholes formula. This is in some sense a reverse phenomenon of what we are used to in the natural sciences: a phenomenon is observed and a theory is built to explain the phenomenon. MacKenzie and Millo [2] seem to indicate that the opposite happened in option pricing: a theory is built and the phenomenon is generated from the theory. To some extent such an event is also noticeable in other areas of social science, but this so-called "reverse theorizing" is certainly most clear in the option pricing theory side of the social sciences, since it is after all the Black–Scholes formula which triggered a trillion dollar market. Hence, in some sense we can thus question to some slight degree the so-called "observability" of option prices.

13.7 The wave function and non-observed state prices

If we want to stay resolutely within a non-observed framework, then we can possibly hint at the use of the so-called state prices, which are used in the theory of non-arbitrage. This theory, within a discrete state space environment, has been treated in different places in this book (for instance see Chapter 4, Section 4.18). We follow here an approach which can connect the wave function in, again, a Bohmian mechanics framework.

Recall the formulation of the non-arbitrage theorem in Chapter 4, Section 4.18.3. We now reproduce[9] the paper by Haven [3] (Sections 4 and 5 (pp. 196–199)). Remark that q indicates the position of the particle (for instance the price of an asset)

The existence of arbitrage is for a large part based upon the existence of information. Hence, changes in information will alter arbitrage opportunities. We can thus imagine a benchmark situation where we start out under no arbitrage with a particular state of information, reflected by a particular functional form of the information wave function. Our goal, in the proposition below, is then to show that if the state of information changes (this can for instance be reflected by a change in the functional form of the wave function) arbitrage can occur. Before we consider our proposition, let us consider again the non-arbitrage theorem (please see Chapter 4, Section 4.18.3). We recall that if there exists a K-dimensional state price vector $\overrightarrow{\Phi} = (\Phi_1, \Phi_2, \Phi_3, \ldots, \Phi_K)$, which solved the system of equations of that theorem, there will be no arbitrage (and vice versa). An associated probability vector $\overrightarrow{\Phi_{prob}} = \left(\frac{\Phi_1}{\Phi_0}, \frac{\Phi_2}{\Phi_0}, \frac{\Phi_3}{\Phi_0}, \ldots, \frac{\Phi_K}{\Phi_0}\right)$ can be defined, where each coordinate is a probability. We have also that $\Phi_0 = \exp(-rT)$, where r is the risk free rate of return and T is time. We must stress that we assume that the probability values making up the probability vector $\overrightarrow{\Phi_{prob}}$ can be drawn from $\|\Psi\|^2 = \int_{\mathbb{R}} |\psi(q)|^2 \, dM(q)$.

[9] Emmanuel Haven (2008). The variation of financial arbitrage via the use of an information wave function. *International Journal of Theoretical Physics*, 47, pp. 196–199.

Proposition 19 *Let there exist an N-dimensional asset price vector $\overrightarrow{p_0}$ and a K-dimensional state price vector $\overrightarrow{\Phi}$. Let there exist a K-dimensional probability vector $\overrightarrow{\Phi_{prob}} = \left(\frac{\Phi_1}{\Phi_0}, \frac{\Phi_2}{\Phi_0}, \frac{\Phi_3}{\Phi_0}, \ldots, \frac{\Phi_K}{\Phi_0}\right)$. Let $N = K$ and let $\Phi_0 = \exp(-rT)$ be fixed. Let there be an information wave function $\Psi(q)$ and a measure M on $\mathbb{R}$. Let each of the probabilities in $\overrightarrow{\Phi_{prob}}$ be drawn from $\|\Psi\|^2 = \int_{\mathbb{R}} |\psi(q)|^2 \, dM(q)$ for each of the respective set of lower and upper bound values of the integral. Let the state prices which are in $\overrightarrow{\Phi_{prob}}$ guarantee no arbitrage. Consider now an information wave function $\gamma(q)$ which has a different functional form from $\psi(q)$ in the following way: (i) we assume that $\gamma(q)$ cannot be the dual wave function of $\psi(q)$;*[10] *(ii) for the same measure M on $\mathbb{R}$ and for the same respective set of lower and upper bound values of the integral we used for $\int_{\mathbb{R}} |\psi(q)|^2 \, dM(q)$, we write $\int_{\mathbb{R}} |\gamma(q)|^2 \, dM(q)$ such that the functions $|\gamma(q)|^2$ and $|\psi(q)|^2$ can be allowed to intersect on different intervals of their domain (of course the functions may not intersect at all) but under the constraint that at least one probability drawn from $\|\gamma\|^2 = \int_{\mathbb{R}} |\gamma(q)|^2 \, dM(q)$ must be different from the probabilities drawn from $\int_{\mathbb{R}} |\psi(q)|^2 \, dM(q)$. Under those conditions, will the change in the information wave function from $\psi(q)$ to $\gamma(q)$ trigger arbitrage.*

Proof. Since $N = K$, there will be a unique state price vector $\overrightarrow{\Phi} = (\Phi_1, \Phi_2, \ldots, \Phi_K)$ solving the system of equations set out by the non-arbitrage theorem. Hence, the probabilities in $\overrightarrow{\Phi_{prob}} = \left(\frac{\Phi_1}{\Phi_0}, \frac{\Phi_2}{\Phi_0}, \frac{\Phi_3}{\Phi_0}, \ldots, \frac{\Phi_K}{\Phi_0}\right)$, where $\Phi_0 = \exp(-rT)$ is fixed, are also unique. Those probabilities are drawn from $\|\psi\|^2 = \int_{\mathbb{R}} |\psi(q)|^2 \, dM(q)$, for each of the respective set of lower and upper bound values of the integral. Now let us consider an information wave function of a different functional form, $\gamma(q)$. This information wave function cannot be the dual of the information wave function $\psi(q)$. As pointed out by one of the referees of the paper, the space of the information wave functions (with the Hermitian inner product) is a Hilbert space. As the referee points out, "Hence (by the Riesz Lemma), the dual and bidual spaces are conjugate isomorphic, respectively isometrically isomorphic to the Hilbert space." In that case, the result will not hold. Keeping this in mind, we do allow both $|\psi(q)|^2$ and $|\gamma(q)|^2$ to overlap on different intervals (of course, the functions may not overlap at all) in their domain. However, the overlap is permissible up to the point where we require, for the same respective set of lower and upper bound values of the integrals, that at least one probability value generated by $\|\gamma\|^2 = \int_{\mathbb{R}} |\gamma(q)|^2 \, dM(q)$ must be different from any of the probability values generated by $\|\psi\|^2 = \int_{\mathbb{R}} |\psi(q)|^2 \, dM(q)$. Without this restriction, the two different functions could possibly overlap on intervals of their domain in such a way to generate (for the same lower and upper bound values of the integrals) exactly the same

[10] Thanks to one of the referees for pointing out this important restriction.

probabilities. We are now sure that the different functional form of the information wave function will induce at least one different probability. Therefore, the emerging probability vector $\overrightarrow{\Phi^{***}_{prob}} = \left(\frac{\Phi_1^{***}}{\Phi_0}, \frac{\Phi_2^{***}}{\Phi_0}, \frac{\Phi_3^{***}}{\Phi_0}, \ldots, \frac{\Phi_K^{***}}{\Phi_0}\right)$, which has now probabilities uniquely drawn from $\|\gamma\|^2 = \int_{\mathbb{R}} |\gamma(q)|^2 \, dM(q)$ will contain at least one state price in $\overrightarrow{\Phi^{***}} = (\Phi_1^{***}, \Phi_2^{***}, \ldots, \Phi_K^{***})$ such that $\overrightarrow{\Phi^{***}} \neq \vec{\Phi}$. Hence, there must be arbitrage. ■

We make the following three remarks.

1. It is feasible to trigger arbitrage by only changing the upper and lower bound values of the integral and keeping the functional form of the wave function unchanged. In this sense, can q be interpreted as the price of information. Since arbitrage is dependent on information, a change in the price of information could trigger arbitrage. We can think of the price of information as the price of proprietary information for instance. A change in the functional form of the information wave function would then indicate a change in the information about the price of information.
2. If we allow a change can occur in the risk free interest rate, r, so that Φ_0 is not fixed anymore, then the change in the functional form of the information wave function from $\psi(q)$ to $\gamma(q)$ could induce a change from the risk free rate of return r to another rate of return, say R, which is non-risk free. The difference between $R - r$ could be denoted as a risk premium (or a risk discount) if, respectively, the difference is positive or negative. Alternatively, we could also change T, although that would have little economic meaning. Clearly, in both cases the state prices will also need to change since the probabilities need to continue to add up to unity.
3. Our proposition will not necessarily hold for the case where $N > K$ since in this case we may obtain more than one set of state price vectors guaranteeing non-arbitrage. Hence, for a different functional form $\gamma(q)$ it could still be possible we find state prices guaranteeing non-arbitrage.

Using the conditions contained in the above proposition, we observe that in the case when $N = K$ a change in the information wave function from $\psi(q)$ to $\gamma(q)$ will change the state price vector $\vec{\Phi} = (\Phi_1, \Phi_2, \ldots, \Phi_K)$ to $\overrightarrow{\Phi^{***}} = (\Phi_1^{***}, \Phi_2^{***}, \ldots, \Phi_K^{***})$ such that $\overrightarrow{\Phi^{***}} \neq \vec{\Phi}$. We need at least one state price in $\overrightarrow{\Phi^{***}}$ which is different from the state prices in $\vec{\Phi}$. We know that the first state price, Φ_1, multiplies all the prices contained in the N-dimensional asset price vector corresponding to state 1. We continue doing this for all K states. Neftci [4] (pp. 19–20) provides for an interesting interpretation of the state prices and likens them to prices used in an insurance policy. As an example, consider the price of asset 2 at time 0, which we denote as p_0^2. An investor could be willing to pay Φ_1

units for an "insurance policy" that offers D_{21} units of currency if state 1 (at time 1) is to occur (but the insurance pays nothing if any other state than state 1 occurs). The investor could be willing to pay Φ_2 units plus Φ_1 units for an "insurance policy" that offers D_{22} units of currency if state 2 (at time 1) is to occur and D_{21} units of currency if state 1 (at time 1) is to occur (but the insurance pays nothing if any other state than states 1 and 2 occur). If the investor wants to insure that he gets a payoff no matter what state occurs, then he will be willing to pay an "insurance policy" of $\Phi_1 + \Phi_2 + \cdots + \Phi_K$.

In order to connect our theory with Bohmian mechanics, we would need to introduce the idea of a continuous state space. We would then, in light of the above, have a continuum of "insurance policy" prices. In a Bohmian mechanics environment, we would have a continuum of information prices, q, and the information wave functions would change following the Schrödinger partial differential equation. So the information about the information prices changes as well as the information prices themselves. Those changes will then affect the "insurance policy" prices.

So we could in physics terms, imagine there exists a smooth information price trajectory which is traced out by the Newton–Bohm equation (which we already covered as equation (6.16), in Chapter 6.

$$m\frac{d^2q(t)}{dt^2} = -\frac{\partial V(q)}{\partial q} - \frac{\partial Q(q)}{\partial q}, \tag{13.1}$$

subject to the initial conditions that $q(t=0) = q_0$ and $q'(t=0) = q_0^{'}$, where q_0 is the information price at $t = 0$ and $q_0^{'}$ is momentum at $t = 0$. We note that m is mass and $\frac{d^2q(t)}{dt^2}$ is acceleration, $-\frac{\partial V(q)}{\partial q}$ is the partial derivative of the real potential towards the information price, and $-\frac{\partial Q(q)}{\partial q}$ is the partial derivative of the quantum potential towards the information price. We note that m, the real potential V, and the quantum potential Q have already been interpreted economically in the work by Khrennikov [5] [6] [7], Choustova [8], and Haven [9] [10] [11]. In Bohmian mechanics, since the quantum potential, Q, depends on the wave function (via the amplitude function of the wave function), we can say the wave function steers the particle. Thus, in the context we have now described, the information about the price of information (i.e. the information wave function) steers the information prices. Moreover, it seems reasonable to claim that the smooth information price trajectory on the prices of information would also trigger a trajectory of "insurance policy" prices. In summary, we have obtained, via the use of the information wave function, a natural device by which we can induce either arbitrage or form a risk premium (or risk discount). Those two financial phenomena are essential in asset pricing and hence we can begin to see the importance of the information wave function in an asset pricing context.

13.8 Price and superposition of values

The above approach looked at a wave function and non-observed state prices. Are there other approaches where non-observed quantities can be rationalized with the help of a wave function? Recall the approach we discussed in Chapter 3, Section 3.1, where Baaquie and Martin [12] (p. 11) describe the psyche of an individual "as being represented by a state vector, denoted as $|P>$ " which is an element of a state space (linear vector space). We could think of the price state of an asset as the superposition of values of the asset. The value of the asset would be a subjective (unobserved measure), while the price state is an observed measure. Those values of assets could be formed from pools of investors who react to information of the market in a variety of ways. However, we must emphasize that the value is unobserved. Thus, the idea would be that the value would be equivalent to the "anti-chamber" stage (or maybe "gestation" stage) before investors decide to buy or sell an asset at a given (observed) price. The same machinery from Baaquie and Martin [12] can now be used. Values may be orthogonal to each other:[11] $<\text{Value Traders Group } 1|\text{Value Traders Group } 2> = 0$. In words, traders in group 1 do not share any common belief as to the value of the asset with traders of group 2. This could be the case if we think of so-called "fundamental traders" (as Group 1 traders) versus "noise traders" (as Group 2 traders). In analogy with Baaquie and Martin's approach [12] (p. 12) of defining the individual's psyche as a superposition of idea states, we can define the price state of an asset as the superposition of different states, which have emerged from the different groups of traders who participate in the buying and selling process. We could then write that the price state of an asset $A = |A\rangle = c_1|VG1\rangle + c_2|VG2\rangle + c_3|VG3\rangle + \cdots + c_N|VGN\rangle$. As before, the squared modulus of the complex number c_i can be interpreted as the probability of each of the values in $|A\rangle$. Remark also that $VG1, VG2, \ldots, VGN$ indicate the values given to the asset from the traders in the N different trading groups. "Free will" (as in Baaquie and Martin [12] (p. 16)) can again be invoked here. But this "free will" now has very much a financial "flavour." The choice of states, in the asset price state formation process is decided by whom? The answer may well be that, depending on the influence various trading groups have, this choice will be auto-determined via the weights of those groups. There is then also the scale dependency which we discussed in Chapter 3, Section 3.1. Such dependency would refer back to the "context" of the price state. The context could be the type of market to which the price state refers to. Maybe the state price is connected to a very local stock market exchange. This could be the case if the asset is a very country specific asset (i.e. for a product which is not sold outside of the country

[11] In Baaquie and Martin [12] , ideas were orthogonal to each other (see p. 12 in that paper).

but produced by a very large company which can be stock exchange traded). The state price's context could be connected to a much larger stock market (i.e. it could be traded virtually simultaneously (assuming comparable time zones) on many different stock market exchanges). This could be the case of banking stocks which trade globally. The information exchange level is indeed very different from one context to the other.

An important issue which needs mentioning is the existence of incompatible observables. In this case, we could mention that one observable variable is "price" and the "time change of price" would be the other observable variable. Both variables cannot be simultaneously observed. See for instance Baaquie [13]. Traders are the "observers" as they cause the price change because of the act of selling and buying. We can also come up with other possibilities. Assume for instance that the financial market could be seen as divided by "greed-driven" traders (i.e. traders who are interested solely in making money) and "fear-driven traders" (traders who are primordially driven by safeguarding invested capital). Assume there is heavy volatility in the financial markets with a definite short to medium outlook that markets will end up in a downward spiral. The practice of so-called "short selling" consists in a broker selling shares at a price, p, on behalf of an investor. It is important to stress that the broker borrows the shares from another investor. Assuming the price of the asset drops, the investor will buy the shares at the price $p_1 < p$ and will make a per share profit (before transaction costs) of $p - p_1$. In case the price rises after the selling, the short seller makes no money. The act of short selling can very much trigger a massive sell off on shares, since it can trigger price drops. This is the main reason why in very volatile markets, with serious downward pressure, short selling can be prohibited by governments.[12] Consider now the "fear-driven" trader. In a heavily volatile market with strong downward pressure, the trader may sell and this may cause further price drops if there exist many "fear-driven" traders. The "greed-driven" traders may sell too, and will opt for the specific trading route which we described above, i.e. short selling. This can only accelerate price drops, if sufficient "greed-driven" traders exist. The two incompatible variables (observables) here could be (i) the price velocity, i.e. the price momentum, and (ii) the price acceleration. If a market were to only exist with "fear-driven" traders, then one could make possibly the argument the acceleration in price drops would be constant. But if "greed-driven" traders are added to the market, this acceleration could become random. If both traders were to be present in the market, then we could make the argument that both observables (price momentum and price acceleration) are not simultaneously observable. Of course, one could also make the argument that even with the sole presence of

[12] During the 2008 crisis, short selling was for a period of time banned by government order in certain countries.

either "fear-driven" or "greed-driven" traders, there could be randomness on the momentum of price and the acceleration of price. Thus, here are possibly two different incompatible observables: price momentum and price acceleration, as opposed to price momentum and price position.

Before concluding, we would also like to mention that there may well be an interesting link between the simultaneous observability of variables and the concept of arbitrage we have discussed before. We could propose that the existence of non-arbitrage on an asset is equivalently translatable as the existence of simultaneous observation (of all possible traders) of this asset's price over all the world financial trading places.[13] Such a type of observation is in reality not feasible, as it would require transmission of information at speeds which are well beyond the optic fibre transmission speeds. But if one were to think of this "equivalence" as being theoretically feasible, then the argument may well invite another possible idea which may merit mentioning. It seems to be the case that the occurrence of arbitrage, via its connection to the existence of a martingale, may indeed have ties to the non-existence of the Hermiticity of an operator. We discussed this in Section 13.1 of this chapter. Hermiticity, we realize is not the sole defining characteristic of quantum mechanics, but it is still a very important aspect. The non-simultaneity of observation of variables is also in itself not a defining characteristic of quantum mechanics. However, again, it is an important characteristic since non-simultaneous observation can be linked back to so-called incompatible observables. Such link back has a historical connotation which can be traced to Bohr. See Khrennikov [14] (p. 48). Here again, only simultaneous observation on an asset's price over all stock exchanges would guarantee the existence of non-arbitrage. We may have therefore a nascent feeling that the existence of arbitrage may be connected to non-simultaneous observation or the existence of Hermiticity.

We have so far considered some possible interpretations of the quantum-like wave function in an economics and finance context. More ideas on how such a wave function may tie into economics and finance will be provided in the next sections of this chapter. Up to this point in the book, we considered the wave function as representing the expectations of financial traders in a market or also as being the result of the wave packet construction (via the Fourier integral). Soros' idea of reflexivity would fit in quite well in such a construction. We also considered the wave function as a put or call function and we also looked into the possibility of using the wave function in the context of superposing values and non-observed state prices.

We can argue that the applications of the quantum formalism to economics, finance, and behavioral science can be considered as a theoretical experiment on the fundamentals of quantum mechanics. The fundamental principles of quantum

[13] For ease of purpose, we assume that transaction costs and currency costs could be instantaneously determined.

mechanics can be modified to match new domains of applications. Such an activity induces a variety of novel quantum-like models. In effect, it can be considered as the testing of the coupling between different quantum postulates and the possibilities of their modification.

13.9 References

[1] Baaquie, B. (2008). *Quantum mechanics and option pricing*. In *Quantum Interaction*. Eds. Bruza, P., Lawless, W., van Rijsbergen, K., Sofge, D. A., Coecke, B., and Clark, S. *Proceedings of the Second Quantum Interaction Symposium (Oxford University)*. College Publications, pp. 49–53.
[2] MacKenzie, D. and Millo, Y. (2005). Negotiating a market, performing a theory: the historical sociology of a financial derivatives exchange. *American Journal of Sociology*, 109, 107–145.
[3] Haven, E. (2008). The variation of financial arbitrage via the use of an information wave function. *International Journal of Theoretical Physics*, 47, 193–199.
[4] Neftci, S. (2000). *An Introduction to the Mathematics of Financial Derivatives*. Academic Press, New York.
[5] Khrennikov, A. Yu. (1999). Classical and quantum mechanics on information spaces with applications to cognitive, psychological, social and anomalous phenomena. *Foundations of Physics*, 29, 1065–1098.
[6] Khrennikov, A. Yu. (2004). *Information Dynamics in Cognitive, Psychological and Anomalous Phenomena. Series in the Fundamental Theories of Physics*. Kluwer, Dordrecht.
[7] Khrennikov, A. Yu. (1999). *Interpretations of Probability*. VSP International Publishers, Utrecht.
[8] Choustova, O. (2006). Quantum Bohmian model for financial markets. *Physica A*, 374, 304–314.
[9] Haven, E. (2005). Pilot-wave theory and financial option pricing. *International Journal of Theoretical Physics*, 44, 1957–1962.
[10] Haven, E. (2005). Analytical solutions to the backward Kolmogorov PDE via an adiabatic approximation to the Schrödinger PDE. *Journal of Mathematical Analysis and Application*, 311, 439–444.
[11] Haven, E. (2008). Private information and the "information function": a survey of possible uses. *Theory and Decision*, 64, 193–228.
[12] Baaquie, B. and Martin, F. (2005). Quantum psyche: quantum field theory of the human psyche. *NeuroQuantology*, 1, 7–42.
[13] Baaquie, B. (2005). *Quantum Finance*. Cambridge University Press.
[14] Khrennikov, A. (2010). *Ubiquitous Quantum Structure: From Psychology to Finance*. Springer Verlag.

13.10 Arbitrage and negative probabilities

Feynman [1] introduced the idea of using negative probabilities in the context of the Young double slit experiment (see Chapter 5, Section 5.3). Scully *et al.* [2] indicate that Feynman used the concept of negative probability (quasiprobability (Wigner-like) distributions) to explain why a particle goes through both holes in the Young double slit experiment (see Chapter 5, Section 5.3). Burgin [3] quotes Dirac

[4] who said that "Negative energies and probabilities should not be considered as nonsense. They are well-defined concepts mathematically, like a negative of money." For more background on probability concepts in quantum mechanics, see Khrennikov [5] and also Suppes [6]. The book by Espen Haug [7], provides for an excellent overview of some of the work of the top academics in finance and probability. There are two relevant chapters: (i) a chapter entitled "Andrei Khrennikov on negative probabilities" (Haug [7], pp. 315–322) and (ii) a chapter entitled "Why so negative about negative probabilities" (Haug [7] (pp. 323–333)). We urge the reader to peruse those two chapters so as to get well acquainted with the notion of negative probability.

In social science, negative probabilities are a taboo subject. It may be "psychologically" more convenient to use the term "negative weights" instead.[14] One of the authors of this book (Khrennikov) published one of his first papers on negative probability (jointly with O. G. Smolyanov under the recommendation of Professor Kolmogorov) in Doklady of the Academy of Sciences of the USSR. Khrennikov also shows the connection between negative probabilities and so-called p-adic probability theory [8] [9]. See Khrennikov [5] for more detail, but see also Khrennikov [10] for an application of p-adic probability theory to quantum physics, to describe negative probability distributions. In the area of social science, p-adic probability theory sheds new light on the St. Petersburg paradox. See Khrennikov [11].

Let us consider the binomial option pricing model which we mentioned in Chapter 2, Section 2.7. We consider a simple example from Haug [7] (p. 325), which provides for a simple extension on the basic model.

Example 20 *Consider the following example. The asset price, S, is \$100. The time to maturity, T, is 6 months: $T = 0.5$. The volatility of the asset is: $\sigma = 2\%$ and the risk free rate of interest is 12%. Assume six time steps and hence $\Delta t = \frac{1}{12} = 0.08$. Assume that the up proportion is given by the formulation $u = e^{\sigma\sqrt{\Delta t}} = e^{0.02(\sqrt{0.08})} = 1.006$, and similarly for the down proportion $d = e^{-\sigma\sqrt{\Delta t}} = e^{-0.02(\sqrt{0.08})} = 0.994\,2$. The risk neutral probability can then be calculated as $p_1 = \frac{e^{0.12(0.08)} - 0.9942}{1.006 - 0.9942} = 1.309$. The complement probability then becomes $p_2 = 1 - p_1 = 1 - 1.309 = -0.309$. Note that to make the probability negative we must impose that $\sigma < \left| r\sqrt{\Delta t} \right| = 0.02 < 0.12\sqrt{0.08} = 0.03$ (see Haug [7] (p. 325)). The value of the asset price in the up state is given by $pSu = 1.309(100(1.006)) = 131.69$ and the value of the asset price in the down state is given by $-0.309(100(0.9942)) = -30.721$.*

[14] Thanks to Karl Gustafson for bringing up this argument.

Clearly, the negative price in the above example is outside the sample space. As Haug [7] (p. 326) proposes: why therefore not call it the "hidden state" of the stock price? Furthermore, since risk neutral probabilities do not measure anything, we may wonder why negative probabilities would be so extraneous in this context.

An alternative formulation is as follows. Consider the risk neutral probability which one obtains out of the binomial option pricing model. If we opt for so-called non-continuous compounding, then the risk neutral probability can be written as $p = \frac{1+r-d}{u-d}$. The relationship between arbitrage and the existence of negative probabilities can be shown to occur immediately if we consider the statement by Björk [12] (p. 8), which says that there does not exist arbitrage if and only if $d \leq 1 + r \leq u$. The proof of this theorem is instructive. We slightly adapt it from Björk [12] (p. 8).

- (I) $\Longrightarrow$: $d \leq 1 + r \leq u$ implies no arbitrage. This part of the proof shows the contrapositive of (I): if there is arbitrage, then $d \leq 1 + r \leq u$ does not hold. Assume a portfolio $h = (x, y)$, where x is the number of bonds in the portfolio and y is the number of units of stock in the portfolio. Assume there exists a stochastic variable Z which takes on either (i) the value u with probability p_u or (ii) the value d with probability p_d. Consider the value of a portfolio at time $t = 0$, $V_0^h = x + ys$, where s is the stock price. Define the value of the portfolio at time $t = 1$, $V_1^h = x(1 + r) + ysZ$. There will be arbitrage if $V_0^h = 0$ and $V_1^h > 0$ (i.e. you get "something ($V_1^h > 0$) for nothing ($V_0^h = 0$)"). Let $V_0^h = 0$; this means $x = -ys$. Substituting this into $V_1^h = -ys(1 + r) + ysZ$. If $Z = u$, $V_1^h = ys\,[u - (1 + r)]$. If $Z = d$, $V_1^h = ys\,[d - (1 + r)]$. Assuming that $y > 0$, then h is an arbitrage strategy if and only if $u > 1 + r$ and $d > 1 + r$, which violates the proposition. For the case $y < 0$, using the same procedure, one obtains $u < 1 + r$ and $d < 1 + r$, which again violates the proposition.
- (II) $\Longleftarrow$:no arbitrage implies $d \leq 1 + r \leq u$. This part of the proof shows the contrapositive of (II): if $d \leq 1 + r \leq u$ does not hold, then there is arbitrage. Assume $s(1 + r) > su$, then surely $s(1 + r) > sd$, and then you invest always in the bond as opposed to the stock. Therefore, let $h = (s, -1)$ (i.e. you sell stock (negative sign) and buy bonds). Then $V_0^h = x + ys = s - s = 0$. Recall $V_1^h = x(1 + r) + ysZ$ and substituting $h = (s, -1)$, one obtains: $V_1^h = s(1 + r) - sZ > 0$ by the assumption we made. Therefore, there exists arbitrage.

In Jammer [13] (p. 256) "Definition I" of a hidden variable is as follows: "In a given theory T about certain physical systems S certain variables v describe the states of S; in a theory T' about S certain variables v' (which may be dynamical quantities or other hypothetical entities) which are not experimentally detectable within the framework of T describe the states of S; if the values of $v\ldots$ can be obtained by some averaging operation over the values of v', (then) v' are called hidden variables with respect to T." If we apply Jammer's statement to the non-arbitrage context, then we can say:

- The "state prices" used in the calculation of risk neutral probabilities (see this chapter,[15] Section 13.7 or see also Chapter 4, Section 4.18.3) are unobservable. Call those v'. The risk free rate of interest, R, is observable. An asset with price $A(t)$ at time t can be carried forward in the future $A(t)(1 + R) = \frac{\Phi_1}{\Phi_0} A_1(t+1) + \frac{\Phi_2}{\Phi_0} A_2(t+1)$, where $\frac{\Phi_i}{\Phi_0}$ is part of the probability vector $\vec{\Phi}$ which was defined before[16] and Φ_i describe the states of nature; A_i are the prices in the future, depending on the state of nature occurring. Call this R, thus v.
- In some sense, we can obtain v by "some averaging"[17] operation over the values v': $1 + R = \frac{\frac{\Phi_1}{\Phi_0} A_1(t+1) + \frac{\Phi_2}{\Phi_0} A_2(t+1)}{A(t)}$.

Before moving to the next section of this chapter, let us reiterate some key issues which may have arisen so far. When considering quantum mechanical applications to social science, one needs to keep carefully in mind that one is not importing quantum physics, as a physical theory, into social science. Khrennikov in the preface of his book (Khrennikov [14]) stresses this quite clearly. Recall the discussion we had on the non-Hermiticity of a very important Hamiltonian, i.e. the Black–Scholes Hamiltonian. This discussion shows us that doing otherwise, i.e. pretending that there is some trace of quantum mechanics in the macroscopic world, can be a very difficult position to take. Any quantum system implies the use of Hermitian operators.

So caution is of importance here. The message remains: we use quantum mechanical principles in social science to potentially better explain certain phenomena in that macroscopic setting. This does not mean that anything quantum mechanical is as such manifest in the macroscopic world.

13.11 References

[1] Feynman, R. (1987). Negative probabilities in quantum mechanics. In *Quantum Implications*. Eds. Hiley, B. and Peat, F. Routledge & Kegan Paul.

[2] Scully, M. O., Walther, H., and Schleich, W. (1994). Feynman's approach to negative probability in quantum mechanics. *Physical Review A*, 49, 3, 1562–1566.

[3] Burgin, M. (2012). Interpretations of negative probabilities. Cornell University Library, http://arxiv.org/ftp/arxiv/papers/1008/1008.1287.pdf

[4] Dirac, P. A. M. (1942). The physical interpretation of quantum mechanics. *Proceedings of the Royal Society London (A180)*, 1–39.

[15] Just above Proposition 19. [16] See this chapter, Section 13.7 or see also Chapter 4, Section 4.18.3.

[17] Indeed "some" averaging operation. . . : i.e. the weighted average is divided by some price $A(t)$. . .

[5] Khrennikov, A. Yu. (2009). *Interpretations of Probability*. 2nd edition. De Gruyter, Berlin.
[6] Suppes, P. (1961). Probability concepts in quantum mechanics. *Philosophy of Science*, 28, 4, 378–389.
[7] Haug, E. (2007). *Derivatives: Models on Models*. J. Wiley.
[8] Khrennikov, A. Yu. (1992). p-adic probability and statistics. *Doklady Akademii Nauk (Proceedings of the Russian Academy of Sciences)*, ser. Math., 322, 1075–1079.
[9] Khrennikov, A. Yu. (1993). Discrete Q_p-valued probabilities. *Doklady Akademii Nauk (Proceedings of the Russian Academy of Sciences)*, ser. Math., 333, 162–164.
[10] Khrennikov, A. Yu. (1997). *Non-Archimedean Analysis: Quantum Paradoxes, Dynamical Systems and Biological Models*. Kluwer, Dordrecht.
[11] Khrennikov, A. Yu. (1994). *p-adic Valued Distributions in Mathematical Physics*. Kluwer, Dordrecht.
[12] Björk, T. (1998). *Arbitrage Theory in Continuous Time*. Oxford University Press.
[13] Jammer, M. (1974). *The Philosophy of Quantum Mechanics*. J. Wiley.
[14] Khrennikov, A. (2010). *Ubiquitous Quantum Structure: From Psychology to Finance*. Springer Verlag, Berlin.

13.12 The Li–Zhang and WKB approach

At the end of Chapter 2 (Section 2.11), we mentioned some of the work by Li and Zhang, which maps option pricing functions into wave functions. We also mentioned related work, using the WKB method. In this section of this chapter, we provide for some details on those two approaches.

Two papers by Li and Zhang argue for the use of the so-called Weyl–Titchmarsh theory in option pricing. They are first the 2004 paper by Li and Zhang [1] and second an update paper by Zhang and Li [2], which appeared in 2012.

In Li and Zhang [1], it is shown that option pricing can be modeled via the use of the Schrödinger PDE.

In a nutshell, Li and Zhang [1] (p. 459), show that the necessary transformations to achieve such a pricing approach are that the call function (which is a function of the stock price, S, and time, t); $C(S,t) = \sqrt{\sigma(S)}g(x,t)$, where $g(x,t)$ is the quantum mechanical wave function and $\sigma(S)$ is some volatility function of the asset price, S. Furthermore, they also set: $x(S) = \delta \int \frac{1}{\sigma(S)} dS$, with $\delta = +/-1$.

Equation (7) in Li and Zhang [1] (p. 459) shows that a heat equation on a call option can be written as $\frac{\partial C}{\partial t} - \sigma^2(S)\frac{\partial^2 C}{\partial S^2} = 0$ with initial condition $C(S,0) = \max(S-K,0)$, where K is the strike price. The authors show that equation (7) can be written as a Schrödinger PDE: $\frac{\partial g}{\partial t} = \frac{\partial^2 g}{\partial x^2} - q(x)g$, with $g(x,t)|_{t=0} = g_0(x)$ (they indicate that $g_0(x)$ is defined from $C(S,0)$).

A beautiful interpretation is obtained of a potential function: the function $q(x)$ is the potential function defined as $\frac{1}{4}\left(\frac{d\sigma(S)}{dS}\right)^2 - \frac{1}{2}\sigma(S)\left(\frac{d^2\sigma(S)}{d\sigma(S)^2}\right)$. See Li and Zhang

[1] (p. 459). The authors also mention that when $q(x)$ is a quadratic function of x, one can then transform the above Schrödinger PDE into a standard heat equation.

The practical utility of the Li and Zhang approach is important. For instance, in their paper they provide for a sophisticated volatility function of the form $\sigma(S) = S\sqrt{\sigma^2 + \epsilon^2\left(\ln\frac{S}{S_0}\right)^2}$, with $\epsilon > 0$ a small parameter and $\sigma > 0$ and $S_0 > 0$ (see Li and Zhang [1] (p. 461)). Using the transformation $x(S) = \delta\int\frac{1}{\sigma(S)}dS$, they then find in this specific case that $x = \frac{1}{\epsilon}\sinh^{-1}\frac{\epsilon}{\sigma}\ln\frac{S}{S_0}$ with $S = S_0\exp\left(\left(\frac{\sigma}{\epsilon}\right)\sinh\epsilon x\right)$. The potential function $q(x) = \frac{1}{4}l(S) + \frac{1}{4}\epsilon^2 - \frac{3}{4}\frac{\sigma^2\epsilon^2}{4l(S)}$, and $l(S) = \sigma^2\cosh^2\epsilon x$. The authors indicate that if $\epsilon \to 0$, then $q(x) \to \frac{\sigma^2}{4}$. A solution is found for $g(x,t)$ in the paper.

Another volatility function, the so-called exponentially decreasing volatility of the form $\sigma(S) = \exp(-S)$ is also treated. A solution for $g(x,t)$ is also presented for this case. See Li and Zhang [1] (p. 462).

Table 1 in Li and Zhang [1] (p. 463), provides for a very interesting overview of potential functions for which solutions for $g(x,t)$ exist.

An update on the 2004 paper is provided for in Zhang and Li [2]. In that paper other interesting potential functions are considered (with associated solutions).

A different approach can be proposed by using the so-called WKB technique in option pricing. We follow here closely, Haven [3]. The so-called WKB approximation, by Bender and Orszag [4], yields a solution, $\psi(x)$, to the time-independent Schrödinger equation.

For that approximation to be senseful, it must satisfy:

$$\frac{\hbar m\left(\frac{dR}{dx}\right)}{\sqrt{(2m(E-R))^3}} \ll 1, \tag{13.2}$$

where R is the potential function (Bohm [5]).[18] Note that m is mass, E is total energy, and $\hbar$ is the rationalized Planck constant. See Haven [3] (pp. 440–441).

In Haven [3], we also mention that when using Bohm [5], the solution for $\psi(x)$ using the WKB approximation:

$$\psi(x) = \vartheta(x)e^{\left[i\int_{x_0}^{x}\sqrt{2m(E-R)}\frac{dx}{\hbar}\right]} + \theta(x)e^{\left[-i\int_{x_0}^{x}\sqrt{2m(E-R)}\frac{dx}{\hbar}\right]}, \tag{13.3}$$

where $\vartheta(x) = \frac{A}{\sqrt[4]{E-R(x)}}$ and $\theta(x) = \frac{B}{\sqrt[4]{E-R(x)}}$ and A and B are constants. Those coefficient A and B can be determined via an initial value problem.

As we have seen in Chapter 2, the Black–Scholes PDE is a so-called backward Kolmogorov PDE. The backward Kolmogorov PDE is of the form $\frac{\partial p}{\partial t} + \frac{b(y,t)^2}{2}\frac{\partial^2 p}{\partial y^2} +$

[18] Although this book is authored by David Bohm, the book does NOT discuss "Bohmian mechanics," as defined in Chapter 6. Hence, the application considered in this section of the book (Section 13.12) does not have a connection with Bohmian mechanics.

$a(y,t)\frac{\partial p}{\partial y} = 0$. Clearly, depending on the functional forms of $b(y,t)$ and $a(y,t)$, the solution approaches (and solutions) will vary substantially. Note that p is a probability density function. As in Haven [3] (p. 441), if we let $a(y,t)$ and $b(y,t)$ be time independent functions and let $a(y,t)$ be a constant function, then the claims we make in that paper (Haven [3] (Claim 2 (p. 441)) is that the essential transformation needed for using the WKB approximation in solving the backward Kolmogorov PDE are: (a) $x = g(b(y))$ and (b) $\psi(x) = f(p(y,t))$. When we apply those transformations to the Black–Scholes PDE, we need to set: (a) $x = g(\sigma(S))$ and (b) $\psi(x) = f(p(S,t))$. Furthermore, we need a semigroup operator, $\widehat{d}$, which relates $p(S,t)$ with the option price $V(S,t) : V(S,t) = \widehat{d}p(S,t)$. In this finance problem, $\frac{\hbar^2}{2m} = 1$.

As expressed in criterion 6 in Haven [3] (p. 442), the WKB approximation is appropriate when, after using $E = \frac{p^2}{\hbar^2}$, the inequality $\frac{\hbar^3}{2}\left(\frac{dR}{dx}\right) \ll \sqrt{(p^2 - \hbar^2 R)^3}$ is respected.

We now reproduce[19] the paper by Haven [3] (latter part of p. 442 and p. 443).

Bender and Orszag [4] provide for an initial value problem where, if they set $\psi(x) = 0$ for $x = 0$ and $\frac{d\psi(x)}{dx}|_{x=0} = 1$, they obtain as WKB solution:

$$\psi(x) = \varsigma(S)\sinh\left[\frac{\sqrt{2m}}{i\hbar}\int_0^S \sqrt{E - R(t)dt}\right], \tag{13.4}$$

with $\varsigma(S) = \frac{i\hbar}{\sqrt{2m}}\frac{1}{\sqrt[4]{E-R(0)}\sqrt[4]{E-R(S)}}$. Note also that t is NOT time, but just the integration variable.

Let us apply Bender and Orszag's initial value problem to our financial problem. A key issue to note is that when using $\psi(x) = f(p(y,t))$, we must make sure that the conditions $\psi(0) = 0$ and $\frac{d\psi(x)}{dx}|_{x=0} = 1$ do make financial sense. Let us set $y = S$ and let us define, as in the paper by Li and Zhang [1], $x(S) = \int \frac{1}{\sigma(S)}dS$ (this takes care of defining $g(.)$).[20] Assume we know what $\widehat{d}$ is and let us have for a given t that $V(S,t) = \psi(x)\sqrt{\sigma(S)}$ (this takes care of defining $f(.)$). As in Li and Zhang [1], we propose a quite interesting (i.e. close to reality) volatility function $\sigma(S) = S\sqrt{\sigma^2 + \epsilon^2\left(\ln\frac{S}{S_0}\right)^2}$, where the volatility $\sigma > 0$ and $S_0 > 0$ and ϵ is small. Following, Li and Zhang [1], one can then write $x = \frac{1}{\epsilon}\sinh^{-1}\frac{\epsilon}{\sigma}\ln\frac{S}{S_0}$. To apply the WKB initial value problem conditions, we need:

[19] Emmanuel Haven (2005). Analytical solutions to the backward Kolmogorov PDE via an adiabatic approximation to the Schrödinger PDE. *Journal of Mathematical Analysis and Applications*, 311, 442–443.

[20] This is Example 7 in Haven [3] (p. 443).

1. to see whether the Bohm condition to use WKB can be satisfied: $\frac{\hbar^3}{2}\left(\frac{dR}{dx}\right) \ll \sqrt{(p^2 - \hbar^2 R)^3}$. In effect, $R(x) = \frac{1}{4}\sigma^2 \cosh^2 \epsilon x + \frac{1}{4}\epsilon^2 - \frac{3}{4}\frac{\sigma^2\epsilon^2}{\sigma^2 \cosh^2 \epsilon x}$ and it is immediate for ϵ very small $\frac{dR}{dx} \to 0$.
2. $x = 0 \Leftrightarrow S = S_0$ and in the WKB equation (13.3), after setting $x_0 = 0$, we get: $\psi(0) = \frac{A+B}{\sqrt[4]{E-R(0)}} = \frac{A+B}{\sqrt[4]{E-\frac{1}{4}\sigma^2+\frac{1}{2}\epsilon^2}} = \frac{V(S=S_0,t)}{\sqrt{\sigma(S_0)}} = 0$. This is definitely possible, from a financial point of view, if $S_0 \ll K$ and hence $A = -B$.
3. Given $A = -B$, we have that $C = \frac{d\psi(x)}{dx}|_{x=0} = (A - B)\sqrt[4]{E - R(0)}\frac{\sqrt{2m}}{i\hbar}$ and let $C = 1$ so we get: $A = \frac{i\hbar}{2\sqrt{2m}\sqrt[4]{E-\frac{1}{4}\sigma^2+\frac{1}{2}\epsilon^2}}$ and we can now use equation (13.4) since $\frac{d\psi(x)}{dx}|_{x=0} = \frac{d\left(\frac{V(S,t)}{\sqrt{S\sqrt{\sigma^2+\epsilon^2\left(\ln\frac{S}{S_0}\right)^2}}}\right)}{d\left(\frac{1}{\epsilon}\sinh^{-1}\frac{\epsilon}{\sigma}\ln\frac{S}{S_0}\right)}|_{S=S_0} = 1$ is financially feasible.

Therefore, using the relationship between $\psi(x)$ and $V(S,t)$ and using the operator $\widehat{d}$ we will have found an analytical solution to the option pricing backward Kolmogorov PDE.

13.13 References

[1] Li, Y. and Zhang, J. E. (2004). Option pricing with Weyl–Titchmarsh theory. *Quantitative Finance*, 4, 4, 457–464.

[2] Zhang, J. E. and Li, Y. (2012). New analytical option pricing models with Weyl–Titchmarsh theory. *Quantitative Finance*, 12, 7, 1003–1010.

[3] Haven, E. (2005). Analytical solutions to the backward Kolmogorov PDE via an adiabatic approximation to the Schrödinger PDE. *Journal of Mathematical Analysis and Applications*, 311, 439–444.

[4] Bender, D. and Orszag, S. (1978). *Advanced Mathematical Methods for Scientists and Engineers*. McGraw Hill.

[5] Bohm, D. (1951). *Quantum Theory*. Prentice-Hall.

13.14 The wave function as a Radon–Nikodym derivative

We can augment our understanding of the wave function in an asset pricing context by considering some of the results which deal with the measurement of the quantity of information. In this section we reproduce[21] (with only slight modifications[22]) the paper by Haven [1] (sections 6 until 9 (pp. 751–755)). To measure the quantity of

[21] Emmanuel Haven (2010). The Blackwell and Dubins theorem and Rényi's amount of information measure: some applications. *Acta Applicandae Mathematicae*, 109, 3, 751–755.

[22] Mainly condition 27 and the brief discussion preceding that condition provides for the modification.

information in the information wave function, we can use a specific theorem which we will highlight below. Before we state the main theorem from the Blackwell and Dubins paper [2], we need to define a so-called predictive probability.

Definition 21 ***(Blackwell and Dubins [2] (p. 882). Predictive probability.*** *Let $\mathcal{B}_i$ be a σ-field of subsets of the set X_i, $i = 1, 2\ldots$ Let $(X, \mathcal{B}) = (X_1 \times X_2 \times \ldots \times \mathcal{B}_1 \times \mathcal{B}_2 \times \ldots)$. Let $(X, \mathcal{B}, P)$ be a probability space and let P_n be the marginal distribution of $(X_1 \times X_2 \times \ldots \times X_n, \mathcal{B}_1 \times \mathcal{B}_2 \times \ldots \times \mathcal{B}_n)$; i.e., $P_n(A) = P(A \times X_{n+1} \times \ldots)$ for all $A \in \mathcal{B}_1 \times \mathcal{B}_2 \times \ldots \times \mathcal{B}_n$. The predictive probability P is defined if, for every $n \geq 1$, there exists a conditional probability P^n for the future $X_{n+1} \times \ldots$ given the past $X_1, \ldots, X_n$. That is, if there exists a function $P^n(x_1, \ldots, x_n)(C)$ where $(x_1, \ldots, x_n)$ ranges over $X_1 \times X_2 \times \ldots \times X_n$ and C ranges over $\mathcal{B}_{n+1} \times \ldots$ with the usual three properties: (i) $P^n(x_1, \ldots, x_n)(C)$ is $\mathcal{B}_1 \times \mathcal{B}_2 \times \ldots \times \mathcal{B}_n$ measurable for fixed C; (ii) a probability distribution on $(X_{n+1} \times \ldots \mathcal{B}_{n+1} \times \ldots)$ for fixed $(x_1, \ldots, x_n)$; and (iii) for bounded $\mathcal{B}$-measurable ϕ we have that $\int \phi dP = \int [\phi(x_1, \ldots, x_n, x_{n+1}, \ldots)$ $dP^n(x_{n+1}, \ldots |x_1, \ldots, x_n)] .dP_n(x_1, \ldots, x_n)$ holds.*

Definition 22 ***(Blackwell and Dubins [2] (p. 883)). Distance between distributions.*** *For any two probabilities μ_1 and μ_2 on the same σ−field $\mathcal{F}$, the distance $\rho(\mu_1, \mu_2)$ is the least upper bound over $D \in \mathcal{F}$ of $|\mu_1(D) - \mu_2(D)|$.*

Theorem 23 ***(Blackwell and Dubins [2] (p. 883)).*** *Let P be a predictive probability on $(X, \mathcal{B})$ and let Q be absolutely continuous with respect to P. Then for each conditional distribution P^n of the future given the past with respect to P, there exists a conditional distribution Q^n of the future given the past with respect to Q such that the distance between $P^n(x_1, \ldots, x_n)$ and $Q^n(x_1, \ldots, x_n)$ converges to zero as $n \to \infty$.*

As is proposed in Nakata [3] (p. 13), we can make the distributions conditional in such a way that we have $P\{.|x_1, x_2, \ldots, x_t\}$ and $Q\{.|x_1, x_2, \ldots, x_t\}$ so that when time $t \to \infty$ one can say that both distributions agree Q–a.s. (almost surely). From an information point of view, the meaning of the theorem indicates that a merging of opinions (amongst agents) occurs if we assume that there exist two agents who have as probability measures P and Q. See Nakata [3] (pp. 1, 28, 13). In this sense, as is well argued in Nakata [3] (p. 13), one can assume mutual absolute continuity so that both distributions agree Q−a.s. and P−a.s. Therefore (Nakata [3] (p. 13)), a special case of this mutual absolute continuity is then that $P = Q$. One possible interpretation of this equality is that agents in an economy have attained a state where all asymmetric information has been removed. See again Nakata [3] (p. 29, 55).

The Radon–Nikodym derivative is a widely used concept in stochastic mathematics and it has applications in many areas, including financial theory where it is used in the theory of financial option pricing. Primbs [4] provides for a dense and excellent background to the measure theoretic issues lying on the back of the Radon–Nikodym derivative. To define the Radon–Nikodym derivative, we need the following ingredients: (1) a set Ω; (2) a set F which is a collection of subsets of Ω (and a σ-algebra); (3) two measures, μ and υ; (4) absolute continuity of one measure with respect to the other measure; and (5) the so-called Radon–Nikodym theorem. Let us first define absolute continuity. The definition is by Primbs [4].

Definition 24 ***Absolute continuity.*** *Let μ and υ be two measures on the measurable space (Ω, F), then υ is absolutely continuous with respect to μ if, for any set A, $\mu(A) = 0 \Rightarrow \upsilon(A) = 0$.*

The Radon–Nikodym theorem is as follows. We follow again Primbs [4].

Theorem 25 ***Radon–Nikodym theorem.*** *If υ is absolutely continuous with respect to μ, then $\upsilon(A) = \int_A g d\mu$.*

The Radon–Nikodym derivative is then defined as follows. We follow again Primbs [4].

Definition 26 ***Radon–Nikodym derivative.*** *The function g is known as the Radon–Nikodym derivative and denoted as $g = \frac{d\upsilon}{d\mu}$.*

In the condition 27, which follows below, we mention there exists a representative agent and a "specific" agent. The representative agent can be thought of as being representative of a trading group. Examples of such trading groups could be "fundamental traders" or also "noise traders." Fundamental traders can be seen as traders who have specific financial information such as market depth (and other measures), which they can objectively use in making a trading decision. Noise traders can be seen as traders who act upon information (which more often than not could be wholly spurious information). We assume that both types of agents have knowledge of a measure, which in effect can be a probability density function. We assume this probability density function can refer to, for instance, rates of return of a stock or also the price of a stock. We note that in the condition below the wave function $\psi(q)$ is defined as $\psi(q) \equiv R(q)\exp((iA(q))$, where q is position, $R(q)$ is the amplitude function, $A(q)$ is the phase, and h is the Planck constant.

Condition 27 $|\psi(q)|^2$ *as a Radon–Nikodym derivative. Assume there exists a representative agent "1" and a specific agent "2" who is not part of the representative's agent trading group. The representative agent "1" has a measure υ (i.e. a probability density function) on the return or the price of an asset. The specific* *agent "2" has a measure μ (i.e. a probability density function) on the return or the price of an asset. Consider a Hilbert state space of quantum states L^2 with respect to υ. Assume that μ is absolutely continuous with respect*[23] *to υ. We assume that the Radon–Nikodym derivative $\frac{d\mu}{d\upsilon} = |\psi(q)|^2$, where $\psi(q)$ is the information wave function of the representative agent "1".*

We remark that $|\psi(q)|^2$ as a probability density function is defined with respect to the measure υ. We could imagine there exist other $\psi(q)$ functions which refer to the specific information of other representative agents.

We are now ready to define the Rényi measure of quantity of information as follows.

Definition 28 ***Rényi's quantity of information measure [5] (pp. 553–554).*** *Consider $\mathcal{P}$ as the unconditional distribution of a random variable and consider $\mathcal{Q}$ as the conditional distribution of the same random variable (obtained from observing an event E). Let $\mathcal{Q}$ be absolutely continuous with respect to $\mathcal{P}$. The amount of information $I(\mathcal{Q}|\mathcal{P})$, concerning the random variable obtained from observing E is given by $I(\mathcal{Q}|\mathcal{P}) = \int_\Omega \log h \, d\mathcal{Q}$, where $h = \frac{d\mathcal{Q}}{d\mathcal{P}}$ is the Radon–Nikodym derivative of $\mathcal{Q}$ with respect to $\mathcal{P}$.*

Definition 29 ***Rényi's information measure [5] and $|\psi(q)|^2$.*** *Given the above condition that $|\psi(q)|^2$ is a Radon–Nikodym derivative, we can apply Rényi's quantity of information measure directly on the information wave function and write that, using the same variables as in the above condition, $I(\mu|\upsilon) = \int_\Omega \log |\psi(q)|^2 \, d\mu$.*

We can thus say that $|\psi(q)|^2$ has in fact a double interpretation: (i) changing $|\psi(q)|^2$ will affect the probability value and (ii) changing $|\psi(q)|^2$ will affect the quantity of information. We thus see here that $|\psi(q)|^2$ has a true information-theoretic interpretation.

We now want to discuss a little more how Rényi's information measure can be connected to the Blackwell and Dubins theorem. Let us first consider the special case of mutual absolute continuity. In this case, $|\psi(q)|^2 = 1$ and, therefore, Rényi's measure yields zero. In the context of the Blackwell and Dubins theorem, we can interpret this result somewhat better. In the context of the special case of mutual absolute continuity, the zero information quantity measure would indicate that all asymmetric information has been removed. This case occurred because, under the economic interpretation of the Blackwell and Dubins theorem, Nakata [3] (p. 28) indicates that there is a merging of opinions. This convergence of measures is obtained because of communication. Communication removes asymmetric information and the highest level of absence of asymmetric information is reached when

[23] Compared to the definition of absolute continuity, we reverse the measures.

$I(\mu|\upsilon) = 0$. In quantum physical terms, we can think of the case $|\psi(q)|^2 = 1$ as the ground state (i.e. the state with minimal total energy). It is not unreasonable to argue that the representative agent "1" can possibly make gainful use of his information, transmitted via $\psi(q)$, if indeed his information is running counter to the information possessed by the specific agent "2." The information wave function would then, in quantum physical terms, have a state with higher than minimal total energy.

Clearly, by slightly modifying the Blackwell and Dubins theorem, we can induce a quantity of information measure which is non-zero. Clearly, if $t \nrightarrow \infty$, then, using Nakata's [3] (p. 13) interpretation, we cannot possibly have convergence of measures. Therefore, it is impossible that $P\{.|x_1, x_2, \ldots, x_t\} = Q\{.|x_1, x_2, \ldots, x_t\}$ using the above terminology. Therefore, it is immediate that, using Rényi's information measure, we have that: $I(Q|P) = \int_\Omega \log \frac{dQ}{dP} dQ \neq 0$.

Definition 30 ***Ambiguity of information.*** *Let* $\psi(q) = \psi_1(q)\psi_2(q)$, *then there is ambiguity of information.*

In this equality, the wave function is factorized into an independent product. As per Bohm and Hiley [6] (p. 61), this could indicate that there exist "independent pools of information." However, the position variable, q, is the same. Hence, this sort of product could indicate ambiguity of information: we have two sources of information on the same position variable. The obvious conditions for this equality to hold is that $R(q) = R_1(q)R_2(q)$ and $A(q) = A_1(q) + A_2(q)$.

Proposition 31 ***Rényi's information measure [5] and ambiguity of information.*** *Given our condition that* $|\psi(q)|^2$ *is a Radon–Nikodym derivative and that* $\psi(q) = \psi_1(q)\psi_2(q)$, *we can apply Rényi's information measure directly on the information wave function and write that, using the same variables as in the above condition ,* $I(\mu|\upsilon) = \int_\Omega \log|\psi|^2 d\mu = \int_\Omega \log|\psi_1|^2 d\mu + \int_\Omega \log|\psi_2|^2 d\mu$.

Proof. Obvious. We can write $|\psi|^2 = |\psi_1 * \psi_2|^2 = \left|R_1 e^{iA_1} * R_2 e^{iA_2}\right|^2 = \left|R_1 R_2 e^{i(A_1+A_2)}\right|^2 = R_1^2 R_2^2$. But $R_1^2 R_2^2$ is nothing else other than $|\psi_1|^2 |\psi_2|^2$. We note that we have omitted Planck's constant. ∎

Definition 32 ***Multi-sourced information.*** *Let* $\psi(q_1, q_2) = \psi_1(q_1)\psi_2(q_2)$, *then there is multi-sourced information.*

Here we have again a factorization into an independent product. Note that each information wave function refers now to a different position variable. We have two sources of information on different position variables. Here the requirements for the equality to hold are again obvious: $A(q_1, q_2) = A_1(q_1) + A_2(q_2)$ and $R(q_1, q_2) = R_1(q_1)R_2(q_2)$.

We now consider postulate 9 in Rényi's 1961 article [5]. This proposition is valid for discrete distributions and is formulated as follows.

Proposition 33 ***Postulate 9 in Rényi [5] (p. 555).*** *Consider $I(\mathcal{Q}_1|\mathcal{P}_1)$ and $I(\mathcal{Q}_2|\mathcal{P}_2)$ and let $\mathcal{P} = \mathcal{P}_1 * \mathcal{P}_2$ and $\mathcal{Q} = \mathcal{Q}_1 * \mathcal{Q}_2$, where "$*$" denoted direct product. Let the correspondence between the elements of $\mathcal{P}$ and $\mathcal{Q}$ be induced by the correspondence between the elements of $\mathcal{P}_1$ and $\mathcal{Q}_1$ and those of $\mathcal{P}_2$ and $\mathcal{Q}_2$. Then we can write that $I(\mathcal{Q}|\mathcal{P}) = I(\mathcal{Q}_1|\mathcal{P}_1) + I(\mathcal{Q}_2|\mathcal{P}_2)$.*

It is interesting to see how this postulate would look if we were to use an information wave function as the basis of the Radon–Nikodym derivative. Assume that we draw from positive rational numbers rather than positive real numbers, so we can keep the correspondence. Therefore, we can use the integral representation for the quantity of information.

We have the following proposition:

Proposition 34 ***Rényi's postulate 9 [5] and the information wave function.*** *If we use $|\psi_1 * \psi_2|^2$, $|\psi_1|^2$, $|\psi_2|^2$ as Radon–Nikodym derivatives in respectively Rényi's quantity of information measures $I(\mathcal{Q}|\mathcal{P})$, $I(\mathcal{Q}_1|\mathcal{P}_1)$, $I(\mathcal{Q}_2|\mathcal{P}_2)$, then $\int_\Omega \log \frac{d\mathcal{Q}_1 * \mathcal{Q}_2}{d\mathcal{P}_1 * \mathcal{P}_2} d\mathcal{Q}_1 * \mathcal{Q}_2 \neq \int_\Omega \log \frac{d\mathcal{Q}_1}{d\mathcal{P}_1} d\mathcal{Q}_1 + \int_\Omega \log \frac{d\mathcal{Q}_2}{d\mathcal{P}_2} d\mathcal{Q}_2$.*

Proof. We observe that $\frac{d\mathcal{Q}_1 * \mathcal{Q}_2}{d\mathcal{P}_1 * \mathcal{P}_2}$ can be written as the information function as follows: $\frac{d\mathcal{Q}_1 * \mathcal{Q}_2}{d\mathcal{P}_1 * \mathcal{P}_2} = |\psi_1 * \psi_2|^2$. The right-hand side can be written as: $\left|R_1 e^{iA_1} * R_2 e^{iA_2}\right|^2 = \left|R_1 R_2 e^{i(A_1 + A_2)}\right|^2 = R_1^2 R_2^2$. But $R_1^2 R_2^2$ is nothing else other than $|\psi_1|^2 |\psi_2|^2$. Therefore, we can write $\int_\Omega \log \frac{d\mathcal{Q}_1 * \mathcal{Q}_2}{d\mathcal{P}_1 * \mathcal{P}_2} d\mathcal{Q}_1 * \mathcal{Q}_2 = \int_\Omega \log \frac{d\mathcal{Q}_1}{d\mathcal{P}_1} \frac{d\mathcal{Q}_2}{d\mathcal{P}_2} d\mathcal{Q}_1 * \mathcal{Q}_2 = \int_\Omega \log |\psi_1|^2 d\mathcal{Q}_1 * \mathcal{Q}_2 + \int_\Omega \log |\psi_2|^2 d\mathcal{Q}_1 * \mathcal{Q}_2$. From this, we can clearly see $I(\mathcal{Q}|\mathcal{P}) \neq I(\mathcal{Q}_1|\mathcal{P}_1) + I(\mathcal{Q}_2|\mathcal{P}_2)$. ■

Proposition 35 ***Rényi's postulate 9 [5] and ambiguity of information.*** *If we have ambiguity of information, then $I(\mathcal{Q}|\mathcal{P}) = \int_\Omega \log |\psi|^2 d\mathcal{Q}_1 * \mathcal{Q}_2 \neq \int_\Omega \log \frac{d\mathcal{Q}_1}{d\mathcal{P}_1} d\mathcal{Q}_1 + \int_\Omega \log \frac{d\mathcal{Q}_2}{d\mathcal{P}_2} d\mathcal{Q}_2$.*

Proof. Obvious. Clearly, since $|\psi_1 * \psi_2|^2 = |\psi_1|^2 |\psi_2|^2$, we can only write $\int_\Omega \log |\psi|^2 d\mathcal{Q}_1 * \mathcal{Q}_2 = \int_\Omega \log |\psi_1|^2 d\mathcal{Q}_1 * \mathcal{Q}_2 + \int_\Omega \log |\psi_2|^2 d\mathcal{Q}_1 * \mathcal{Q}_2$. ■

Proposition 36 ***Rényi's postulate 9 [5] and multi-sourced information.*** *If we have multi-sourced information, then $I(\mathcal{Q}|\mathcal{P}) = \int_\Omega \log |\psi|^2 d\mathcal{Q}_1 * \mathcal{Q}_2 \neq \int_\Omega \log \frac{d\mathcal{Q}_1}{d\mathcal{P}_1} d\mathcal{Q}_1 + \int_\Omega \log \frac{d\mathcal{Q}_2}{d\mathcal{P}_2} d\mathcal{Q}_2$.*

Proof. Obvious. Since $\psi(q_1, q_2) = \psi_1(q_1)\psi_2(q_2)$, the left-hand side of the non-equality is integrating a two-dimensional function under $d\mathcal{Q}_1 * \mathcal{Q}_2$, while the right-hand side is integrating a one-dimensional function again towards $d\mathcal{Q}_1 * \mathcal{Q}_2$. ■

In conclusion, we have connected the information wave function with the concept of a Radon–Nikodym derivative. We have then tried to show how this derivative could be used in Rényi's quantity of information measure [5]. We also briefly alluded to how the derivative could be interpreted in the context of the Blackwell and Dubins theorem [2]. We also introduced the possible notions of ambiguity of information and multi-sourced information. Finally, we considered Rényi's postulate 9 [5] for the case when we considered an information wave function as the basis of the Radon–Nikodym derivative.

13.15 References

[1] Haven, E. (2010). The Blackwell and Dubins theorem and Rényi's amount of information measure: some applications. *Acta Applicandae Mathematicae*, 109, 3, 743–757.
[2] Blackwell, D. and Dubins, L. (1962). Merging of opinions with increasing information. *Annals of Mathematical Statistics*, 33, 882–886.
[3] Nakata, H. (2000). On the Dynamics of Endogenous Correlations of Beliefs. Ph.D. Thesis; Stanford University.
[4] Primbs, J. (2006). *Measure Theory in a Lecture.* http://www.stanford.edu/~japrimbs/MeasureTheory5.ppt. Cited September 28, 2006.
[5] Rényi, A. (1961). *On Measures of Entropy and Information.* In Neyman, J. (ed.) Proceedings of the Fourth Berkeley Symposium on Mathematical Statistics and Probability, California.
[6] Bohm, D. and Hiley, B. (2005). *The Undivided Universe: An Ontological Interpretation of Quantum Theory*. Routledge.

13.16 Universal Brownian motion: definition and discussion

13.16.1 Basic steps

The hypothesis of Universal Brownian motion was proposed by Edward Nelson. The key references for this important work can be found in Nelson [1] [2] [3]. Paul and Baschnagel [4] provide for an interesting overview of the essential steps in this hypothesis. Elbaz [5] also gives a good account of Nelson's stochastic mechanics.

Recall the Brownian motion (see Chapter 4, Section 4.18.2): $dx(t) = b(x, t)dt + \sigma dW(t)$, where $dW(t)$ is Wiener process. Nelson [1] (p. 1080) defines a mean forward derivative:

$$Dx(t) = \lim_{\Delta t \to 0} E\left[\frac{x(t + \Delta t) - x(t)}{\Delta t}\right]. \tag{13.5}$$

The mean backward derivative is then (Nelson [1] (p. 1080)):

$$D_* x(t) = \lim_{\Delta t \to 0} E\left[\frac{x(t) - x(t - \Delta t)}{\Delta t}\right]. \tag{13.6}$$

One can write, using the above Brownian motion, what Nelson calls [1] (p. 1081) "mean forward velocity" as:

$$Dx(t) = b(x, t). \tag{13.7}$$

Similarly (mean backward velocity):

$$D_* x(t) = b_*(x, t). \tag{13.8}$$

Nelson [1] (p. 1081) then defines a current velocity as:

$$v(x, t) = \frac{1}{2}(b(x, t) + b_*(x, t)). \tag{13.9}$$

As Paul and Baschnagel [4] remark (p. 84) for Newtonian motion, the mean forward and mean backward derivative will be equal, i.e. $D_* x(t) = Dx(t)$ and hence $v(x, t) = D_* x(t) = Dx(t) = b_*(x, t) = b(x, t)$.

The osmotic velocity is defined as (see Paul and Baschnagel [4] remark (p. 84)):

$$u(x, t) = \frac{1}{2}(D - D_*)x(t). \tag{13.10}$$

Recall the Fokker–Planck PDE we mentioned in Chapter 2, Section 2.3. Paul and Baschnagel [4] (p. 84) show (see also Nelson [1] (p. 1081)) that by using the Fokker–Planck PDE (using the Brownian motion above), one finds that:

$$u(x, t) = \frac{\sigma^2}{2} \nabla \ln p(x, t) = \frac{\sigma^2}{2} 1/p(x, t) \frac{\partial p}{\partial x}, \tag{13.11}$$

where p is a probability density function.

13.16.2 The Universal Brownian motion

Consider the drift in the Brownian motion $dx(t) = b(x, t)dt + \sigma dW(t)$. Paul and Baschnagel [4] (p. 86) indicate (see also Nelson [1] (p. 1083)) that, when using the polar form of the quantum mechanical wave function, one obtains:

$$b(x, t) = \frac{\hbar}{m} \nabla \ln R + \frac{1}{m} \nabla S, \tag{13.12}$$

where R is the amplitude function, S is the phase, m is mass, and $\hbar$ is the rationalized Planck constant. Substituting the above equation in to $dx(t) = b(x, t)dt + \sigma dW(t)$

yields:

$$dx(t) = \left(\frac{1}{m}\nabla S + \frac{\hbar}{2m}\frac{\nabla |\psi|^2}{|\psi|^2}\right) dt + dW(t). \tag{13.13}$$

We assume here also that the Wiener process has the constraints $\overline{dW} = 0$ and $\overline{(dW)^2} = \frac{\hbar}{m}dt$. Bacciagaluppi [6] also formulated a stochastic differential equation like the one above. We note that Bohm and Hiley [7] indicate that $\frac{\nabla|\psi|^2}{|\psi|^2}$ is osmotic velocity.

Wallstrom [8] is also an interesting source for more detail on the link between the Schrödinger equation and stochastic mechanics.

13.17 References

[1] Nelson, E. (1966). Derivation of the Schrödinger equation from Newtonian mechanics. *Physical Review*, 150, 1079–1085.

[2] Nelson, E. (1967). *Dynamical Theories of Brownian Motion*. Princeton University Press, NJ.

[3] Nelson, E. (2005). *The Mystery of Stochastic Mechanics*. http://www.math.princeton.edu/~nelson/papers/talk.pdf.

[4] Paul, W. and Baschnagel, J. (2000). *Stochastic Processes: From Physics to Finance*. Springer Verlag, Berlin.

[5] Elbaz, E. (2006). *The Quantum Theory of Particle, Fields, and Cosmology*. Springer Verlag, Berlin.

[6] Bacciagaluppi, G. (1999). Nelsonian mechanics revisited. *Foundations of Physics Letters*, 12, 1–16.

[7] Bohm, D. and Hiley, B. (1998). Non-locality and locality in the stochastic interpretation of quantum mechanics. *Physics Reports*, 172, 93–122.

[8] Wallstrom, T. C. (1989). On the derivation of the Schrödinger equation from stochastic mechanics. *Foundations of Physics Letters*, 2, 2, 113–126.

13.18 Universal Brownian motion and option pricing

13.18.1 Pricing of information risk

In Chapter 6, Section 6.3, equations (6.14) and (6.15) were the two key equations which emerged from putting in the polar form of the quantum mechanical wave function into the Schrödinger equation. In Ishio and Haven [1], it is argued that there is a connection between Bohmian mechanics and the Nelson approach. However, in Ishio and Haven [1] it is also argued that the stochastic model does not necessarily imply Bohmian mechanics.

Recall our discussion in Chapter 2, Section 2.7, where we set up the basics of Black–Scholes option pricing. The famous Black–Scholes partial differential equation was equation (2.12). We recall the equation again:

$$\frac{\partial F}{\partial t} + \frac{\sigma^2 S^2}{2} \frac{\partial^2 F}{\partial S^2} + rS\frac{\partial F}{\partial S} = rF, \tag{13.14}$$

where F is the option price, t is time, σ is the volatility, S is the stock price (not to be confused with the phase S of (13.13)!), r is the risk free rate of interest.

In Ishio and Haven [1], we propose to define a so-called "information risk parameter." We argued there that information carried a degree of erroneous information and the parameter we propose would measure this type of information. In that same paper, we proposed two Brownian motions, one which is (13.13) (but adapted to become a Brownian motion on the return, μ) and the other stochastic differential equation which is the Brownian motion we used in the original Black–Scholes set up, i.e. equation (2.4) (in Chapter 2, Section 2.7.3), i.e. $dS = \mu S dt + \sigma S dW$. In Ishio and Haven [1] (p. 40), the stochastic differential equation on μ is introduced as:

$$d\mu = \left(\nabla Ph(\mu) + \frac{1}{2} \frac{\nabla \left|\psi^1\right|^2}{\left|\psi^1\right|^2} \right) dt + dW_1, \tag{13.15}$$

where $\nabla Ph(\mu)$ is the gradient of the phase of the wave function, ψ^1, μ is the asset (or stock) return, and dW_1 is a Wiener process. As in Wilmott [2], a correlation is assumed to exist between dW_1 and the Wiener process dW, which is part of the (geometric) Brownian motion of the asset (or stock) price. Ishio and Haven [1] follow Wilmott's [2] (pp. 299–302) set up, but use equation (13.15) instead of the stochastic differential equation on volatility.

We reproduce[24] the paper by Ishio and Haven [1] (Section 4 (pp. 40–42)).

We now follow Wilmott's approach [2] (p. 300). Consider the option price F to be a function of the drift μ, the stock price S, and time t: $F(\mu, S, t)$. The infinitesimal change in the option price dF can be expanded as follows:

$$dF = \frac{\partial F}{\partial S}dS + \frac{\partial F}{\partial \mu}d\mu + \frac{\partial F}{\partial t}dt + \frac{1}{2}\frac{\partial^2 F}{\partial S^2}(dS)^2 + \frac{1}{2}\frac{\partial^2 F}{\partial \mu^2}(d\mu)^2 + \frac{\partial^2 F}{\partial \mu \partial S}dSd\mu. \tag{13.16}$$

In contrast to the Black–Scholes portfolio, which was $\Pi = -F + \frac{\partial F}{\partial S}S$, where $F(S, t)$ is the option price, the portfolio we have now is somewhat different (Wilmott [2]):

$$\Pi = F - \Delta S - \Delta_1 F_1. \tag{13.17}$$

[24] Ishio, H. and Haven, E. (2009). Information in asset pricing: a wave function approach. *Annalen der Physik*, 18, 1, 40–42.

We remark that F and F_1 are two different option prices. In full analogy with Wilmott [2], the two option prices F and F_1 are needed in this set up since one needs to hedge against two sources of randomness: (i) stock price randomness (as in Black–Scholes) (as expressed by the geometric Brownian motion on the stock price) and now also (ii) information risk randomness (as expressed by (13.15)). We are ultimately interested in finding the functional form of F, which can be found by solving the PDE in (13.24) below. The variables Δ and Δ_1 indicate, respectively, the quantities of stock and options which are needed to make sure (as in Black–Scholes) that the portfolio change per unit of time is risk free. At this stage of the derivation, the quantities Δ and Δ_1 are open. However, as we will see following from (13.20) below, those two quantities will be fully determined.

As is the case in the Black–Scholes derivation, the infinitesimal change in the value of the portfolio given by (13.17) can be written as:

$$d\Pi = dF - \Delta dS - \Delta_1 dF_1, \tag{13.18}$$

where dF (and dF_1) is given by (13.16).[25] Remark that dS is given by the geometric Brownian motion on stock.

Substituting those expressions in (13.18) above, one obtains:

$$\begin{aligned} d\Pi = &\left(\frac{\partial F}{\partial S} - \Delta_1 \frac{\partial F_1}{\partial S} - \Delta\right) dS + \left(\frac{\partial F}{\partial \mu} - \Delta_1 \frac{\partial F_1}{\partial \mu}\right) d\mu \\ &+ \left(\frac{\partial F}{\partial t} + \frac{1}{2}\sigma^2 S^2 \frac{\partial^2 F}{\partial S^2} + \rho\sigma S \frac{\partial^2 F}{\partial S \partial \mu} + \frac{1}{2}\frac{\partial^2 F}{\partial \mu^2}\right) dt \\ &- \Delta_1 \left(\frac{\partial F_1}{\partial t} + \frac{1}{2}\sigma^2 S^2 \frac{\partial^2 F_1}{\partial S^2} + \rho\sigma S \frac{\partial^2 F_1}{\partial S \partial \mu} + \frac{1}{2}\frac{\partial^2 F_1}{\partial \mu^2}\right) dt. \end{aligned} \tag{13.19}$$

In the Black–Scholes set up, we had one source of risk given by the unique Wiener process, dW. Now, we have two sources of risk, dW (the dS term of (13.19)) and dW_1 (the $d\mu$ term of (13.19)), which are the Wiener processes corresponding respectively to the geometric Brownian motion on stock and (13.15). To eliminate those two sources of risk, one now sets:

$$\left(\frac{\partial F}{\partial S} - \Delta_1 \frac{\partial F_1}{\partial S} - \Delta\right) = \left(\frac{\partial F}{\partial \mu} - \Delta_1 \frac{\partial F_1}{\partial \mu}\right) = 0. \tag{13.20}$$

Clearly, from (13.20) we find immediately that Δ and Δ_1 are now defined:

$$\Delta_1 = \frac{\partial F}{\partial \mu} \Big/ \frac{\partial F_1}{\partial \mu} \tag{13.21}$$

[25] We note that to obtain dF_1, F is to be replaced by F_1 in (13.16).

and

$$\Delta = \frac{\partial F}{\partial S} - \left(\frac{\partial F}{\partial \mu} / \frac{\partial F_1}{\partial \mu} \right) \frac{\partial F_1}{\partial S}. \tag{13.22}$$

At this stage of the derivation, one can invoke the same important arguments which were used in the Black–Scholes derivation (see Chapter 2, Section 2.7.3): (i) the sources of risk (dW and dW_1) are eliminated by imposing (13.20), and (ii) we assume there is no arbitrage. Given (i) and (ii) we can write again that $d\Pi = r\Pi dt$. Thus:

$$\begin{aligned} d\Pi &= \left(\frac{\partial F}{\partial t} + \frac{1}{2}\sigma^2 S^2 \frac{\partial^2 F}{\partial S^2} + \rho\sigma S \frac{\partial^2 F}{\partial S \partial \mu} + \frac{1}{2}\frac{\partial^2 F}{\partial \mu^2} \right) dt \\ &- \Delta_1 \left(\frac{\partial F_1}{\partial t} + \frac{1}{2}\sigma^2 S^2 \frac{\partial^2 F_1}{\partial S^2} + \rho\sigma S \frac{\partial^2 F_1}{\partial S \partial \mu} + \frac{1}{2}\frac{\partial^2 F_1}{\partial \mu^2} \right) dt \\ &= r(F - \Delta S - \Delta_1 F_1) dt = r\Pi dt. \end{aligned} \tag{13.23}$$

Then the final PDE becomes [2] (p. 301):

$$\begin{aligned} &\frac{\partial F}{\partial t} + \frac{1}{2}\sigma^2 S^2 \frac{\partial^2 F}{\partial S^2} + \rho\sigma S \frac{\partial^2 F}{\partial S \partial \mu} + \frac{1}{2}\frac{\partial^2 F}{\partial \mu^2} + rS\frac{\partial F}{\partial S} \\ &+ \left(\nabla Ph(\mu) + \frac{1}{2}\frac{\nabla \left|\psi^1\right|^2}{\left|\psi^1\right|^2} - \lambda \right) \frac{\partial F}{\partial \mu} \\ &= rF, \end{aligned} \tag{13.24}$$

where we could claim that λ is the market price of **information risk**. In Wilmott [2] (p. 301), this λ, within the stochastic volatility model, is called the market price of volatility risk. We now claim that dW_1 is a source of information risk. We remark that the phase Ph and amplitude $\left|\psi^1\right|$ of the wave function, which appear in the above PDE, will have a functional form which will have been determined within the Bohmian mechanics setting. As was pointed out at the beginning of this section, those functional forms carry over only in one direction: from Bohmian mechanics to the Universal Brownian motion, but not in the opposite direction.

We write, for the case of information risk, in complete analogy with Wilmott [2] (p. 302):

$$d\Pi - r\Pi dt = \frac{\partial F}{\partial \mu}(\lambda dt + dW_1). \tag{13.25}$$

The right-hand side of the above equation should vanish in Black–Scholes because of $d\Pi = r\Pi dt$.

13.19 References

[1] Ishio, H. and Haven, E. (2009). Information in asset pricing: a wave function approach. *Annalen der Physik*, 18, 1, 33–44.
[2] Wilmott, P. (1999). *Derivatives: The Theory and Practice of Financial Engineering*. J. Wiley, Chichester.

13.20 Wave functions in drift-dependent option pricing

In Chapter 2, Section 2.7.3, we briefly mentioned an alternative way of deriving the option pricing formula by Bouchaud [1], Bouchaud and Potters [2], Bouchaud and Sornette [3], Schweizer [4], and Schäl [5]. Bouchaud [1] (p. 241) writes the so-called "global wealth balance" as $\Delta W = ce^{rT} - \max(x(T) - x_s, 0) + \sum_i \phi(x_i, t_i) \exp(r(T - t_i)) [x_{i+1} - x_i - rx_i\tau]$, with c the price of a call, $x(T)$ the underlying asset price at maturity time T, x_s the exercise price, r the interest rate, $\phi(x_i, t_i)$ (as per Bouchaud [1] (p. 241)) "the trading strategy, i.e. the number of stocks per option in the portfolio of the option writer,"$x_i = x(t_i)$ the underlying asset price, and τ is what Bouchaud [1] (p. 241) terms the "elementary time interval."

The first two terms are self explanatory in the above equation. The price of the contract (c) is what the writer of the contract receives and is used to buy bonds with interest r. The term with the summation indicates what Bouchaud [1] (p. 241) calls "the profit or loss . . . incurred due to the trading strategy ϕ."

When imposing the condition $E(\Delta W) = 0$ and a so called "risk minimization condition"[26] (see Bouchaud and Sornette [3], Schweizer [4], and Schäl [5]); the call value is then found to be: $\int_{x_s}^{\infty} (x' - x_s) P(x', T|x_0, 0) dx'$, with P a conditional probability.

Tan [6] introduces the following condition:

$$E\left(\int \phi(x,t)\frac{dx}{dt}dt\right) = \mu x_0 \int_0^T \int_0^\infty \phi^*(x,t) P(x_i, t; \mu|x_0, 0) dx dt, \qquad (13.26)$$

where now $P(x_i, t; \mu|x_0, 0)$, is in the words of Tan [6] (p. 114) "the biased (μ-dependent) distribution of the stock." Note that $\phi^*(x,t) = \int_{x_s}^{\infty} (x(T) - x_s) \frac{\partial P}{\partial x} dx(T)$. Tan [6] then shows (p. 116) that, when using (13.26), the call option price can be rewritten in the following form:

$$c_{new} = c_{Black} - \frac{\mu x_0 T}{\sigma^2 x_0^2} \sum_{n=3}^{\infty} \frac{a_n}{(n-1)!} \frac{\partial^{n-3} P(x_s, T|x_0, 0)}{\partial x_0^{n-3}}, \qquad (13.27)$$

[26] This condition is essentially requiring that the first partial derivative of ΔW towards ϕ is zero.

where $a_n = 0$ for all $n \geq 3$ when the Gaussian probability density is used. Hence, in that case one recuperates c_{Black}, which is the Black–Scholes option price.

Equation (13.27) is a highly interesting result. In Haven [7], we substitute the drift term we imposed in equation (13.15) (see Section 13.18 of this chapter) into the above correction term to the Black–Scholes call price. This term then reads as:

$$-\frac{\left(\nabla Ph + \frac{1}{2}\frac{\nabla|\psi|^2}{|\psi|^2}\right) S_0 T}{\sigma^2 S_0^2} \sum_{n=3}^{\infty} \frac{a_n}{(n-1)!} \frac{\partial^{n-3} P(X, T|S_0, 0)}{\partial S_0^{n-3}}. \qquad (13.28)$$

13.21 References

[1] Bouchaud, J. P. (2002). An introduction to statistical finance. *Physica A*, 313, 238–251.

[2] Bouchaud, J. P. and Potters, M. (2000). *Theory of Financial Risks*. Cambridge University Press.

[3] Bouchaud, J. P. and Sornette, D. (1994). The Black–Scholes option pricing problem in mathematical finance: generalization and extensions for a large class of stochastic processes. *Journal de Physique I*, 4, 863–881.

[4] Schweizer, M. (1994b). Approximating random variables by stochastic integrals. *Annals of Probability*, 22, 1536–1575.

[5] Schäl, M. (1994). On quadratic cost criteria for option hedging. *Mathematics of Operations Research*, 19, 121–131.

[6] Tan, A. (2005). Long memory stochastic volatility and a risk minimisation approach for derivative pricing and hedging. Doctoral Thesis, School of Mathematics, Faculty of Engineering and Physical Sciences, University of Manchester (UK).

[7] Haven, E. (2009). The use of the information wave function in a drift dependent option price: a simple example. *Foundations of Probability and Physics*, 5, Sweden, 1101, 74–77. American Institute of Physics Conference Proceedings.

13.22 Generalizations of Itô stochastics: path integration and other tools

Path integration was developed by Feynman. It has very wide applications in science, but has only trickled-in, in very narrow ways, into social science (mostly in mathematical finance).

Baaquie [1] in a recent paper which appeared in *Physical Review E* proposes the use of quantum fields in the establishment of a theory of forward rates. Any advanced undergraduate and graduate student in finance will have heard of the so-called HJM theory [2] of forward interest rates which was developed by Heath, Jarrow, and Morton. In the context of LIBOR markets, the Baaquie approach generalizes white noise, a characteristic of stochastics, to what Baaquie [1] (p. 046119-4) calls a "two dimensional quantum field." The key formulation is given by equation (9) in Baaquie's paper [1] (p. 046119-4): $\frac{\partial f(x,t)}{\partial t} = \mu(x,t) + v(t,x)\mathcal{A}(t,x)$, where $f(t,x)$ is the LIBOR forward interest rate on position x and time t, $v(t,x)$ is the volatility, and $\mathcal{A}(t, x)$ is the two-dimensional quantum field. In Baaquie [1] (p. 046119-4), $f(t, x)$ is built up as $f(t, x) = f(t_0, x) + \int_{t_0}^{t} \mu(t, x)dt + \int_{t_0}^{t} v(t, x)\mathcal{A}(t, x)dt$ and the quantum field dynamics are given by what Baaquie [1] (p. 046119-5) calls a "stiff Lagrangian." The so-called Wilson expansion [3] in that paper is central because it allows for considering the product of non-Gaussian quantum fields. Furthermore, this very same expansion for Gaussian quantum fields also provides for a generalization of Itô calculus.

13.23 References

[1] Baaquie, B. (2009). Interest rates in quantum finance: the Wilson expansion. *Physical Review E*, 80, 046119.

[2] Heath, D., Jarrow, R., and Morton, A. (1992). Bond pricing and the term structure of interest rates: a new methodology for contingent claims valuation. *Econometrica*, 60, 1, 77–105.

[3] Wilson, K. G. (1969). Non-Lagrangian models in current algebra. *Physical Review*, 179, 1499–1512.

13.24 q-calculus and finance

q-calculus is also known under the name of "quantum calculus." Key references to this type of calculus can be found in Kac and Cheung [1] and Andrews [2] [3]. The Heisenberg uncertainty principle is to be slightly adapted when such calculus is used. We would also like to show how this type of calculus can be connected to some elementary financial principles. We apply it here to option pricing. We now reproduce[27] the paper by Haven [4] (Sections 2–4 (pp. 530–536)).

13.24.1 q-derivatives, Hilger delta time derivatives and links between them

For definitions 37, 38, 39, and 40, below,[28] we follow Kac and Cheung [1]. We also use their notation.

Definition 37 *For an arbitrary function $f(x)$, the q-differential is $d_q f(x) = f(qx) - f(x)$.*

Definition 38 *For an arbitrary function $f(x)$, the h-differential is $d_h f(x) = f(x + h) - f(x)$.*

[27] Haven, E. (2010). Itô's Lemma with quantum calculus: some implications, *Foundations of Physics*, 41, 3, 530–536.

[28] The definitions in this section are also given in Haven [5] except for definitions 42, 43, 44, and 45, and Theorem 46.

The relationship between q-calculus and h-calculus follows via either $q = e^h$ or $q = e^{ih}$, where $i \in \mathbb{C}$. We also have the following definitions.

Definition 39 *The q-derivative of a function $f(x)$ is* $D_q f(x) = \frac{d_q f(x)}{d_q(x)} = \frac{f(qx)-f(x)}{qx-x}$.

Definition 40 *The h-derivative of a function $f(x)$ is* $D_h f(x) = \frac{d_h f(x)}{d_h(x)} = \frac{f(x+h)-f(x)}{(x+h)-x}$.

Note that $\lim_{q\to 1} D_q f(x) = \lim_{h\to 0} D_h f(x) = \frac{df(x)}{dx}$.

What is the q-derivative of a function $f(x) = x^n$? Since $D_q f(x) = \frac{x^n(q^n-1)}{x(q-1)}$ and $\frac{q^n-1}{q-1} = q^{n-1} + \cdots + 1 = [n]$ $(n \in \mathbb{Z}_0^+)$ as in Kac and Cheung [1].

The so-called Hilger delta time derivative follows the calculus of Hilger [6] [7]. See also Bohner and Peterson [8] [9] and Davis *et al.* [10] for excellent sources. In order to define this derivative, we need to provide for some background first. Please see for instance Hilger [7] or also Bohner and Peterson [8]. In Definitions 41, 42, 43, 44, and 45, and Theorem 46 below, we follow Bohner and Peterson [8].

Let us first define the most basic object: a time scale $\mathbb{T}$.

Definition 41 *A time scale $\mathbb{T}$ is an arbitrary non-empty closed subset of $\mathbb{R}$.*

As Bohner and Peterson [8] indicate, the sets $]0, 1[$, $\mathbb{Q}$, $\mathbb{R}\backslash\mathbb{Q}$, and $\mathbb{C}$ do contradict Definition 41 and hence are not time scales.

Definition 42 *For $t \in \mathbb{T}$, the forward jump operator $\eta : \mathbb{T} \to \mathbb{T}$ is defined as* $\eta(t) \equiv \inf\{s \in \mathbb{T} : s > t\}$.

Definition 43 *For $t \in \mathbb{T}$, the backward jump operator $\lambda : \mathbb{T} \to \mathbb{T}$ is defined as* $\lambda(t) \equiv \sup\{s \in \mathbb{T} : s < t\}$.

Definition 44 *The graininess function $\theta : \mathbb{T} \to [0, \infty[$ is defined as $\theta(t) \equiv \eta(t) - t$.*

Definition 45 *If $t < \sup \mathbb{T}$ and $\eta(t) = t$, then t is right dense. If $t > \inf \mathbb{T}$ and $\lambda(t) = t$, then t is left dense.*

Still following Bohner and Peterson [8], t is said to be right-scattered, when $\eta(t) > t$. When $\lambda(t) < t$, then t is said to be left-scattered. We have the usual derivative when $\mathbb{T} = \mathbb{R}$. As an example, the usual time derivative, $\lim_{t\to s} \frac{f(t)-f(s)}{t-s}$ is obtained when $\mathbb{T} = \mathbb{R}$. When $\mathbb{T} = \mathbb{Z}$, one uses the forward difference operator $f(t+1) - f(t)$. When $\mathbb{T} = \mathbb{R}$, then for all $t \in \mathbb{T}$, $\theta(t) = 0$. When $\mathbb{T} = \mathbb{Z}$, then for all $t \in \mathbb{T}$, $\theta(t) = 1$.

An interesting question is now: "How can one define a 'derivative' when $\mathbb{T}$ is arbitrary?" Theorem 1.16 in Bohner and Peterson [8] provides for an answer. Before we proceed to enumerate part of that theorem, we need the definition of $\mathbb{T}^\kappa$. If $\sup \mathbb{T} < \infty$, then $\mathbb{T}^\kappa = \mathbb{T} \setminus\,]\lambda(\sup \mathbb{T}), \sup \mathbb{T}]$. If $\sup \mathbb{T} = \infty$, then $\mathbb{T}^\kappa = \mathbb{T}$.

Theorem 46 *Bohner and Peterson [8] (p. 13). Assume the function* $f : \mathbb{T} \rightarrow \mathbb{R}$ *and let* $t \in \mathbb{T}^{\kappa}$. *If* f *is continuous at* t *and* t *is right-scattered, then* f *is differentiable at* t *with* $f^{\Delta}(t) = \frac{f(\eta(t))-f(t)}{\eta(t)-t}$, *where for* $t \in \mathbb{T}$, $\eta(t) \equiv \inf\{s \in \mathbb{T} : s > t\}$.

We call $f^{\Delta}(t)$ the Hilger delta time derivative. The link between $D_q f(x)$ and $f^{\Delta}(t)$ is quite straightforward: if one sets $\eta(t) = qt$ in $f^{\Delta}(t)$, then a q-derivative in t is obtained. We need the graininess $\theta(t) = \eta(t) - t = qt - t > 0$ for t to be right-scattered. For $\mathbb{T} = \mathbb{Z}$, we know $\mathbb{T}^{\kappa} = \mathbb{T}$. We require that $qt - t = 1$, where $q = \frac{1}{t} + 1$. For the case when $\mathbb{T} = \mathbb{R}$, we have $\mathbb{T}^{\kappa} = \mathbb{T}$ and $\theta(t) = \eta(t) - t = qt - t = 0$. Therefore, $q = 1$ in that case.

13.24.2 How can we formulate Itô's Lemma with q-derivatives?

Itô's Lemma [11] is a crucial mathematical result, which allows for the effective use of stochastic calculus in financial derivative (or also option) pricing. The title of this sub-section suggests investigating whether q-derivatives could be used in Itô's Lemma. Hence, we need to have a closer look at the proof of Itô's Lemma to check the feasibility of this query. We follow Øksendal's [12] treatment of the proof (in one dimension) of Itô's Lemma. Itô's Lemma can be expressed in integral form or in differential notation. Assume that X_t is a stochastic integral[29] $dX_t = udt + vdB_t$, where B_t is a one-dimensional Brownian motion, u and v are respectively drift and diffusion factors, and t is time. Let a function $g(t, x)$ be twice continuously differentiable on the product space $[0, \infty[\times\mathbb{R}$, then $g(t, X_t)$ is also a stochastic integral and Itô's Lemma indicates that:[30]

$$g(t, X_t) - g(0, X_0) = \int_0^t \left(\frac{\partial g}{\partial s} + u\frac{\partial g}{\partial x} + \frac{1}{2}v^2\frac{\partial^2 g}{\partial x^2} \right) ds + \int_0^t v\frac{\partial g}{\partial x} dB_s. \quad (13.29)$$

It can be assumed, as is indicated in Øksendal [12], that g, $\frac{\partial g}{\partial t}$, $\frac{\partial g}{\partial x}$, and $\frac{\partial^2 g}{\partial x^2}$ are bounded. But the Lemma can also be proven without this restriction. See, for instance, Brzeźniak and Zastawniak [13].

While still following Øksendal [12], Taylor's theorem now reads as:

$$\begin{aligned} g(t, X_t) - g(0, X_0) = & \sum_j \frac{\partial g}{\partial t}\Delta t_j + \sum_j \frac{\partial g}{\partial x}\Delta X_j + \frac{1}{2}\sum_j \frac{\partial^2 g}{\partial t^2}(\Delta t_j)^2 \\ & + \sum_j \frac{\partial^2 g}{\partial t \partial x}(\Delta t_j)(\Delta X_j) + \frac{1}{2}\sum_j \frac{\partial^2 g}{\partial x^2}(\Delta X_j)^2 + \sum_j R_j, \end{aligned} \quad (13.30)$$

[29] Please see Theorem 4.2, p. 33 in Øksendal [12].

[30] Please see Theorem 4.5, p. 35 in Øksendal [12].

where $\Delta t_j = t_{j+1} - t_j$; $\Delta X_j = X_{t_{j+1}} - X_{t_j}$ and $R_j = o\left(\left|\Delta t_j\right|^2 + \left|\Delta X_j\right|^2\right)$, $\forall j$. The partial derivatives are evaluated at $\left(t_j, X_{t_j}\right)$.

Øksendal [12] proposes, so as to prove Itô's Lemma, that the following evaluations are made. Each term in (13.30) is evaluated as follows:

(a) If $\Delta t_j \to 0$, then $\sum_j \frac{\partial g}{\partial t} \Delta t_j \to \int_0^t \frac{\partial g}{\partial s}(s, X_s)ds$.

(b) If $\Delta t_j \to 0$, then $\sum_j \frac{\partial g}{\partial x} \Delta X_j \to \int_0^t \frac{\partial g}{\partial x}(s, X_s)dX_s$.

This means the two first-order terms will remain in the Taylor expansion. However, the situation is different for some of the second-order terms.

(c) If $\Delta t_j \to 0$, then in $\sum_j \frac{\partial^2 g}{\partial x^2}(\Delta X_j)^2 = \sum_j \frac{\partial^2 g}{\partial x^2} u_j^2 (\Delta t_j)^2 + 2\sum_j \frac{\partial^2 g}{\partial x^2} u_j v_j (\Delta t_j) \times (\Delta B_j) + \sum_j \frac{\partial^2 g}{\partial x^2} v_j^2 (\Delta B_j)^2$ the first two terms are zero. As an example, Øksendal [12], shows, the limit $\Delta t_j \to 0$ in the mean square sense of $\sum_j \frac{\partial^2 g}{\partial x^2} u_j v_j \times (\Delta t_j)(\Delta B_j)$, is $\lim_{\Delta t_j \to 0} E\left[\left(\sum_j \frac{\partial^2 g}{\partial x^2} u_j v_j (\Delta t_j)(\Delta B_j)\right)^2\right] = \lim_{\Delta t_j \to 0} \sum_j E\left[\left(\frac{\partial^2 g}{\partial x^2} u_j v_j\right)^2 (\Delta t_j)^3\right] = 0$.

The situation is different with the last term in (c). In effect, Øksendal [12] shows that $\sum_j \frac{\partial^2 g}{\partial x^2} v_j^2 \left(\Delta B_j\right)^2$, when $\Delta t_j \to 0$, tends to $\int_0^t \frac{\partial^2 g}{\partial x^2} v^2 ds$.

(d) The other terms in (13.30), $\sum_j \frac{\partial^2 g}{\partial t \partial x}\left(\Delta t_j\right)\left(\Delta X_j\right)$ and $\frac{1}{2}\sum_j \frac{\partial^2 g}{\partial t^2}\left(\Delta t_j\right)^2$, can equally be shown to vanish.

(e) The remainder term $\sum_j R_j^2$ can also be shown to tend to zero, when $\Delta t_j \to 0$.

To stay faithful to the title of this section, we now query how (a), (b), (c), (d), and (e) behave if we consider q-derivatives. A key remark can be as follows. Let us denote q-derivatives as $\frac{\partial_q \cdot}{\partial_q \cdot}$. If we consider (a), then we can write:

$$\lim_{\Delta t_j \to 0} \sum_j \frac{\partial_q g}{\partial_q t} \Delta t_j \neq \frac{\partial_q g}{\partial_q t} \lim_{\Delta t_j \to 0} \sum_j \Delta t_j. \tag{13.31}$$

This non-equality will appear for the following reason.[31] Let us use the q-derivative definition on time $f^{\Delta}(t) = \frac{f(\eta(t)) - f(t)}{\eta(t) - t}$, where for $t \in \mathbb{T}$, $\eta(t) \equiv$

[31] Of course, $\lim_{\Delta t_j \to 0} \sum_j \frac{\partial g}{\partial t} \Delta t_j = \frac{\partial g}{\partial t} \lim_{\Delta t_j \to 0} \sum_j \Delta t_j$.

$\inf\{s \in \mathbb{T} : s > t\}$ and let us set $\eta(t) = qt$. Set $qt = t_2$ and $t = t_1$ and $t_1 < t_2$ (therefore $q > 0$, which is fine when we assume that $q = \exp(h)$). Let $\Delta t_j = t_2 - t_1$. Thus, $\lim_{\Delta t_j \to 0} \sum_j \frac{\partial_q g}{\partial_q t} \Delta t_j = \lim_{\Delta t_j \to 0} \sum_j \frac{g(t_2)-g(t_1)}{t_2-t_1} \Delta t_j$, but $\lim_{\Delta t_j \to 0} \frac{g(t_2)-g(t_1)}{t_2-t_1} = \frac{\partial g}{\partial t} \neq \frac{\partial_q g}{\partial_q t}$. This then means that, if we apply the delta time limit as in (a) on a q-derivative on time, we plainly cannot keep the q-derivative on time. The q-derivative converts itself to a regular derivative. So we cannot keep the q-derivative on time in (a). We must use an ordinary time derivative. This argument does not repeat itself for (b), where the delta time limit is applied to a position derivative. Hence, we can write in (b) that $\lim_{\Delta t_j \to 0} \sum_j \frac{\partial_q g}{\partial_q x} \Delta X_j \to \int_0^t \frac{\partial_q g}{\partial_q x}(s, X_s) dX_s$. The same argument goes for (c). In (d) and (e), we must again be prudent to use regular derivatives and no q-derivatives.

In non-summation form, we can write (13.30), with the blend of ordinary derivatives on time and the q-derivatives on position, simply as follows:

$$\Delta g = \frac{\partial_q g}{\partial_q X} \Delta X_j + \frac{\partial g}{\partial t} \Delta t_j + \frac{1}{2} \frac{\partial_q^2 g}{\partial_q X^2} \left(\Delta X_j^2\right) + \frac{1}{2} \frac{\partial^2 g}{\partial t^2} \left(\Delta t_j\right)^2 + \frac{\partial^2 g}{\partial X \partial t} (\Delta t_j)(\Delta X_j) + R_j. \tag{13.32}$$

If we consider $\Delta X_j = u_j \Delta t_j + v_j \Delta B_j$, then using (a), (b), (c), (d), and (e) from above, we can write:

$$\Delta_q g = \frac{\partial_q g}{\partial_q X} \left(u_j \Delta t_j + v_j \Delta B_j\right) + \frac{\partial g}{\partial t} (\Delta t_j) + \frac{1}{2} \frac{\partial_q^2 g}{\partial_q X^2} v_j^2 \Delta t_j. \tag{13.33}$$

Note the mixed use of derivatives in this expansion.

13.24.3 Financial interpretation of q

In Haven [5], we provide for an option pricing application using the Itô's Lemma with mixed q-derivatives.[32] We summarize this application as follows. Before we do so, let us consider an example.

Example 47 *Consider a small generalization of $dX_t = udt + vdB_t$ which is $dX = u(X, t)dt + v(X, t)dB$. Consider a stock price, S. Let $X = S$ and $u(X, t) = \mu S$, where μ is a rate of return on the stock. Let $v(X, t) = \sigma S$, where σ is a constant stock price volatility. Let $g(S) = S^n$. We want to express*

[32] Haven [5] does not show why we can only have mixed derivatives in Ito's Lemma. We attempt to show this in this section.

$dg(S) = dS^n$ when we use q-derivatives. We use therefore (13.33) in differential form. We obtain $dS^n = \left(\frac{d_q g(S)}{d_q(S)}\mu S + \frac{1}{2}\frac{d_q^2 g(S)}{d_q(S)^2}\sigma^2 S^2\right)dt + \frac{d_q g(S)}{d_q(S)}\sigma S dB = \left([n]\mu S^n + \frac{1}{2}[n][n-1]\sigma^2 S^n\right)dt + [n]\sigma S^n dB$.

This takes on a quite complicated form if we replace $[n]$. We get: $dS^n = \left((1+q+..q^{n-1})\mu S^n + \frac{1}{2}(1+q+..q^{n-1})(1+q+..q^{n-2})\sigma^2 S^n\right)dt + (1+q+ ..q^{n-1})\sigma S^n dB$.

A financial option is a contract entitling the buyer of such contract to either have the right to buy or sell an asset (from the seller of that contract) at a certain price at a certain date in the future. The value at maturity of a "call" option is $\max\{S-K, 0\}$, where K indicates the price at which the holder of a call option has the right to buy the underlying asset (a stock with price S in this case). The "put" option has as intrinsic value: $\max\{K-S, 0\}$. K indicates the price at which the holder of a put option has the right to sell the underlying asset (a stock with price S in this case). See Wilmott [14]. See also Baaquie [15]. Black–Scholes [16] option pricing theory yields a partial differential equation, which when solved yields the value of a (European) call and put contract.

As in Haven [5], we consider an option with intrinsic value, $\max\{S^2-K, 0\}$ for the call and $\max\{K-S^2, 0\}$ for the put.[33] Consider, as in example 47, $g(S) = S^n$, but now for $n = 2$. The expression, with q-derivatives, for dS^2 is then $dS^2 = \left([2]\mu S^2 + \frac{1}{2}[2]\sigma^2 S^2\right)dt + [2]\sigma S^2 dB$. The infinitesimal change of the value of the option price function, $df(S^2, t)$ can be shown to be equal to:

$$df(S^2,t) = \frac{\partial_q f}{\partial_q S^2}\left(\left([2]\mu S^2 + \frac{1}{2}[2]\sigma^2 S^2\right)dt + [2]\sigma S^2 dB\right) + \frac{\partial f}{\partial t}dt + \frac{1}{2}\frac{\partial_q^2 f}{\partial_q S^{2^2}}[2]^2\sigma^2 S^4 dt. \tag{13.34}$$

Note that we made use of $(dB)^2 = dt$. The above equation (13.34) simply uses the result (13.33). The traditional Black–Scholes [16] set up defines the value of a portfolio Π which consists of the buying and selling of two positions: an option and $\frac{\partial f}{\partial S}$ units of a stock. In analogy with the Black–Scholes [16] set up, the infinitesimal change in the value of this portfolio, $d\Pi$, is defined as $d\Pi = -df + \frac{\partial_q f}{\partial_q S^2}dS^2$. Substituting (13.34) and the expression for dS^2 in $d\Pi$, one obtains:

$$\frac{d\Pi}{dt} = -\frac{\partial f}{\partial t} - \frac{1}{2}\frac{\partial_q^2 f}{\partial_q S^{2^2}}[1+q]^2\sigma^2 S^4. \tag{13.35}$$

Note that we replaced $[2] = 1 + q$, using the notation of Kac and Cheung [1].

[33] McDonald [18] provides for a discussion, albeit without reference to q-derivatives, on claims other than S.

13.24.4 Non-use of the macroscopic q-derivative version of the Heisenberg uncertainty principle

A main argument consists now in debating whether equation (13.35) is reasonable when q-derivatives are used. We make a different argument from Haven [5]. We note that the term $\frac{\partial_q^2 f}{\partial_q S^{2^2}}$ is indicative of the so-called "hedging error," or also the "gamma" of the option. This term is thus nothing else other than a measure of the curvature of the option pricing function, f. This measure of the hedging error is becoming more imprecise when using $q \neq 1$, since no limits are being used. But what interpretation could we give to the imprecision of the curvature measurement? What does this mean? We could claim that the value of q indicates uncertainty as to the precision of f to measure the option price. Under the use of $\frac{\partial^2 f}{\partial S^{2^2}}$, we effectively advance that the curvature of the option pricing function is precisely measurable. Under the use of $\frac{\partial_q^2 f}{\partial_q S^{2^2}}$, we effectively advance that the measurement of the curvature is not precisely measurable, and this could be *equivalently translated* into saying that there is *uncertainty* as to the precision of the option pricing function, f. This could lead to non-uniqueness of option prices.

On prima facie, this imprecision in measurement of f, given by the parameter q, could also be possibly quantified in some sort of macroscopic Heisenberg uncertainty principle. The q-derivative version of the Heisenberg uncertainty principle, at the quantum level, can be written as:

$$P.Q - q.Q.P = -i\hbar, \tag{13.36}$$

where P and Q are respectively the position and momentum operators. See Baez [17]. See also Haven [5], where we discuss the q-derivatives version of the macroscopic Heisenberg uncertainty principle in different terms. At the macroscopic level (with no operators thus), we could write:

$$\Delta f.\Delta\frac{d_q f}{d_q t} - q.\Delta\frac{d_q f}{d_q t}.\Delta f = -k, \tag{13.37}$$

where Δf is the uncertainty on the measurement of the option price f and $\Delta\frac{d_q f}{d_q t}$ measures the uncertainty on the momentum of the option price. The parameter $k > 0$ would be some macroscopic "equivalent" of the Planck constant. We note that a macroscopic version of the Heisenberg uncertainty principle (albeit not in q-derivative form) is discussed, in a stock price context, by Baaquie [15]. For related work, see also Segal and Segal [19] and Accardi and Boukas [20].

We can claim that for $q = 1$ (in (13.37)), we indeed would have no macroscopic uncertainty principle: there would be no imprecision as to the measurement of the option pricing function (or the option price momentum for that matter). This conclusion could tie in with the argument we made above that for $q = 1$ there would be no uncertainty as to the imprecision of the option pricing function.

It is important to observe that from the above development, we found that we could not use q-derivatives on time in the context of the application of Itô's Lemma. Clearly, in (13.37) we must use such derivative, $\frac{d_q f}{d_q t}$. Therefore, we can logically conclude that in the $q = \exp(k)$ world, the uncertainty argument we tied to the existence of $\frac{\partial_q^2 f}{\partial_q S^{2^2}}$ is still valid, but the non-commutativity, as captured by (13.37), would not hold. The specific use of q-derivatives in Itô's Lemma prohibits us from doing this.

In conclusion, if we assume there should exist a macroscopic uncertainty relation which exhibits non-commutativity at the level of option pricing, then is the q-derivative world the correct generalization for financial option pricing? Would we have to work instead with a Langevin-based approach on option pricing, where the $dt \to 0$ is not an issue?

13.25 References

[1] Kac, V. and Cheung, P. (2002). *Quantum Calculus*. Springer Verlag.

[2] Andrews, G. E., Askey, R., and Roy, R. (1999). *Special Functions*. Cambridge University Press.

[3] Andrews, G. E. (1986). *q-Series: Their Development and Application in Analysis, Number Theory, Combinatorics, Physics and Computer Algebra*. In *CBMS Regional Conference Lecture Series in Mathematics*, 66, American Mathematical Society.

[4] Haven, E. (2010). Itô's Lemma with quantum calculus: some implications. *Foundations of Physics*, 41, 3, 529–537.

[5] Haven, E. (2009). *Quantum calculus (q-calculus) and option pricing: a brief introduction*. In *Quantum Interaction*. Eds. Bruza, P., Sofge, D., Lawless, W., and van Rijsbergen, K., *Lecture Notes in Artificial Intelligence* (LNAI), 5494, 308–314, Springer-Verlag.

[6] Hilger, S. (1988). Ein Maßkettenkalkül mit Anwendung auf Zentrumsmannigfaltigkeiten. Ph.D. thesis, Universität Würzburg (Germany).

[7] Hilger, S. (1990). Analysis on measure chains – a unified approach to continuous and discrete calculus. *Results in Mathematics*, 18, 18–56.

[8] Bohner, M. and Peterson, A. (2001). *Dynamic Equations on Time Scales: An Introduction with Applications*. Birkhäuser Verlag.

[9] Bohner, M. and Peterson, A. (2003). *Advances in Dynamic Equations on Time Scales*. Birkhäuser, Basel.

[10] Davis, J. M., Gravagne, I. A., Jackson, B. J., Marks II, R. J., and Ramos, A. A. (2007). The Laplace transform on time scales revisited. *Journal of Mathematical Analysis and Applications*, 332, 1291–1307.

[11] Itô, K. (1951). On stochastic differential equations. *Memoirs of the American Mathematical Society*, 4, 1–51.

[12] Øksendal, B. (1992). *Stochastic Differential Equations*. Springer Verlag.

[13] Brzeźniak, Z. and Zastawniak, T. (1999). *Basic Stochastic Processes*. Springer Verlag.

[14] Wilmott, P. (1999). *Derivatives: The Theory and Practice of Financial Engineering*. J. Wiley.

[15] Baaquie, B. (2004). *Quantum Finance*. Cambridge University Press.
[16] Black, F. and Scholes, M. (1973). The pricing of options and corporate liabilities. *Journal of Political Economy*, 81, 637–654.
[17] Baez, J. (2003). This week's finds in mathematical physics *(Week 183)*. http://math.ucr.edu/home/baez/week183.html.
[18] McDonald, R. L. (2003). *Derivatives Markets*. Addison-Wesley.
[19] Segal, W. and Segal, I. E. (1998). The Black–Scholes pricing formula in the quantum context. *Proceedings of the National Academy of Sciences of the USA*, 95, 4072–4075.
[20] Accardi, L. and Boukas, A. (2007). The quantum Black–Scholes equation. *Global Journal of Pure and Applied Mathematics*, 2, 155–170.

14

The neurophysiological sources of quantum-like processing in the brain

14.1 Introduction

In previous chapters, we discussed the conjecture of quantum-like (QL) processing of information in the brain. See, for instance, Chapter 1 (Section 1.16) and Chapter 3 (Section 3.1). In general, such processing need not be based on the physical quantum brain (see, for instance, Hameroff [1] [2] and Penrose [3] [4]), which is the quantum physical carrier of information. As we remarked in Khrennikov [5], the brain created the QL representation (QLR) of information in Hilbert space. Such representation uses quantum information rules in decision making. The existence of such QLR was (at least preliminary) confirmed by experimental data from cognitive psychology (please see Chapter 8). The violation of the law of total probability in these experiments is an important sign of the non-classicality of data. In the constructive wave function approach (see Section 8.16 of Chapter 8), such data can be represented by complex amplitudes. We also presented the QL model of decision making in Chapter 9.

In this chapter we reproduce[1] (with only slight modifications) the paper by Khrennikov [6]. The next natural step is to try to find possible physical realizations of the QLR in the brain. One of the possibilities is to appeal to the quantum physics of the microprocesses in the brain – the quantum brain. However, quite surprisingly, it is possible to proceed within the classical field framework, i.e. to create a classical wave model producing the QLR.

Neurophysiologically, our model is based on a *variety of time scales in the brain*. Each pair of scales – a fine scale (the background fluctuations of the electromagnetic field) and a rough scale (the mental image scale) – induces the QL representation. The background field plays the crucial role in the creation of "superstrong QL correlations" in the brain.

[1] Andrei Khrennikov (2011). Quantum-like model of processing of information in the brain based on classical electromagnetic field. *Biosystems*, 105, 251–259.

We propose a classical (!) wave model, which reproduces probabilistic effects of quantum information theory. Why do we appeal to classical electromagnetic fields in the brain and not to quantum phenomena? In neurophysiological and cognitive studies, we see numerous classical electromagnetic waves in the brain. Our conjecture is that these waves are carriers of mental information which are processed in the framework of quantum information theory.

In the quantum community, there is a general opinion that quantum effects cannot be described by classical wave models. Even those who agree that the classical and quantum interferences are similar will emphasize the role of quantum entanglement and its irreducibility to classical correlations (but recall Einstein–Podolsky–Rosen). It is well known that entanglement is crucial in quantum information theory. Although some authors emphasize the role of quantum parallelism in quantum computing, i.e. superposition and interference, experts know well that without entanglement the quantum computer is not able to beat the classical digital computer.

Recently, one of the authors of this book proposed a classical wave model which reproduces all the probabilistic predictions of quantum mechanics, including correlations of entangled systems. This model was called *prequantum classical statistical field theory* (PCSFT) [7]–[9] and see paper [10] for the recent model for composite systems. It seems that, in spite of the mentioned common opinion, the classical wave description of quantum phenomena is still possible.

In this chapter, we apply PCSFT to model the QL processing of information in the brain on the basis of classical electromagnetic fields. This model is based on the presence of various *time scales in the brain.* Roughly speaking, as mentioned already above, each pair of time scales (fine[2], rough[3]) can be used for the creation of the QLR in the brain. The background field (background rhythms in the brain), which is an important part of our model, plays the crucial role in the creation of "superstrong QL correlations" in the brain. These mental correlations are nonlocal due to the background field. These correlations might provide a solution to the so-called *binding problem.*

Each of such pairs of time scales (fine, rough) induces the QLR of information. As a consequence of the variety of time scales in the brain, we get a variety of QL representations serving various mental functions. This QL model of the brain's functioning was originated in Khrennikov [11]. The main improvement of the "old model" is due to a new possibility achieved recently by the PCSFT: to represent the quantum correlations for entangled systems as the correlations of the classical random field, i.e. the prequantum field. This recent development also enlightened the role of the *background field*, i.e. the vacuum fluctuations. We now transfer

[2] Fine: the background fluctuation of the electromagnetic (classical) field in the brain.
[3] Rough: the mental image scale.

this mathematical construction designed for quantum physics to the area of brain sciences. Of course, in some sense it is a slightly naive model, since we do not know the "QL code" used by the brain, i.e. the correspondence between images and probability distributions of random electromagnetic fields in the brain.

We speculate that decision making through the non-classical law of total probability (LTP) is based on a *wave representation of information* in the brain. The brain is full of classical electromagnetic radiation. It could be that the brain is able to create the QLR of information via classical electromagnetic signals, cf. Fichtner *et al.* [12].

It is well known that classical waves produce superposition and, hence, violate the LTP. However, quantum information processing is based not only on superposition, but also on ENTANGLEMENT. It is the source of superstrong non-local correlations. Correlations are really superstrong (violations of Bell's inequality).

Can entanglement be produced by classical signals? Can quantum information processing be reproduced by using classical waves? Surprisingly, the answer is positive.

The crucial element of our classical wave model in explaining the functioning of the brain is the presence of the random background field. In physics, this background field consists of the fluctuations of the vacuum. However, in the cognitive model this field consists of the background fluctuations of the brain. Such a random background increases essentially the correlations between different mental functions and it generates the non-local presentation of information. As was already remarked, we might couple this non-local representation of information to the binding problem:

"How is the unity of conscious perception brought about by the distributed activities of the central nervous system?"

14.2 Why could the brain use the quantum-like representation of information which is based on classical electromagnetic waves?

As we have emphasized already, profound neurophysiological studies have demonstrated that the brain definitely processes information by using classical electromagnetic signals. We would like to apply the results of these studies and propose a classical signal model of the functioning brain. However, we do not plan to just solely explore the standard classical signal theory. We speculate that information processing in the brain should be described by the mathematical formalism of quantum mechanics and that classical electromagnetic waves are used by the brain to create the QLR.

"Why is the brain not satisfied with classical signal processing?"

"What are the advantages of the QL processing of information (even with a classical field)?"

14.2.1 The incomplete processing of information

If we speculate (as we do in this chapter) that in physics quantum probabilistic behavior can be expressed through classical random waves, then we definitely reject Bohr's thesis on the completeness of quantum mechanics (QM). Please note that this "Einsteinian" attitude is characteristic only for this chapter. In previous chapters, we were able to proceed even with the orthodox Copenhagen interpretation. Here we have to assume that the QM formalism provides only an approximate description of processes in the micro-world. The main difference between quantum processing of information and classical processing is that the quantum approach provides for a possibility to ignore consistently a part of information, i.e. to make a consistent cutoff on information (which is described by the mathematical formalism of QM).

The operation through incomplete information processing is very profitable for cognitive systems. An important part of cognition is the extraction of a part of information from huge information flows coming to the brain. By operating in the QL framework, for example, on the basis of the wave representation, the brain gets a possibility to work harmonically with incomplete information. This is one of the sources of creation of QL processing in the brain.

By ignoring a part of information, the brain is able to create abstract mental images, ideas, concepts, categories. Another advantage is the incredible increase in the speed of computations. Here we speak about computations based on classical electromagnetic signals, but performed on the basis of quantum formalism.

14.2.2 Background noise: how can the worst enemy become the best friend?

By the PCSFT, the standard QM formalism provides a possibility to extract signals (in fact, their averages) from the noisy background. QM can be interpreted as a kind of renormalization theory which is applicable to signals with irreducible noise. In quantum physics, this is the noise of vacuum fluctuations, the background field. It seems that this noise is a fundamental feature of space. One cannot hope to isolate a signal from these fluctuations. The only possibility is to take them into account in a consistent way.

Let us now project this situation to the functioning of the brain. The brain is a complex electric system. Its functioning definitely induces noise. The more complex brains are, the more noisy they are. The higher the brain activity, the noisier it is. Of course, in the process of evolution, cognitive systems might try to reduce the impact of the background noise in the brain. However, it is clear that it would be a very complicated task: it seems impossible to isolate signals in the brain from the background fluctuations (induced by a huge number of neurons).

We speculate that the brain has chosen another way to evolve, namely to elaborate a procedure to extract mental images from signals of the noisy background. Hence, the brain proceeds with the noise, but it has a procedure (which is consistent for different signals) for extracting images. By using the PCSFT, the processing of information on the basis of the QLR provides for such a possibility.

Moreover, in physics (by the PCSFT) the background field is a source of super-strong non-local correlations between entangled systems. Those are "entangled classical waves" in the PCSFT framework. By projecting this situation to the brain functioning, we see that the brain can obtain a great advantage from the presence of background fluctuations. They produce entanglement between processes in different (including spatially separated) domains of the brain, between different mental functions.

14.2.3 The joint processing of an abstract image and its concrete realizations

By the PCSFT, a random electromagnetic field is represented in the QL way by its covariance operator (="density operator"). In the cognitive model, the QL representation corresponds to abstract mental images. Thus, they are given by covariance operators. To process such QL images, the brain is not interested in the complex structure of random fluctuations of classical signals. The brain operates with images encoded by operators – covariance operators (matrices) of classical random signals. However, in this process the brain might need to proceed from the abstract image to its concrete realization. For example, the brain can operate with abstract notions, e.g. house and tree, but it can switch to the concrete house and the concrete tree. In the processing of the first type (the QL processing), the brain operates with operators D_{house} and D_{tree} and in the processing of the second type (the classical processing) the brain has to switch to the classical signals encoding this house and this tree.

We remark that two different realizations of the random signal with the fixed covariance operator can differ essentially, i.e. they really can encode two different houses. Take a coin. Consider a series of tossings, e.g. a few thousand: $x = (x_1, x_2, \ldots, x_n)$, where $x_j = 0, 1$ are labels of coin's sides. After this tossing is finished, start a new series: $y = (y_1, y_2, \ldots, y_n)$. Although both sequences are samples of the same random law, they can differ essentially bit-wise. Thus, we can encode "house" (as the abstract notion) by the probability law of this coin, but two concrete houses are encoded by sequences x and y.

If the brain works in the QL (abstract) regime, it recognizes only encoding of the corresponding probability laws. If it works in the classical regime, it has to use essentially more processing resources, since it operates not with the codes of probabilistic laws, but with the real data streams.

In the QL model induced by PCSFT, we consider only *Gaussian random fields* with zero averages (symmetric rhythms). Such random fields are uniquely determined by covariance operators. The former remark on encoding by using the coin and its tossings can be modified in the following way. We have a Gaussian random generator producing vector data. Each vector (realization) has dimension m. Starting with some input given by the vector x_0 the generator produces the stream of vectors x encoding the concrete house. Starting with another input x_0', the generator produces another stream of random data y encoding another concrete house. The concept of house is represented by the covariance operator $D = D_{\text{house}}$ of this random generator, the density operator in the QL formalism.

14.3 Prequantum classical statistical field theory: non-composite systems

Quantum mechanics (QM) is a statistical theory. It cannot tell us anything about an individual quantum system, e.g. an electron or photon. It predicts only probabilities for results of measurements of ensembles of quantum systems. Classical statistical mechanics (CSM) does the same. Why are QM and CSM based on different probability models?

In CSM, averages are given by integrals with respect to probability measures and in QM they are given by traces. In CSM, we have:

$$\langle f \rangle_\mu = \int_M f(\phi) d\mu(\phi), \tag{14.1}$$

where M is the state space. In probabilistic terms, there is a random vector $\phi(\omega)$ given, which is taking values in M. Then $\langle f \rangle_\phi = Ef(\phi(\omega)) = \langle f \rangle_\mu$. In QM, the average is given by the operator trace-formula:

$$\langle \widehat{A} \rangle_\rho = \text{Tr}\rho\widehat{\text{A}}. \tag{14.2}$$

This formal mathematical difference induces a prejudice on the fundamental difference between classical and quantum worlds. Our aim is to show that, in spite of common opinion, quantum averages can be easily represented as classical averages and, moreover, even correlations between entangled systems can be expressed as classical correlations (with respect to fluctuations of classical random fields).

14.3.1 Einstein's dreams:

Albert Einstein did not believe in irreducible randomness, i.e. the completeness of QM. He dreamed of a better, so to say "prequantum," model [13]:

(1) **Dream 1.** A mathematical model reducing quantum randomness to classical randomness.

(2) **Dream 2.** Renaissance of the causal description.
(3) **Dream 3.** Instead of particles, classical fields will provide the complete description of reality: the reality of fields (Einstein and Infeld [13] (p. 257)):

"But the division into matter and field is, after the recognition of the equivalence of mass and energy, something artificial and not clearly defined. Could we not reject the concept of matter and build a pure field physics? What impresses our senses as matter is really a great concentration of energy into a comparatively small space. We could regard matter as the regions in space where the field is extremely strong. In this way a new philosophical background could be created."

The real trouble with the prequantum wave model (in the spirit of early Schrödinger) is not the various no-go theorems[4] (e.g., the Bell inequality [14]), but the problem which had already been recognized by Schrödinger. In fact, he gave up with his wave quantum mechanics because of this problem: *A composite quantum system cannot be described by waves on physical space!* Two electrons are described by the wave function on $\mathbb{R}^6$ and not by two waves on $\mathbb{R}^3$.

Einstein also recognized this problem (Einstein and Infeld [13] (p. 305)):

"For one elementary particle, electron or photon, we have probability waves in a three-dimensional continuum, characterizing the statistical behavior of the system if the experiments are often repeated. But what about the case of not one but two interacting particles, for instance, two electrons, electron and photon, or electron and nucleus? We cannot treat them separately and describe each of them through a probability wave in three dimensions ..."

14.3.2 Quantum system = classical random field

Einstein's Dreams 1 and 3 came true in PCSFT (but not Dream 2!), which is a version of CSM in which fields play the role of particles.[5] In particular, composite systems can be described by vector random fields, i.e. by the Cartesian product of state spaces of subsystems and not the tensor product. The basic postulate of PCSFT can be formulated in the following way:

A quantum particle is the symbolic representation of a "prequantum" classical field fluctuating on the time scale which is essentially finer than the time scale of measurements.

In the prequantum state space $M = L_2(\mathbb{R}^3)$, states are fields $\phi : \mathbb{R}^3 \to \mathbb{R}$; "electronic field," "neutronic field," "photonic field" – classical electromagnetic field.

[4] See Chapter 1, Section 1.22 for more on no-go theorems
[5] It seems surprising that, although Dream 1 came true, Dream 2 did not. The situation differs essentially from CSM where dynamics of probability distributions given by the Liouville equation can be reduced to the deterministic Hamiltonian dynamics. The main difference is due to the presence of the background fluctuations – irreducible noise.

An ensemble of "quantum particles" is represented by an ensemble of classical fields. The probability measure μ on $M = L_2(\mathbb{R}^3)$, or the random field $\phi(x, \omega)$ is taking values in $M = L_2(\mathbb{R}^3)$. For each fixed value of the random parameter $\omega = \omega_0$, $x \to \phi(x, \omega_0)$ is a classical field on physical space.

14.3.3 Density operator = covariance operator

Each measure (or random field) has the covariance operator, say D. It describes correlations between various degrees of freedom. The map $\rho \mapsto D = \rho$ is one-to-one between density operators and the covariance operators of the corresponding prequantum random fields, in the case of non-composite quantum systems. In the case of composite systems, this correspondence is very tricky.

Thus, each quantum state (an element of the QM formalism) is represented by the classical random field in PCSFT. The covariance operator of this field is determined by the density operator. We also postulate that the prequantum random field has *zero mean value.* These two conditions determine uniquely Gaussian random fields. We restrict our model to such fields. Thus, by PCSFT, quantum systems are Gaussian random fields.

Finally, we recall that the covariance operator D of a random field ϕ is defined by its bilinear form ($u, v \in H$):

$$\langle Du, v\rangle = E\rangle\phi, u\langle\langle v, \phi\rangle = E\left(\int_O \phi(x,\omega)\overline{u(x)}dx\right)\left(\int_O v(x)\overline{\phi(x,\omega)}dx\right) \quad (14.3)$$

or by using the probability distribution μ of the random field:

$$\langle Du, v\rangle = \int_H \left(\int_O \phi(x)\overline{u(x)}dx\right)\left(\int_O v(x)\overline{\phi(x)}dx\right) d\mu(\phi). \quad (14.4)$$

14.3.4 Quantum observable = quadratic form

The map $\widehat{A} \to f_A(\phi) = (\widehat{A}\phi, \phi)$ establishes a one-to-one correspondence between quantum observables (self-adjoint operators) and classical physical variables (quadratic functionals of the prequantum field).

It is easy to prove that the following equality holds:

$$Ef_A(\phi_(\omega)) = \int_M f_A(\phi)d\mu(\phi) = \mathrm{Tr}\rho\widehat{\mathrm{A}}. \quad (14.5)$$

In particular, for a pure quantum state ψ, consider the Gaussian measure with zero mean value and the covariance operator $\rho = \psi \otimes \psi$ (the orthogonal projector on the vector ψ), then

$$\int_M f_A(\phi) d\mu(\phi) = (\widehat{A}\psi, \psi).$$

This mathematical formula coupling the integral of a quadratic form and the corresponding trace is well known in measure theory. Our main contribution is the coupling of this mathematical formula with quantum physics.

This is the end of the story for quantum non-composite systems, e.g. a single electron or photon [7] [9].

14.3.5 Beyond QM

In fact, PCSFT not only reproduces quantum averages, but it also provides a possibility to go beyond QM. Suppose that not all prequantum physical variables are given by QUADRATIC forms. Consider a more general model, such as all the smooth functionals $f(\phi)$ of classical fields. We only have the illusion of the representation of all quantum observables by self-adjoint operators.

The map

$$f \mapsto \widehat{A} = f''(0)/2 \tag{14.6}$$

projects smooth functionals of the prequantum field (physical variables in PCSFT) on self-adjoint operators (quantum observables). Then quantum and classical (prequantum) averages do not coincide precisely, but only approximately:

$$\int_M f_A(\phi) d\mu(\phi) = \mathrm{Tr}\rho\widehat{A} + O(\tau/T), \tag{14.7}$$

where T is the time scale of measurements and τ is the time scale of fluctuations of the prequantum field. The main problem is that PCSFT does not provide a quantitative estimate of the time scale of fluctuations of the prequantum field. If this scale is too fine, e.g. the Planck scale, then QM is "too good an approximation of PCSFT," i.e. it would be really impossible to distinguish them experimentally. However, even a possibility to represent QM as the classical wave mechanics can have important theoretical and practical applications. In the present chapter, we shall use the mathematical formalism of PCSFT to model the functioning of the brain. Although even in this case the choice of the scale of fluctuations is a complicated problem, we know that it is not extremely fine. Hence, the model can be experimentally verified (in contrast to Roger Penrose, we are not looking for cognition at the Planck scale!).

14.4 Cognitive model: two regimes of brain's functioning

We now turn to considerations which were exhibited in Section 14.2.3 and proceed on the basis of the short presentation of PCSFT given in Section 14.3. We first consider a one fixed mental function of the brain, say F, which is physically concentrated in some spatial domain $O \subset \mathbb{R}^3$ of the brain. We shall then come to the model of QL cooperation of a few mental functions ("entanglement of mental functions") after presentation of the PCSFT for composite systems, Section 14.8.

Classical regime

By getting an input ϕ_0 (from an environment or another mental function), the mental function F produces a random signal $\phi(x, \omega)$, a classical electromagnetic field resulting from neuronal activity.[6] It is a random signal depending on the chance parameter ω. For each ω_0, this electromagnetic field, $x \to \phi(x, \omega_0)$, is distributed on the domain O.

We use the complex representation for the electromagnetic field, the Riemann–Silberstein representation:

$$\phi(x) = E(x) + iB(x),$$

where $E(x) = (E_1(x), E_2(x), E_3(x))$ and $B(x) = (B_1(x), B_2(x), B_3(x))$ are the electric and magnetic components, respectively.

In our model, *each concrete mental image is associated with a random signal.* Its mental features, *"qualia,"* are given by functionals of this signal. In the simplest case, these are quadratic forms of the signal.

The main problem is to create the classical signal code, i.e. to establish a correspondence between random signals and mental images as well as between field functionals and qualia of images. We speculate that at least some field functionals represent *emotions* related to the mental image (which is represented by the classical electromagnetic signal). Consider a number of emotions, say $\mathcal{E}_1, \ldots, \mathcal{E}_k$, related to some image, say MI_ϕ (associated with the signal ϕ). Then, the mental function F physically operates with the corresponding field functionals. In the simplest case, these are quadratic functionals and they can be represented by integral kernels:

$$f_{\mathcal{E}_j}(\phi) = \int_{O \times O} K_j(x, y)\phi(x)\overline{\phi(y)}dxdy. \tag{14.8}$$

[6] Thus, we consider the ensemble of neurons which are located in O. By our model the brain is not so much interested in the "private life" of the individual neurons, i.e. the frequency of spikes and so on, it is only interested in the electromagnetic field induced by the activity of these neurons.

However, in the classical regime *non-quadratic functionals* are also in use, e.g.:

$$f(\phi) = f_{\mathcal{E}}(\phi) + \int_{O\times O\times O} K(x, y, z)\phi(x)\phi(y)\overline{\phi(z)}dxdydz,$$

where $f_{\mathcal{E}}$ is the functional of the form (14.8).

Remark 1. (Spatial distribution of qualia) By considering integral functionals of the classical electromagnetic field, we suppose that the F-function performs integration of a signal over its domain of spatial concentration. Thus, we consider only spatially concentrated mental functions. If a mental function F is concentrated in a domain $O = \cup_k O_k$, where the O_k are located in the brain far away from each other, then we represent F as a collection of "elementary mental functions" F_k concentrated in domains O_k. Some qualia of F are associated with elementary functions F_k. However, there are also exist global qualia which are obtained by the summation of local ones (so integration on each O_k and then collection and summation in a "special center").

Since signals are random, field functionals are fluctuating quantities – random variables $\xi_\phi(\omega) = f(\phi(\omega))$. It is clear that the brain cannot operate with such unstable mental entities. Thus, it has to produce averages of emotions and operate with them. It will be especially clear in the time-representation of random signals, see Section 14.5. Let μ be the probability distribution of a random signal ϕ. Then emotions are quantified by averages:

$$\langle f_{\mathcal{E}_j}\rangle = \int_H f_{\mathcal{E}_j}(\phi)d\mu(\phi), \tag{14.9}$$

where $H = L_2(O)$. In our model, not only emotions but all qualia are quantified by averages.

It is clear that quantification of each qualia consumes the brain's resources. Therefore, only a special class of qualia (in particular, emotions) is associated with each mental image. How the mental function F selects them is the open question! The crucial point is that in principle any two emotions $\mathcal{E}_1$ and $\mathcal{E}_2$ or other qualia can be associated with the image MI_ϕ and quantified. This *total compatibility of emotions and qualia in general* may induce some problems. For example, it is not always profitable for survival to combine some emotions. We shall see that the situation is totally different in the QL processing of information.

Functionals of classical electromagnetic signals represent not only emotions, but even other qualia of the image MI_ϕ. For example, in the PCSFT we have the energy variable (representing the intensity of a signal):

$$f_I(\phi) = \int_O |\phi(x)|^2 dx = \int_O (E^2(x) + B^2(x))dx. \tag{14.10}$$

We relate this functional to the *intensity of feeling* of the image MI_ϕ. This intensity is quantified as:

$$\langle f_I \rangle = \int_H f_I(\phi) d\mu(\phi) = \int_H \left(\int_O |\phi(x)|^2 dx \right) d\mu(\phi), \tag{14.11}$$
$$= \int_H \left(\int_O (E^2(x) + B^2(x)) dx \right) d\mu(\phi).$$

In QM, the position observable is given by the multiplication operator:

$$\widehat{x}\phi(x) = x\phi(x)$$

and in the PCSFT it is represented by the field functional:

$$f_x(\phi) = (\widehat{x}\phi, \phi) = \int_O x|\phi(x)|^2 dx = \int_O x(E^2(x) + B^2(x))dx. \tag{14.12}$$

What quale can be coupled to this functional?

Already here, on the level of classical mental processing, mathematics led us to the notion of *conjugate qualia of a mental image.* For example, consider the momentum functional in the PCSFT:

$$f_p(\phi) = (\widehat{p}\phi, \phi) = \int_{\mathbb{R}^3} p|\tilde{\phi}(p)|^2 dp, \tag{14.13}$$

where $\tilde{\phi}(p)$ is the Fourier transform of the signal $\phi(x)$. What is a cognitive interpretation of conjugation between qualia given by functionals f_x and f_p?

It seems that in the classical regime the brain can process conjugate qualia simultaneously. In the case of "position and momentum" functionals, this means simultaneous processing in the spatial and frequency representations.

Classical mental coding: How does the brain associate the mental image MI_ϕ with a classical signal ϕ? In our model, this is done through calculation of its covariance operator $D = D(\phi)$. This association, mental image – covariance operator, is especially natural in the time representation of random signals, see Section 14.5.

Quantum-like regime

We are now interested in the QLR (quantum-like representation) of information. In the QLR, the brain operates with density operators which represent not only concrete mental images (coming from the classical regime of mental processing), but also abstract concepts (of different levels of abstraction, see Section 14.7 for details) which do not correspond to classically produced images.

In the QLR, the brain's state space is a space of density operators $\mathcal{D}(H)$, $H = L_2(O)$. In principle, each density operator can be used as a QL state of the brain.

However, it is natural to assume that each mental function F operates in its own subspace $\mathcal{D}_F(H)$ of $\mathcal{D}(H)$.

In standard QM a system has not only the state, but also the "properties" or (depending on the interpretation) there are defined observables on this system (in this state) – e.g. the energy observable, the coordinate observable, and so on. In QM, they are represented by self-adjoint operators.

To simplify the mathematics, we shall consider only bounded (continuous) operators. Denote the space of all bounded self-adjoint operators by the symbol $\mathcal{L}_s(H)$. For a given quantum state $\rho \in \mathcal{D}(H)$ and observable $\widehat{A} \in \mathcal{L}_s(H)$, the QM formalism gives the average of this observable in this state, see (14.2).

We encode qualia of a QL cognitive image MI_ρ (which is encoded by a density operator ρ) by self-adjoint operators. Thus, qualia of a mental image (in the QLR of information) are described by the space $\mathcal{L}_s(H)$. They are quantified by their averages, via (14.2).

Contrary to classical processing, in the QLR the brain cannot quantify all qualia simultaneously. There exist *incompatible qualia*; in particular, incompatible emotions. The QL brain can select different representations of the mental image MI_ρ and different collections of compatible qualia of the image. The QL brain escapes the simultaneous use of e.g. some emotions ("incompatible emotions"). In this way the QLR-processing differs essentially from classical processing. We can speculate that in the process of evolution, the brain created (on the basis of experience) commutative algebras corresponding to compatible qualia.

Coupling between classical and quantum-like representations

The crucial point of the QLR of information is that this "operator-thinking" is naturally coupled with the processing of classical electromagnetic signals. On the level of mental images, we have:

(aMI). From the classical regime to QL: a classical signal ϕ induces the mental image MI_ϕ given by its covariance operator $D = D(\phi)$ and it is transferred to the QLR through normalization by the trace $D \to \rho = D/\mathrm{Tr}D$. Thus, there is a map from classical mental images to QL mental images $MI_\phi \to MI_\rho$.

(bMI). From QL to classical: a QL mental image MI_ρ can be represented by a classical (Gaussian) signal $\phi = \phi_\rho$ with the covariance operator ρ, i.e. by the image MI_ϕ.

On the level of qualia we have:

(aQU). From the classical regime to QL: each functional of classical field, $f(\phi)$, is represented by its second derivative – self-adjoint operator, see (14.6).

(bQU). From QL to classical: each quantum qualia (given by a self-adjoint operator) is represented by its quadratic form.

Since (aQU) is not one-to-one, i.e. since a huge class of different classical qualia given by various functionals of the electromagnetic field (all functionals with the same second derivative) is mapped into the same self-adjoint operator, the QLR makes the mental picture less rich than it was in the classical representation. The same can be said about (aMI). Various classical signals, concrete mental images, can have the same covariance operator. Please note that we do not claim that input signals are obligatory Gaussian, so there is no one-to-one correspondence.

Take a classical mental image MI_ϕ. It is represented by the covariance operator $D = D(\phi)$. Of course, it can be mapped to a QL image MI_ρ, see aMI. Values of all QL qualia can be obtained by scaling from values of corresponding classical qualia, since, for any operator $\widehat{A} \in \mathcal{L}_s(H)$:

$$\begin{aligned}\langle \widehat{A} \rangle_\rho &= \mathrm{Tr}\rho\widehat{A} = \frac{1}{\mathrm{Tr}D}\mathrm{Tr}D\widehat{A}\\ &= \frac{1}{\mathrm{Tr}D}\int_H f_A(\phi)d\mu(\phi) = \frac{1}{\mathrm{Tr}D}\langle f_A \rangle_\phi,\end{aligned}$$

where μ is the probability distribution of the signal. We remark that:

$$\mathrm{Tr}D = \int_H \left(\int_O |\phi(x)|^2 dx \right) d\mu(\phi) = \langle f_I \rangle$$

is the intensity of the signal, or in the cognitive model the intensity of feeling of the mental image MI_ϕ. Thus, QL qualia are normalized by the intensity of feeling:

$$\langle \widehat{A} \rangle_\rho = \frac{\int_H f_A(\phi)d\mu(\phi)}{\int_H \left(\int_O |\phi(x)|^2 dx\right) d\mu(\phi)}.$$

The QL processing of all mental images is performed on the same level of intensity of feeling, so it is *"calm thinking."* We remark once again that some classical qualia do not have a quantum counterpart.

14.5 Classical regime: time representation

As usual in signal theory, we can switch from the ensemble representation for averages to the time representation (under the standard assumption of *ergodicity*). Thus, instead of a random field $\phi(x, \omega)$, which is distributed with some probability distribution $d\mu(\phi)$ on H, we consider a time-dependent signal:

$$\phi(s) \equiv \phi(s, x),$$

where $x \in O, s \in [0, +\infty)$. Then, for each functional $f(\phi)$ such that $\int_H |f(\phi)|d\mu(\phi) < \infty$, we have (by ergodicity):

$$\langle f \rangle_\mu \equiv \int_H f(\phi)d\mu(\phi) = \lim_{T\to\infty} \frac{1}{T}\int_0^T f(\phi(s))ds \equiv \langle f \rangle_\phi. \qquad (14.14)$$

Consider two time scales: τ is a fine scale and $T >> \tau$ is a rough time scale. In QM the latter is the scale of measurements and τ is the scale of fluctuations of the prequantum field.[7] In cognitive science, we use the following interpretation of time scales: T is the scale of the QLR and τ is the scale of the real physical processing of the electromagnetic signal in the brain. Thus

$$\langle f \rangle_\phi \approx \frac{1}{T}\int_0^T f(\phi(s))ds, \qquad (14.15)$$

where s denotes the time variable at the τ-scale. We call the T-scale the *mental time scale*. We can also speak about *psychological time*. The T-scale is the scale of creation of mental images by the brain. The τ-scale is the physical processing scale or *premental time scale*. Theoretical and experimental studies in neurophysiology provide the following estimate of relative magnitudes of these time scales. If we select $\tau = 1$ mls., then $T \approx 80$ mls.

For each signal $\phi(s, x)$, $x \in O$, the brain can find its qualia, e.g. the strength of feeling of this image:

$$\langle f \rangle_\mu \approx \frac{1}{T}\int_0^T \left(\int_O (E^2(s,x) + B^2(s,x))dx\right) ds. \qquad (14.16)$$

In particular, emotions (special qualia) are given by such functionals, e.g., $f_{\text{anger}}, f_{\text{sadness}} \ldots$ Our formal mathematical model cannot provide the form of concrete emotion functionals. We hope that in the future it will be described as the result of neurophysiological and cognitive studies.

In the time representation, the covariance operator (its bilinear form) of the signal $\phi(s, x)$ is given by

$$(Du, v) = \lim_{T\to\infty} \frac{1}{T}\int_0^T \left(\int_O u(x)\overline{\phi(s,x)}dx \int_O \phi(s,x)\overline{v(x)}dx\right) ds,$$
$$\approx \frac{1}{T}\int_0^T \left(\int_O u(x)\overline{\phi(s,x)}dx \int_O \phi(s,x)\overline{v(x)}dx\right) ds, \qquad (14.17)$$

where $u(x)$, $v(x)$ are two "test signals," $u, v \in L_2(O)$.

[7] The PCSFT does not predict the magnitude of the scale of prequantum field fluctuations. One may speculate (motivated in particular by cosmology and string theory), cf. G. 't Hooft [15] [16] and T. Elze [17]), that it has to be the Planck scale $\tau_P \approx 10^{-44}$ s. If it were really the case, then prequantum fluctuations have only a theoretical value: they will never be approached experimentally. However, in Khrennikov [18] we discussed a possibility that the prequantum scale may have a larger magnitude and, hence, fluctuations will be sooner or later approached experimentally.

14.6 Classical signal processing of mental images

This section contains a detailed presentation of the classical processing of information, see Section 14.4. We proceed in the time representation of random signals.

CSP1. **Electromagnetic field basis of mental images.** Inputs from external and internal worlds induce electromagnetic signals in the brain. Each signal has a variety of qualia, in particular emotions associated with the signal $\phi(s, x)$. Qualia are realized by various functionals, $\phi \mapsto f(\phi)$, of the signal. They are quantified by averages of these functionals, $f \mapsto \langle f \rangle_\phi$, see (14.14) and (14.16). In principle, all possible qualia (e.g., emotions) can be jointly associated with $\phi(s, x)$.

The physical dynamics of a signal is in general *non-linear* and very complicated. It depends essentially on the context of the signal processing:

$$\phi(s, x) = \phi(s_0, x) + \int_{s_0}^{s} d\alpha \left(\int_O K(x, y, \alpha; \phi(\alpha, y); \phi(s_0, y)) dy \right), \quad (14.18)$$

where the kernel K depends on (i) the spatial variables $x, y \in O$ (the distribution of the signal in the brain), (ii) the time variable α, (iii) the previous dynamics of the signal $\phi(\alpha, y)$, $\alpha \in [s_0, s)$, and (iv) the signal at $s = s_0$, the input. We remark that the dynamics $\phi(s, x)$ depend on input not only additively, i.e. as the initial state which then will evolve in accordance with some integral equation, but even the kernel of the equation depends non-trivially on the input. Thus, the dynamics are different for different inputs.

CSP2. **Calculation of correlations, creation of mental images.** For each signal $\phi(s, x)$, the brain calculates the corresponding covariance operator $D \equiv D(\phi)$, see (14.17). The completion of this process, i.e. calculation of D, is associated with the creation of the *mental image*, MI_ϕ, induced by the signal, ϕ. Thus, on the cognitive level, the brain is not interested in the dynamics of the physical signal (14.18). It is only interested in the dynamics of the covariance operator

$$t \mapsto D(t). \quad (14.19)$$

We remark that the dynamics (14.18) and (14.19) have different time scales. The first one is performed on the physical time scale and the second one on the mental time scale. Thus, it is very important that the "physical brain" and the "cognitive brain" work on two different time scales: the scale of physical signal – τ, and the scale of the QLR – T. The interval of time $T >> \tau$, so its size justifies the ergodic interplay between ensemble and time representations of random signals.

CSP3. **Memory of correlations.** The density operator D is recorded in the brain. The PCSFT basis of the model in combination with the ergodic argument makes the following model of memory very attractive.

The operator D determines uniquely the Gaussian probability distribution μ_D (with zero mean value). The brain records this probability distribution. How can it do this? We speculate that, to encode μ_D, the brain uses the statistical distribution by assigning statistical weights to elements of some ensemble Ω. What are elements of Ω? They might be neurons or even distributions of chemical components in the brain. We emphasize once again that such a model of memory for probabilistic laws is based on the ergodicity of processes in the brain: from a signal $\phi(s) \equiv \phi(s, x)$ (the time representation) to its covariance, and from the covariance to the probability distribution on an ensemble.[8]

CSP4. **Recollection of images.** Recollection is the process of activation of a special mental image. We keep to the model of statistical (ensemble) representation of the probability distribution encoding the image, see CSP3. We obtain the following procedure of recollection.

Suppose that a mental image MI_ϕ was recoded in the memory, $\phi \mapsto D = D(\phi) \mapsto \mu = \mu_D$.

The process of recollection proceeds as follows. Starting with the probability distribution μ, the brain computes the covariance operator D of this probability distribution by using the ensemble averaging, see (14.3), (14.4).

On the basis of this covariance operator, it produces a signal $\phi_{\text{recall}}(s, x)$, $x \in O$, a trajectory of the corresponding Gaussian process.

In this situation, the brain does not reproduce the original signal $\phi(s, x)$, see CSP1. The graphs of $\phi(s, x)$ and $\phi_{\text{recall}}(s, x)$ can differ essentially point wise. Moreover, the original signal $\phi(s, x)$ need not be Gaussian at all. However, correlations inside both signals approximately coincide and, hence, their qualia:

$$\langle f \rangle_\phi = \lim_{T\to\infty} \frac{1}{T} \int_0^T f(\phi(s))ds \approx \lim_{T\to\infty} \frac{1}{T} \int_0^T f(\phi_{\text{recall}}(s))ds = \langle f \rangle_{\phi_{\text{recall}}}.$$

We also remark that even two different recollections $\phi^1_{\text{recall}}(s, x)$ and $\phi^2_{\text{recall}}(s, x)$ of the same image can be very different as physical signals – two different realizations of the same Gaussian process. However, their qualia coincide:

$$\langle f \rangle_{\mu_D} = \lim_{T\to\infty} \frac{1}{T} \int_0^T f(\phi^1_{\text{recall}}(s))ds = \lim_{T\to\infty} \frac{1}{T} \int_0^T f(\phi^2_{\text{recall}}(s))ds. \quad (14.20)$$

To be more precise, we say that they coincide approximately, since in reality the brain does not calculate the limit for $T \to \infty$, but it uses the finite T. Thus:

$$\langle f \rangle_{\mu_D} \approx \frac{1}{T} \int_0^T f(\phi^1_{\text{recall}}(s))ds \approx \frac{1}{T} \int_0^T f(\phi^2_{\text{recall}}(s))ds. \quad (14.21)$$

[8] The choice of a Gaussian probability law can be debated. But at least mathematically it works well, because of the one-to-one correspondence between covariance matrices and Gaussian probability distributions (with zero mean values).

CSP5. **Recognition of images.** Suppose now that some mental image was saved in the memory: starting with the input signal $\phi(s, x)$ and through its covariance operator $D(\phi)$. As an example, consider the Moscow Kremlin. I revisit Moscow once again and I look at the Kremlin. This visual input induces a signal $q(s, x)$. Its covariance operator $D(q)$ is produced, see CSP2. It is compared with covariance operators in the memory to match with the operator $D(\phi)$. Note that the model under consideration does not describe the mechanism of this comparing process. However, see CSP5n (Section 14.6.1). Finally, matching of the operators, $D(\phi) \approx D(q)$, is approached. This activates in the memory the statistical probability distribution $\mu_{D(\phi)}$ in the form of a Gaussian signal $\phi_{\text{recall}}(s, x)$. Its triggers the feeling of recognition of the image, which was encoded by $D(\phi)$.

14.6.1 Classical signal processing of mental images: finite-dimensional approximations

In principle, the brain can calculate the complete covariance operator (14.17), especially if it works as an analogous computational device. However, it consumes a lot of computational resources. We might speculate that the brain selects a finite number of test functions. It is always possible to assume that they are orthogonal in $L_2(O)$:

$$u_1(x), \ldots, u_n(x). \tag{14.22}$$

Instead of the complete covariance operator $D = D(\phi)$, the brain calculates its cutoff, the covariance $n \times n$ matrix $D_n = D_n(\phi)$. Thus, instead of an infinite dimensional L_2-space, the brain works (for a given mental function) in a fixed finite dimensional subspace H_n. We modify CSP1–CSP5. The first step CPS1 is not changed. We have:

CSP2n. The signal $\phi(s, x)$ induces the mental image $MI_{\phi;n}$ encoded by the covariance matrix D_n.

CSP3n. The $MI_{\phi;n}$ is recorded in the memory through the probability distribution μ_{D_n} on the finite-dimensional Hilbert space H_n.

CSP4n. On the basis of D_n, the brain produces a signal $\phi_{\text{recall}}(t)$ in H_n. Its activation is a recollection of the memory on $MI_{\phi;n}$.

CSP5n. The memory contains the image $MI_{\phi;n}$ in the form of the matrix $D_n(\phi)$. The new signal produces $MI_{q;n}$ with the covariance matrix $D_n(q)$. These matrices must be compared. Since a lot revolves around covariance matrices, so $n \times n$ matrices, it is natural to expect a comparing algorithm which compares cutoffs of these $n \times n$ covariance matrices. First, the matrix is of dimension two, then three, and so on. Thus, first the $D_2(q)$ is compared with 2×2 matrices obtained through the projection of mental images on H_2 until the cluster of matrices with the left-up block $D_2(q)(= D_2(\phi))$ is found. Then inside this cluster the brain is looking for the sub-cluster with the left-up block $D_3(q)(= D_3(\phi))$ and so on.

How does the brain select the subspace H_n with the basis (14.22)?

The most natural way is to assume that it just selects a band of frequencies. It is also natural that different mental functions may use different bands, i.e. different Hilbert spaces. For mental functions F and G, two Hilbert spaces H_F and H_G are used.

14.7 Quantum-like processing of mental images

We still proceed with the functioning of one fixed mental function, say F.

QLP1. **Density operator code.** At some stage in its growth, a cognitive system creates a sufficiently extended database of classical mental images. They are encoded by covariance operators which are transferred in density operators by normalization, see (aMI), Section 14.4. Thus, the brain created a collection of QL mental states borrowed from the classical processing, $\rho \in \mathcal{D}_{\text{data}}(H)$ which is a subspace of $\mathcal{D}(H)$. At this stage the brain can be fine by working inside $\mathcal{D}(H)$, i.e., even without contact with the physical and mental environment.

QLP2. **Unitary thinking.** The processing of information inside $\mathcal{D}(H)$ is the process of QL thinking. Starting with the operator ρ_0 the brain induces the evolution $\rho(s)$ of the mental QL state. The simplest dynamics correspond to the process of thinking in the absence of inputs from the environment (which includes the body). It is given by the von Neumann equation:

$$i\frac{d\rho(t)}{dt} = [\widehat{\mathcal{H}}, \rho(t)], \;\; \rho(0) = \rho_0, \tag{14.23}$$

where $\widehat{\mathcal{H}} : H \to H$ is the "mental Hamiltonian" (given by a self-adjoint operator). It describes the functioning of the mental function F under consideration.

In the simplest case, the Hamiltonian $\widehat{\mathcal{H}}$ is completely determined by the mental function F, so $\widehat{\mathcal{H}} \equiv \widehat{\mathcal{H}}_F$. However, even more complex dynamics seem to be reasonable – with $\widehat{\mathcal{H}}$ which also depends on the initial state $\rho_0 : \widehat{\mathcal{H}} \equiv \widehat{\mathcal{H}}_{F,\rho_0}$ (cf. with the QL model of decision making, Chapter 9).

We remark that by starting with e.g. the QL version of a concrete image, i.e., $\rho_0 \in \mathcal{D}_{\text{data}}(H)$, the QL dynamics can go away from this subspace of $\mathcal{D}(H)$. New "really QL" images are created. They can be visualized through the production of corresponding classical signals, see (bMI), Section 14.4.

We emphasize that the QL dynamics of mental images is performed on the T-scale which is rough compared with the τ-scale of physical processing of signals in the brain. Each instant of time t of the T-scale is the (large) interval T of the s-time.

It may be more illustrative to consider the discrete dynamics, the mental time t is considered as the discrete parameter: $t_n = nT$. Then

$$i\rho(t_{n+1}) = T[\widehat{\mathcal{H}}, \rho(t_n)], \ \rho(0) = \rho_0. \tag{14.24}$$

QLP3. **Dynamics of QL qualia.** In QL processing qualia are reduced to *quadratic functionals* of premental (physical) signals. These functionals are represented by their QL counterparts – corresponding self-adjoint operators. The evolution of QL qualia is described by the Heisenberg equation:

$$-i\frac{d\widehat{A}(t)}{dt} = [\widehat{\mathcal{H}}, \widehat{A}], \ \widehat{A}(0) = \widehat{A}_0, \tag{14.25}$$

or in the discrete representation of the mental time:

$$i\widehat{A}(t_n) = T[\widehat{\mathcal{H}}, \widehat{A}(t_n)], \ \widehat{A}(0) = \widehat{A}_0. \tag{14.26}$$

The qualia (encoded by $\widehat{A}$) of a mental image MI_ρ (encoded by the density operator ρ) is quantified by its average, given by the quantum formula (14.2).

Of course, the transition from the class of classical qualia (given by arbitrary functionals of signals) to QL mental features corresponding to only quadratic functionals simplifies the mental representation of an image. However, this reduction can be justified by (14.7) in the framework of the QLR (14.6) of classical functionals of signals.

QLP4. **Thinking via operator algebra.** This is the crucial point. In QL thinking, the brain switches from the classical physical signal processing, i.e. non-linear equations of the type (14.18) to the *linear processing* of mental images represented by density operators (14.26). The representation of qualia is also essentially simplified and it can be done in the linear operator form. Our conjecture is that the brain is really able to realize such linear operator processing of mental entities. This type of processing is especially profitable for "abstract thinking," i.e. thinking which has a high degree of independence from inputs.

How does the brain realize the QL (operator) processing on the physical level?

We do not know yet. However, we hope that our model may stimulate neurophysiologists to look for the corresponding neuronal representation of QL processing. We can present the following scheme of mental operator processing.

Since the brain has no other computational resources different from neural electric activity, it seems reasonable to assume that the QL mental dynamics ((14.26), (14.29)) also has to be performed through this activity. The production of density operators can be done similarly to the production of covariance operators in the classical regime. The only difference is that the brain wants to escape the complicated non-linear evolution (14.18). We consider the following stochastic linear dynamics in Hilbert space H (of classical electromagnetic fields):

$$\frac{\partial \phi}{\partial t}(t, x, \omega) = \widehat{\mathcal{H}}\phi(t, x, \omega), \ \phi(t_0, x, \omega) = \phi_0(x, \omega), \tag{14.27}$$

where the random variable $\phi_0(x, \omega)$ is the Gaussian field with zero mean value and the covariance operator ρ_0. Hence, $\langle \rho_0 u, v \rangle = E\langle \phi_0, u \rangle \langle v, \phi_0 \rangle, u, v \in H$. The solution of the Cauchy problem (14.27) is the random field:

$$\phi(t, x, \omega) = U_t \phi_0(x, \omega), \tag{14.28}$$

where $u_t = e^{-it\widehat{\mathcal{H}}}$ is the standard for QM one parametric group of unitary operators. The covariance operator $\rho(t) \equiv \rho_{\phi(t)}$ can be easily found: $\langle \rho(t)u, v \rangle = E\langle U_t\phi_0, u \rangle \langle v, U_t\phi_0 \rangle = E\langle \phi_0, U_t^* u \rangle \langle U_t^* v, \phi_0 \rangle = \langle \rho(t) U_t^* u, U_t^* v \rangle$. Thus:

$$\rho(t) = U_t \rho_0 U_t^*.$$

This operator valued function $\rho(t)$ satisfies the von Neumann equation (14.29). Thus, the von Neumann evolution of the mental state can be induced by the linear dynamics with random initial condition (14.27). As was mentioned, the crucial point is that those dynamics are much simpler than the "classical signal dynamics" (14.18).

Finally, we have the following model of physical realization of the evolution (14.29). In fact, the brain produces the Gaussian random signal by realizing on the neuronal level the linear Schrödinger type evolution. At each moment of mental time t by calculating its covariance operator, the brain creates the mental image given by the covariance density operator $\rho(t)$.

In quantum information theory it is well known that in general, i.e. in the presence of interaction with the environment, the von Neumann equation should be modified to the Gorini–Kossakowski–Sudarshan–Lindblad equation:

$$i\frac{d\rho(t)}{dt} = \widehat{L}(\rho(t)), \ \rho(0) = \rho_0, \tag{14.29}$$

where $\widehat{L} : \mathcal{L}(H) \to \mathcal{L}(H)$ is a linear map with special properties. All previous considerations can be easily generalized to such mental dynamics.

QLP5. **Concepts.** Consider a subspace L of H and the orthogonal projector $\pi \equiv \pi_{HL} : H \to L$. It induces the map $\pi : \mathcal{D}(H) \to \mathcal{D}(L)$,

$$\rho_L \equiv \pi(\rho) = \frac{\pi\rho\pi}{\mathrm{Tr}\pi\rho\pi}.$$

Mental images corresponding to elements of $\mathcal{D}(L)$ can be considered as abstractions of mental images corresponding to elements of $\mathcal{D}(H)$. We call them L concepts or simply concepts. Take some $\rho_L \in \mathcal{D}(L)$. The mental image MI_{ρ_L} can be interpreted as an abstract concept induced by the cluster of mental images:

$$W_{\rho_L} = \{\rho \in \mathcal{D}(H) : \pi(\rho) = \rho_L\}.$$

Each concept is based on common correlations of a cluster of mental images. It is especially interesting to consider the case dim L $= m$, and m is quite small. These are very abstract concepts which contain only the basic common correlations in a huge cluster of mental images. It is extremely profitable for the brain to think on the conceptual level, especially to operate in a finite-dimensional L. The operator unitary dynamics (14.29) is reduced to the matrix dynamics. Conceptually, the Hamiltonian is given by a symmetric $m \times m$ matrix. For small m, the dynamics of such a type are very simple and processing is very rapid.

For example, consider the QLR for the concept "house." In the classical regime, the brain created a collection of images of concrete houses $MI_1, \ldots, MI_k$. They were classically encoded by covariance operators $D_1, \ldots, D_k$. These operators contain some common correlations. In the matrix representation, they have a common block. For simplicity, suppose that this block is of the diagonal type. Consider a subspace L of H related to this block. Then this block can be represented as a self-adjoint and positive operator D_L in L. We have:

$$D_L = \pi D_j \pi, \; j = 1, 2, \ldots, k,$$

where $\pi : H \to L$ is the orthogonal projector. Its QL image is given by

$$Q_L = D_L/\mathrm{Tr} D_L = \frac{\pi D_j \pi}{\mathrm{Tr} \pi D_j \pi} = \frac{\pi (D_j/\mathrm{Tr} D_j) \pi}{\mathrm{Tr} \pi (D_j/\mathrm{Tr} D_j) \pi} = \frac{\pi \rho_j \pi}{\mathrm{Tr} \pi \rho_j \pi} = \rho_L.$$

Here ρ_j are QL representations of the covariance operators D_j. The density operator ρ_L gives the abstract concept of house.

We now describe the process of creation of a concept of a higher level of abstractness from a cluster of concepts. Consider a subspace Z of L. The brain can create new concepts belonging to $\mathcal{D}(Z)$ starting with clusters of L concepts. Of course, it can proceed directly starting with mental images from $\mathcal{D}(H)$. However, such step-by-step increasing of the level of abstraction is very natural.

QLP6. **Neuronal location of the QL processor.** From the general viewpoint there are no reasons to assume that the QLR is realized in the same physical domain, i.e. there is the same ensemble of neurons, as in classical processing. It may be that there is a special domain O_{QL} which is used for dynamics (14.29). Our model induced an interesting problem of experimental neurophysiology, i.e. to find domains of the brain coupled to the QLR. If the hypothesis that the dynamics (14.29) of mental images is based on the physical dynamics (14.27) is correct, then domains of the QLR can be identified by the presence of Gaussian stochastic dynamics.

Unfortunately, at the present level of measurements it is impossible to measure directly the electromagnetic field inside the brain (at least to make measurements in a sufficiently dense set of points). However, even the measurement technology based on EEG provides a possibility of reconstruction of the field inside the brain by using the methods of the inverse problem.

QLP7. **Quantum-like consciousness.** We may speculate that consciousness can be associated with QL processing in the brain. The von Neumann equation (or more generally the Gorini–Kossakowski–Sudarshan–Lindblad equation) represents the "continuous flow of consciousness." The feeling of continuity is generated through the averaging of physical signals with respect to the mental time scale, i.e. the representation of mental images by covariance density operators. In fact, on the physical time scale the dynamics are discrete, see (14.26).

QLP8. **Correspondence between classical and quantum qualia.** Consider a classical quale given by a functional $f(\phi)$. The corresponding quantum quale is given by the self-adjoint operator, the second derivative of $f(\phi)$ at the point $\phi = 0$. In general, the behavior of the functional $f(\phi)$ differs essentially from the behavior of its quadratic part. However, on the level of averages the difference is not so large: in the limit $\tau/T \to 0$ they coincide, see (14.7).

14.8 Composite systems

We turn again to physics. As remarked in Khrennikov [19] (pp. 2–3), in CSM, a composite system $S = (S_1, S_2)$ is mathematically described by the Cartesian product of state spaces of its parts S_1 and S_2. In QM, it is described by the tensor product. The majority of researchers working in quantum foundations and, especially quantum information theory, consider this difference in the mathematical representation as crucial. In particular, the entanglement which is a consequence of the tensor space representation is considered as a totally non-classical phenomenon. However, we recall that Einstein considered the EPR-states as exhibitions of classical correlations due to the common preparation. The PCSFT will realize Einstein's dream on entanglement.

Let $S = (S_1, S_2)$, where S_i has the state space H_i – complex Hilbert space. Then, by the CSM, the state space of S is $H_1 \times H_2$. By extending the PCSFT to composite systems, we should describe ensembles of composite systems by probability distributions on this Cartesian product, or by a random field $\phi(x, \omega) = (\phi_1(x, \omega), \phi_1(x, \omega)) \in H_1 \times H_2$.

In our approach, each quantum system is described by its own random field: S_i by $\phi_i(x, \omega)$, $i = 1, 2$. However, these fields are CORRELATED – in a completely classical sense. Correlation at the initial instant of time $s = s_0$ propagates in time in complete accordance with the laws of QM. There is no action at a distance. We have purely classical dynamics of two stochastic processes which were correlated at the beginning. In fact, the situation is more complex: there is also the common random background, i.e. vacuum fluctuations. We shall come back to this question in future research.

14.8.1 Operator realization of wave function

Consider now the QM model. Take a pure state case $\Psi \in H_1 \otimes H_2$. Can one peacefully connect the QM and the PCSFT formalisms? Yes! But Ψ should be interpreted in a completely different way than in conventional QM.

The main mathematical point is: Ψ is not a vector! It is an operator! It is, in fact, the non-diagonal block of the covariance operator of the corresponding prequantum random field: $\phi(x, \omega) \in H_1 \times H_2$. The wave function $\Psi(x, y)$ of a composite system determines the integral operator:

$$\widehat{\Psi}\phi(x) = \int \Psi(x, y)\phi(y)dy.$$

We keep now to the finite-dimensional case. Any vector $\Psi \in H_1 \otimes H_2$ can be represented in the form $\Psi = \sum_{j=1}^{m} \psi_j \otimes \chi_j, \;\; \psi_j \in H_1, \chi_j \in H_2$, and it determines a linear operator from H_2 to H_1

$$\widehat{\Psi}\phi = \sum_{j=1}^{m} (\phi, \chi_j)\psi_j, \;\; \phi \in H_2. \tag{14.30}$$

Its adjoint operator Ψ^* acts from H_1 to $H_2 : \widehat{\Psi}^*\psi = \sum_{j=1}^{m}(\psi, \psi_j)\chi_j, \psi \in H_1$. Of course, $\widehat{\Psi}\widehat{\Psi}^* : H_1 \to H_1$ and $\widehat{\Psi}^*\widehat{\Psi} : H_2 \to H_2$ and these operators are self-adjoint and positively defined. Consider the density operator corresponding to a pure quantum state, $\rho = \Psi \otimes \Psi$. Then, the operators of the partial traces $\rho^{(1)} \equiv \mathrm{Tr}_{H_2}\rho = \widehat{\Psi}\widehat{\Psi}^*$ and $\rho^{(2)} \equiv \mathrm{Tr}_{H_1}\rho = \widehat{\Psi}^*\widehat{\Psi}$.

14.8.2 Basic equality

Let $\Psi \in H_1 \otimes H_2$ be normalized by 1. Then, for any pair of linear bounded operators $\widehat{A}_j : H_j \to H_j, j = 1, 2$, we have:

$$\mathrm{Tr}\widehat{\Psi}\widehat{A}_2\widehat{\Psi}^*\widehat{A}_1 = \langle \widehat{A}_1 \otimes \widehat{A}_2 \rangle_\Psi \equiv (\widehat{A}_1 \otimes \widehat{A}_2\Psi, \Psi). \tag{14.31}$$

This is a mathematical theorem (Khrennikov [10]). It will play a fundamental role in further considerations.

14.8.3 Coupling of classical and quantum correlations

In the PCSFT a composite system $S = (S_1, S_2)$ is mathematically represented by the random field $\phi(\omega) = (\phi_1(\omega), \phi_2(\omega)) \in H_1 \times H_2$. Its covariance operator D has the block structure

$$D = \begin{pmatrix} D_{11} & D_{12} \\ D_{21} & D_{22} \end{pmatrix},$$

where $D_{ii} : H_i \to H_i$, $D_{ij} : H_j \to H_i$. The covariance operator is self-adjoint. Hence, $D_{ii}^* = D_{ii}$, and $D_{12}^* = D_{21}$.

Here by the definition: $(D_{ij}u_j, v_i) = E(u_j, \phi_j(\omega))(v_i, \phi_i(\omega))$, $u_i \in H_i$, $v_j \in H_j$. For any Gaussian random vector $\phi(\omega) = (\phi_1(\omega), \phi_2(\omega))$ having zero average and any pair of operators $\widehat{A}_i \in \mathcal{L}_s(H_i)$, $i = 1, 2,$ the following equality takes place: $\langle f_{A_1}, f_{A_2}\rangle_\phi \equiv Ef_{A_1}(\phi_1(\omega))f_{A_2}(\phi_2(\omega)) = (\mathrm{Tr}D_{11}\widehat{A}_1)(\mathrm{Tr}D_{22}\widehat{A}_2) + \mathrm{Tr}D_{12}\widehat{A}_2D_{21}\widehat{A}_1$. We remark that $\mathrm{Tr}D_{ii}\widehat{A}_i = Ef_{A_i}(\phi_i(\omega))$, $i = 1, 2$. Thus, we have $f_{A_1}f_{A_2} = Ef_{A_1}Ef_{A_2} + \mathrm{Tr}D_{12}\widehat{A}_2D_{21}\widehat{A}_1$. Consider a Gaussian vector random field such that $D_{12} = \widehat{\Psi}$:

$$E(f_{A_1} - Ef_{A_1})(f_{A_2} - Ef_{A_2}) = (\widehat{A}_1 \otimes \widehat{A}_2\Psi, \Psi) \equiv \langle \widehat{A}_1 \otimes \widehat{A}_2\rangle_\Psi, \quad (14.32)$$

or, for the covariance of two classical random vectors f_{A_1}, f_{A_2}, we have: $\mathrm{cov}\,(f_{A_1}, f_{A_2}) = \langle \widehat{A}_1 \otimes \widehat{A}_2\rangle_\Psi$.

14.9 References

[1] Hameroff, S. (1994). Quantum coherence in microtubules: a neural basis for emergent consciousness? *Journal of Consiousness Studies*, 1, 91–118.

[2] Hameroff, S. (1994). Quantum computing in brain microtubules? The Penrose–Hameroff Orch Or model of consciousness. *Philosophical Transactions of the Royal Society*; London A 1-28.

[3] Penrose, R. (1989). *The Emperor's New Mind.* Oxford University Press.

[4] Penrose, R. (1994). *Shadows of the Mind.* Oxford University Press.

[5] Khrennikov, A. Yu (2010). On the physical basis of the theory of "mental waves." *Neuroquantology*, 8, 4 (Supp. 1), S71–S80.

[6] Khrennikov, A. Yu (2011). Quantum-like model of processing of information in the brain based on classical electromagnetic field. *Biosystems* 105, 250–262.

[7] Khrennikov, A. Yu. (2005). Prequantum classical statistical model with infinite dimensional phase-space. *Journal of Physics A: Mathematical and General*, 38, 9051–9073.

[8] Khrennikov, A. Yu. (2005). Generalizations of quantum mechanics induced by classical statistical field theory. *Foundations of Physics Letters*, 18, 637–650.

[9] Khrennikov, A. Yu. (2007). Nonlinear Schrödinger equations from prequantum classical statistical field theory. *Physics Letters A*, 357, 171–176.

[10] Khrennikov, A. Yu. (2009). Entanglement's dynamics from classical stochastic process. *Europhysics Letters*, 88, 40005.

[11] Khrennikov, A. Yu. (2008). The quantum-like brain on the cognitive and subcognitive time scales. *Journal of Consciousness Studies*, 15, 39–77.
[12] Fichtner, K. H., Fichtner, L., Freudenberg, W., and Ohya, M. (2008). On a quantum model of the recognition process. In *Quantum Probability and White Noise Analysis*. Eds. Accardi, L., Freudenberg, W., and Ohya, M.; 21, 64–84; World Scientific.
[13] Einstein, A. and Infeld, L. (1961). *Evolution of Physics: The Growth of Ideas from Early Concepts to Relativity and Quanta*. Simon & Schuster.
[14] Khrennikov, A. Yu. (2009). *Contextual Approach to Quantum Formalism*. Springer Verlag, Berlin.
[15] 't Hooft, G. (2002). *Determinism beneath quantum mechanics*. Conference Proceedings "Quo Vadis Quantum Mechanics"; Philadelphia (see also http://arxiv.org/abs/quant-ph/0212095; Cornell University.
[16] 't Hooft, G. (2007). *The free-will postulate in quantum mechanics*. Please see http://arxiv.org/abs/quant-ph/0701097; Cornell University.
[17] Elze, T. (2008). *The attractor and the quantum states*. Please see http://arxiv.org/abs/0806.3408; Cornell University.
[18] Khrennikov, A. Yu. (2008). Quantum randomness as a result of random fluctuations at the Planck time scale? *International Journal of Theoretical Physics*, 47, 114–124.
[19] Khrennikov, A. Yu. (2011). A classical field theory comeback? The classical field viewpoint on triparticle entanglement. *Physica Scripta*, T143, 014013 (6pp).

15
Conclusion

This book had as its main objective to discuss how (quantum) physics concepts could be used in a relevant way in the modeling of a variety of economics/finance and psychology-based processes.

For those readers who have come to read this conclusion, without skipping its "backbone," we would like to say "thank you" for keeping with us until the end.

The possible claim, which says that social science driven dynamics would need analyzing with techniques from the natural sciences, may well have been tempered with this book. Although quantum physics as a theory of prediction is one of the most successful theories humankind ever devised, it is not unreasonable to suggest that the number of foundational interpretations of quantum physics is sizable.

Consider now the following natural question: "How can a theory with such a wide diversity of foundational interpretations serve to inform and elucidate a non-cognate area of research (like economics or psychology)?" Indeed, we should want to stand upright and write, without hesitation, the words "non-cognate."

No theory in social science can come close to the predictive power of quantum physics. However, very few social science theories are "plagued" by foundational issues of such depth as quantum physics. As an example, consider the issue of why quantum physical wave functions do have to be complex valued. Do physicists have a satisfactory answer to this very basic question? A theory of such predictive power has a partial differential equation, the Schrödinger equation, which cannot be rigorously derived. However, in economics and finance we strive to build theoretical edifices with the utmost respect for rigor.

Consider the recent financial crisis. Are we comfortable to propose that physics should now lend a helping hand to the social sciences? This is indeed an extremely difficult question. There are good examples of theories in finance which can be reformulated in (quantum) physics-based terms. Option pricing is one such theory. In economics, expected utility and the violation of the sure-thing principle can be explained with the concept of probability interference. However, it is unclear whether those novel viewpoints have added to our understanding of the pricing or decision making processes expressed in the above two examples.

We hope that this book has given more clues to aid in our understanding of how certain foundational aspects in the theories of finance, economics, and psychology can be re-interpreted. We may have opened up new avenues for research but it is still far too early to ascertain whether this bears any promise. However, there may be "rays of light" on the horizon. As an example, we can observe that several important research funding agencies have now provided financial aid to projects which involve the use of quantum physical techniques in psychology, finance, and economics. This is a healthy positive step forward and we can only hope that this continued support can muster sufficient enthusiasm and wit around the world so that true breakthroughs can be expected in the near future. We may soon be in dire need for those.

Glossary of mathematics, physics, and economics/finance terms

Some mathematics based definitions (in alphabetical order)

- Bayes' formula: the conditional probability $\mathbf{P}(B|A)$ is defined by using the probability $\mathbf{P}(B \cap A)$ of the intersection of events A and B (their joint occurrence): $\mathbf{P}(B|A) = \frac{\mathbf{P}(B\cap A)}{\mathbf{P}(A)}$; $\mathbf{P}(A) > 0$.
- Bra vector: elements of the dual space H^*, the space of linear continuous functionals on H, are called bra vectors. They are denoted as $\langle\phi|$ (see also "Ket vector" below).
- Complex numbers: $z = x + iy$; $x, y \in \mathbb{R}$ and $i^2 = -1$.
- Dirac braket: it can be defined as $\langle\psi_1|\widehat{w}\psi_2\rangle \equiv \int \psi_1^*(\widehat{w}\psi_2)dv$, where ψ_1^* denotes the complex conjugate of the state function ψ_1, and the operator $\widehat{w}$ acts on the state function ψ_2.
- Eigenstate equation – see "Eigenvalue equation" below: when the function f_l is replaced by a quantum mechanical wave function, ψ, the eigenvalue equation is often re-named as an eigenstate equation: $\widehat{w}\psi = l\psi$, and ψ can be seen as an eigenstate.
- Eigenvalue equation: this equation for an operator $\widehat{w}$, with eigenfunction f_l and eigenvalue l, is given by $\widehat{w}f_l = lf_l$ (see "eigenstate equation" above).
- Fourier integral of a function f: $f(x) = \frac{1}{\sqrt{2\pi}}\int_{-\infty}^{\infty}\widehat{f}(w)\exp(iwx)dw$, where i is a complex number (see "Fourier transform" below).
- Fourier transform of a function f: $\widehat{f}(w) = \frac{1}{\sqrt{2\pi}}\int_{-\infty}^{\infty} f(x)\exp(-iwx)dx$, where i is a complex number. Note that $f(x)$ (i) needs to be piecewise continuous on every finite interval and (ii) $f(x)$ is absolutely integrable on the X-axis.
- h-derivative of a function f: $D_h f(x) = \frac{d_h f(x)}{d_h(x)} = \frac{f(x+h)-f(x)}{(x+h)-x}$.
- h-differential of a function f: $d_h f(x) = f(x+h) - f(x)$.
- Hermitian Hamiltonian: the Hamiltonian, $\widehat{\mathcal{H}}$, is Hermitian when $\langle\widehat{\mathcal{H}}\psi_1|\psi_2\rangle = \langle\psi_1|\widehat{\mathcal{H}}\psi_2\rangle$, where $\langle.|.\rangle$ is a Dirac braket.

- Hilbert space: see Chapter 4: Vector calculus and other mathematical preliminaries.
- Infinitesimal transformation: see entry "Dirac brakets and bras and kets" in Chapter 4: Vector calculus and other mathematical preliminaries.
- Hyperbolic numbers: $z = x + jy$; $x, y \in \mathbb{R}$, and $j^2 = +1$.
- Kantian prejudice: physical space has to be identified with its Euclidean model (later modified by Lobachevsky and Einstein) – see also "Laplace's prejudice" in Chapter 1: Classical, statistical, and quantum mechanics: all in one.
- Ket vector: elements of the Hilbert space H are called ket vectors and they are denoted as $|\psi\rangle$ (see also "bra vector" above).
- Linear space: see Chapter 4: Vector calculus and other mathematical preliminaries.
- Operator: an operator, $\widehat{w}$, is defined relative to a function $f(x)$ in the following way: $g(x) = \widehat{w} f(x)$.
- Operator (adjoint): see entry "Dirac brakets and bras and kets" in the Chapter 4: Vector calculus and other mathematical preliminaries.
- Operator (bounded): an operator $\widehat{w}$ is bounded if: $\|\widehat{w}\| = \sup_{\{\Psi \in H : \|\Psi\| = 1\}} \|\widehat{w}\psi\| < \infty$.
- Operator (Hermitian): an operator, $\widehat{w}$, is Hermitian if $\langle \widehat{w}\psi_1 | \psi_2 \rangle = \langle \psi_1 | \widehat{w}\psi_2 \rangle$, where $\langle . | . \rangle$ is a Dirac braket.
- Operator (idempotent): an operator $\widehat{w}$ is *idempotent* if $\widehat{w}^2 = \widehat{w}$.
- Operator (identity): such an operator leaves a quantum state unchanged. As an example, in the Heisenberg uncertainty principle, one can write the identity operator $\widehat{q}\widehat{p} - \widehat{p}\widehat{q} = i\hbar\widehat{1}$, where $\widehat{1}$ is the identity operator.
- Operator (inverse): an inverse operator is such that $\widehat{w}\widehat{w}^{-1} = \widehat{w}^{-1}\widehat{w} =$ identity operator.
- Operator (linear): i) $\widehat{w}\alpha\psi = \alpha\widehat{w}\psi$ and ii) $\widehat{w}\,[\alpha\psi_1 + \beta\psi_2] = \alpha\widehat{w}\psi_1 + \beta\widehat{w}\psi_2$, where α and β are arbitrary complex constants, and $\widehat{w}$ and $\widehat{v}$ are operators. See also entry "Operators" in Chapter 4: Vector calculus and other mathematical preliminaries.
- Operator (momentum): $\widehat{p} = -i\hbar\frac{\partial}{\partial q}$, where q is position, i is a complex number, and $\hbar$ is the rationalized Planck constant.
- Operator (non-commutative): consider a function $f(x)$, operators $\widehat{q}$, $\widehat{p}$ do not commute if $\widehat{q}\widehat{p} f(x) \neq \widehat{p}\widehat{q} f(x)$.
- Operator (projection): an operator which is idempotent and Hermitian.
- Operator (unitary): see entry "Dirac brakets and bras and kets" in Chapter 4: Vector calculus and other mathematical preliminaries.
- Ordinary differential equation (ODE): $F(x, y, y' \ldots y^{(n)}) = 0$, where y are functions of x and y' indicates the first ordinary derivative towards x; similarly for $y^{(n)}$, which indicates the nth derivative towards x.

- Ordinary differential equation (ODE) (linear): An ODE will be linear if F is a linear function of the variables $y, y' \dots y^{(n)}$.
- q-derivative of a function f: $D_q f(x) = \frac{d_q f(x)}{d_q(x)} = \frac{f(qx)-f(x)}{qx-x}$.
- q-differential of a function f: $d_q f(x) = f(qx) - f(x)$.
- Stochastic matrix (double): a square matrix of non-negative real numbers, where each row and each column sums to 1.
- Stochastic matrix (right): a square matrix where each of its rows consists of non-negative real numbers, with each row summing to 1.
- Stochastic matrix (left): a square matrix whose columns consist of non-negative real numbers whose sum is 1.
- Unitary transformation: see entry "Dirac brakets and bras and kets" in Chapter 4: Vector calculus and other mathematical preliminaries.

Some physics-based definitions (in alphabetical order)

- Bohr's postulate: an electron can move only on a discrete set of orbits and hence its energy can take only a discrete series of values.
- Bohr–Heisenberg Copenhagen interpretation of quantum mechanics: this interpretation says that the wave function of a quantum system provides the most complete description of its state, i.e. no finer description could be created.
- Born's rule (discrete version): the probability to find a particle at the point x of physical space is given by the square of the absolute value of the ψ-function (complex probability amplitude).
- de Broglie relation: $p = \hbar k$; p is momentum and $\hbar$ is the rationalized Planck constant; k is the wave number.
- Conditioning (context): conditioning which is not only based on the Bayes' formula and Boolean algebra, since a context C need not (although can in some cases) be associated with any event.
- Conditioning (event): conditioning which is based on the Bayes' formula and Boolean algebra.
- Conservative force: $f(x) = -\frac{dV}{dx}$, where f is force, V is the real potential, x is position.
- Context: a context C is a complex of conditions: for instance physical, social, financial.
- Continuity equation: $\frac{\partial R^2}{\partial t} + \frac{1}{m}\frac{\partial}{\partial q}\left(R^2 \frac{\partial S}{\partial q}\right) = 0$, where R is the amplitude of the wave function, S is the phase of the wave function, m is mass, q is position, t is time.
- Correspondence principle: the principle which establishes a coupling between classical and quantum mechanics.

- Double slit experiment: the basic experiment in quantum mechanics which indicates wave-particle duality.
- Energy preservation in the process of motion: $\mathcal{H}(q(t), p(t)) = \mathcal{H}(q(t_0), p(t_0))$, where $\mathcal{H}$ is the Hamiltonian function (please see below for "Hamiltonian function").
- Factorizability of the density: the probabilistic independence of particles (this does not imply factorization of the probability amplitude).
- Hamiltonian function: $\mathcal{H}(q, p) = \frac{p^2}{2m} + V(q)$, where p is momentum, m is mass, $V(.)$ is the real potential, $\frac{p^2}{2m}$ is kinetic energy, q is position.
- Hamilton–Jacobi equation: consider the equation $\frac{\partial S}{\partial t} + \frac{1}{2m}\left(\frac{\partial S}{\partial q}\right)^2 + \left(V - \frac{h^2}{2mR}\frac{\partial^2 R}{\partial q^2}\right) = 0$, where S is the phase of the wave function, t is time, m is mass, q is position, h is the Planck constant, V is the real potential, R is the amplitude function. When $\frac{h^2}{2m} << 1$ and let $\frac{h^2}{2mR}\frac{\partial^2 R}{\partial q^2}$ be negligibly small, then the above equation is a Hamilton–Jacobi equation.
- Heisenberg's postulate: operators on position and momentum should satisfy a peculiar commutation relation (the Heisenberg uncertainty principle).
- Heisenberg uncertainty relation (first proven by Kennard (1927) – see Chapter 1: Classical, statistical, and quantum mechanics: all in one): $\sigma_x \sigma_p \geq \frac{\hbar}{2}$, where $\hbar = h/2\pi$, and σ_x, σ_p are the standard deviations of respectively position and momentum.
- Information: $i = -\sum_j p_j \ln p_j$, where p_j is a probability distribution of states and the index j can describe a feature of a system (or a particle for instance).
- Information gain: $K = \sum_j p_j \ln \frac{p_j}{p'_j}$, where p_j and p'_j are different probability distributions (see also "Information" above).
- Maps (quantum information theory): a map is important for describing an information transmission such as a measurement process or a signal transmission.
- Mean forward derivative of a function f: $D_+ f(x,t) = \lim_{\Delta t \to 0} E\left[\frac{f(x(t+\Delta t), t+\Delta t) - f(x(t), t)}{\Delta t}\right]$, where t is time.
- Mean backward derivative of a function f: $D_- f(x,t) = \lim_{\Delta t \to 0} E\left[\frac{f(x(t), t) - f(x(t-\Delta t), t-\Delta t)}{\Delta t}\right]$, where t is time.
- Mixed state: such a state is obtained when a statistical mixture of wave functions is used.
- Newton's second law: $m.a = f$, m is mass, a is acceleration, f is force.
- Newton–Bohm equation: $m\frac{d^2 q(t)}{dt^2} = -\frac{\partial V(q,t)}{\partial q} - \frac{\partial Q(q,t)}{\partial q}$, where m is mass, q is position, V is the real potential, and Q is the quantum potential. $Q(q,t)$ depends on the wave function, and the wave function evolves according to the Schrödinger equation. The initial conditions are $q(t_0) = q_0$ and $q'(t_0) = q'_0$ (momentum).

- No-go theorems: theorems about the impossibility to construct "prequantum models" reproducing the probabilistic predictions of quantum mechanics and operating with classical probabilities.
- Non-local realism: a condition of the world where determinism holds true. There is instantaneous action (with transmission speed faster than speed of light) at a distance.
- Non-stationary states: when the wave function is not separable, then we must obtain a non-stationary state.
- Operators on position and momentum (respectively): $\hat{q}\psi(q) = q\psi(q)$, $\hat{p}\psi(q) = -i\hbar\frac{\partial\psi(q)}{\partial q}$, where $\psi(q)$ is a square integrable function, i is a complex number.
- Planck constant: $h \approx 6.6260693(11) \times 10^{-34}$ J $\times$ sec.
- Polar form of wave function: $\psi(q,t) = R(q,t)e^{i\frac{S(q,t)}{h}}$, where $R(q,t) = |\psi(q,t)|$, $S(q,t)/h$ is the phase of the wave function (divided by the Planck constant), q is position, t is time.
- Prequantum approach: approach by which classical-like models reproduce results of quantum experiments have been created.
- Prequantum classical statistical field theory (PCSFT): in this model quantum non-locality has been reduced to classical correlations in the background field (please see below for "quantum non-locality").
- Probability interference term: $2\,|\psi_1(x)|\,|\psi_2(x)|\cos(S_1 - S_2)$, where $S_i(x)$ is the phase of the wave function i and $|\psi_i(x)|$ is the amplitude of the wave function i.
- Pure state: when a system is described by a single wave function then it is called a pure state.
- Quantization: a transition from functions on the phase space, $f(q,p)$, to Hermitian matrices or more generally operators, by using operators of position and momentum, instead of corresponding classical variables.
- Quantum entanglement: the existence of wave functions of the form $\psi(q_1,q_2) = \psi_1(q_1)\psi_2(q_2) + \psi_2(q_1)\psi_1(q_2)$. The correlation of phases (and not the coordinates) is the essence of quantum entanglement.
- Quantum-like approach: the approach by which we do not start with a classical model and then quantize it, but we directly mimic the quantum approach.
- Quantum non-locality: a consequence of quantum entanglement (please see above for "quantum entanglement").
- Quantum potential: $-\frac{h^2}{2mR}\frac{\partial^2 R}{\partial q^2}$, where R is the amplitude of the wave function, m is mass and q is position, h is the Planck constant.
- Quantum state representation: only normalizable state functions can represent a quantum state. The state function, before complex conjugation is applied, is indicative of the probability wave or the probability amplitude.
- Rationalized Planck constant: $\hbar = h/2\pi$.

- Schrödinger partial differential equation (time dependent – for one particle system): $i\hbar\frac{\partial\psi(q,t)}{\partial t} = -\frac{\hbar^2}{2m}\frac{\partial^2\psi(q,t)}{\partial q^2} + V(q,t)\psi(q,t)$, where i is a complex number, $\hbar$ is the rationalized Planck constant, ψ is the wave function, m is mass, V is the real potential, q is position, and t is time.
- Stationary states (existence): in the case we have a conservative system, i.e. with a real potential which is time independent, stationary states will exist.
- Wave packet construction: $\psi(x,t) = \frac{1}{\sqrt{2\pi}}\int_{-\infty}^{\infty} A(k)\exp(i(kx-\omega t))dk$, where k is a wave number, $A(k)$ is the amplitude function, t is time, x is position, and ω is the angular frequency.

Some economics/finance and psychology-based definitions (in alphabetical order)

- Absolute risk aversion (ARA): ARA$= \frac{-u''(W)}{u'(W)}$; where $u''(W)$ indicates the second derivative of the utility function towards wealth (and similarly for the first derivative).
- Absolute risk aversion (increasing): if wealth increases ARA increases (often denoted as IARA).
- Absolute risk aversion (decreasing): if wealth increases ARA decreases (often denoted as DARA).
- Anscombe–Aumann expected utility model: a mixture of objective and subjective probabilities are used in the calculation of expected utility.
- Arbitrage (no-): the absence of making a riskless profit. Example: if two deposit accounts, A and B, are equally default free (ceteris paribus), then if the interest rate on deposit account A is higher than on the deposit account B, an arbitrage profit situation occurs; the excess interest rate between accounts A and B reflects no additional risk.
- Binomial option pricing: a model which aids in setting the non-arbitrage price of a financial option (see "Black–Scholes option pricing model" below).
- Black–Scholes partial differential equation: $\frac{\partial F}{\partial t} + r_f S\frac{\partial F}{\partial S} + \frac{1}{2}\sigma^2 S^2\frac{\partial^2 F}{\partial S^2} = r_f F$, where F is the option contract and $\frac{\partial F}{\partial S}$ is the "delta" of the option, r_f is the risk free rate of interest, σ is the volatility of the price of the underlying asset S, t is time.
- Black–Scholes option pricing: solutions to the Black–Scholes partial differential equation (with the proper end conditions (please see "Black–Scholes partial differential equation" above)) provide for the pricing equations of the Black–Scholes model.
- Black–Scholes portfolio: $\Pi = -F + \frac{\partial F}{\partial S}S$, where Π, F is the option contract and $\frac{\partial F}{\partial S}$ is the "delta" of the option, S is the price of the underlying asset.
- Brownian motion on the price of an asset: a stochastic differential equation $dS = \alpha(S)dt + \beta(S)dX$, where dS denotes the infinitesimal change in the

asset price S, $\alpha(S)$ is some drift function of the asset price, dt denotes the infinitesimal change in time, $\beta(S)$ is some diffusion function of the asset price, dX is a Wiener process.

- Call option (plain vanilla): a security that gives its owner the right, but not the obligation, to purchase a specified asset for a specified price, known as the exercise price or the strike price.
- Capital asset pricing model (CAPM): the return of asset i, r_i, is written as $r_i = r_f + \beta_i(r_m - r_f) + \epsilon_i$, where ϵ_i is the error term, β_i is the CAPM beta (see below for "CAPM – β_i"), r_f is the risk free rate of interest (please see below for "risk free rate of interest"), r_m is the rate of return on the market portfolio (see below for "market portfolio").
- CAPM – β_i: for a security i, β_i, is defined as $\beta_i = \frac{cov(r_i, r_m)}{\sigma_m^2}$, where r_i is the return of asset i and r_m is the return of the market portfolio and $cov(r_i, r_m)$ is the covariance of the return of asset i and the return of the market portfolio, m (see also "market portfolio" below).
- Efficiency: efficiency expresses what type of information (for instance public information) prices reflect.
- Equivalence between preference relation and utility function: there exists an equivalence between the preference relation of an object x over an object y, denoted as $x \succ y$, if and only if there exists a utility function $u(.)$, which maps a set of objects into $\mathbb{R}$, such that $u(x) > u(y)$.
- Ellsberg paradox: this paradox indicates that the "sure-thing principle," a key assumption of the Savage model can be refuted on experimental grounds.
- Incompatible observables (economics/finance interpretation): non-simultaneous observables. As an example, one observable variable is "price" and the "time change of price" would be the other observable variable. Both variables cannot be simultaneously observed.
- Information price trajectory: a Newton–Bohm path (see above "Newton–Bohm equation") of non-observed prices which affect non-observed risk neutral probabilities (see below "Risk neutral probabilities") in the non-arbitrage theorem.
- Information (quantum-like representation): in the quantum-like representation of information, the brain operates with density operators which represent not only concrete mental images (coming from the classical regime of mental processing), but also abstract concepts (of different levels of abstraction which do not correspond to classically produced images).
- Kinetic energy in economics: $\frac{1}{2}\sum_{j=1}^{n} m_j v_j^2$, where m_j is the number of shares of stock j and $v_j(t) = \dot{q}_j(t) = \lim_{\Delta t \to 0} \frac{q_j(t+\Delta t) - q_j(t)}{\Delta t}$, where t is time and q_j is the price of asset j.
- Market capitalization: $T_j(t) = m_j q_j(t)$: market capitalization of trader j at time t, where m_j is the number of stock of asset j, and q_j is the price of asset j.

- Market portfolio: a theoretical portfolio used in the capital asset pricing model. A close proxy of such portfolio can be an index portfolio like the Dow Jones 30 portfolio.
- Martingale of a stock price: the conditional expectation of a stock price, S_T at time $T > t$, given the information at time $t : I_t$, is $E[S_T|I_t] = S_t$ (note: the mathematical definition is more involved than this heuristic definition).
- Phase space of prices: a product space $Q \times V$; where $V \equiv \mathbb{R}^n$ and $\overrightarrow{v} = (v_1, v_2, \ldots, v_n) \in V$, $v_j(t) = \dot{q}_j(t) = \lim_{\Delta t \to 0} \frac{q_j(t+\Delta t) - q_j(t)}{\Delta t}$; t is time and $\overrightarrow{q} = (q_1, q_2, \ldots, q_n) \in Q$, where q_j is the price of the share of the jth corporation.
- Potential energy in economics: describes the interactions between traders as well as interactions from other factors such as macro-economic factors.
- Pricing rule: $\frac{-\partial Q}{\partial q}$, where Q is the quantum potential (adopted for a financial environment) and q is price.
- Psychological time scale (T-scale): the T-scale is called the mental time scale or psychological time scale. The T-scale is the scale of creation of mental images by the brain.
- Put option (plain vanilla): a security that gives its owner the right, but not the obligation, to sell a specified asset for a specified price, known as the exercise price or the strike price.
- Qualia: in prequantum classical statistical field theory, each concrete mental image is associated with a random signal. Its mental features, " qualia," are given by functionals of this signal. In the simplest case, these are quadratic forms of the signal.
- Quantum-like wave function (economics/finance interpretation): as an input in the Radon–Nikodym derivative (see Chapter 13).
- Quantum-like wave function (economics/finance interpretation): as an analogue of the put and call functions (see Chapter 13).
- Quantum-like pilot wave function (economics/finance interpretation): pilot waves as carriers of erroneous information (see Chapter 13).
- Quantum-like pilot wave function (economics/finance interpretation): pilot wave in a drift-dependent option (see Chapter 13).
- Risk averting: $u(E(W)) > E(u(W))$, i.e. if the utility of expected wealth is larger than the expectation of the utility of wealth, then the individual is risk averse.
- Risk averting (degree of): if agent 1 is more risk averse than agent 2, then agent's 1 utility function is "more concave" than agent's 2 utility function; we get for agent's 1 utility function that $u(w) = h(v(w))$ where $h(.)$ is an increasing/strictly concave function.
- Risk free rate of interest: the rate of interest on an asset which carries no risk at all. Example: a deposit account upon which no default can occur should carry a risk free rate of return.

- Risk loving: $u(E(W)) < E(u(W))$, i.e. if the utility of expected wealth is smaller than the expectation of the utility of wealth, then the individual is risk loving.
- Risk neutral probability: a probability measure which allows a risky asset to be discounted at the risk free rate of interest.
- Savage expected utility model: subjective probabilities are used in the calculation of expected utility.
- Short selling: a broker sells shares at a price, p, on behalf of an investor. The broker borrows the shares from another investor. Assuming the price of the asset drops, the investor will buy the shares at the price $p_1 < p$ and will make a per share profit (before transaction costs) of $p - p_1$.
- von Neumann–Morgenstern expected utility model: objective probabilities are used in the calculation of expected utility.

Index